D0844749

The University of Wisconsin

*Publications
in
Medieval Science*

Mechanics in Sixteenth-Century Italy

Mechanics in Sixteenth-Century Italy

Selections from
*Tartaglia, Benedetti,
Guido Ubaldo, & Galileo*

Translated & Annotated by

Stillman Drake
&
I.E. Drabkin

The University of Wisconsin Press
Madison, Milwaukee, & London
1969

Published by
The University of Wisconsin Press
Box 1379, Madison, Wisconsin 53701

The University of Wisconsin Press, Ltd.
27–29 Whitfield Street, London, W.1

Copyright © 1969
by the Regents of the University of Wisconsin
Printed in the United States of America by
Kingsport Press, Inc., Kingsport, Tennessee

Standard Book Number 299–05100–5
Library of Congress Catalog Card Number 68–9015

For
Miriam, Bill
and
Sue Drabkin

Preface

The volumes already published by the University of Wisconsin Press in this series on medieval science have made available in English translation much source material for the history of physics that had previously been known only to specialists. That material was drawn largely from medieval manuscripts and from scarce printed editions to be found only in a few large libraries. An exception in both regards was the volume containing the principal writings of Galileo on motion and mechanics, prepared by Professor Drabkin and myself. Its purpose was to supply the student of the history of medieval science with a convenient means of judging its role in the transition to the physical science of the modern era.

Having supplied the Galileo volume, we were conscious of the need for English translations of mechanical science in the sixteenth century. So far as theories of motion and of simple machines are concerned, the period from Leonardo da Vinci to Galileo has been relatively neglected. Yet the sixteenth century is of critical importance to the thesis of continuity in the history of science, a thesis which has been uppermost in the minds of most researchers in the field during the present century. And though the advent of printing reduced the dependence of scholars on manuscript material for that epoch, the books which seem to be of central importance are by no means common today. Accordingly, we projected an anthology of translations which would bridge the gap, to some degree, for students of the history of thought in physical science.

In making our selection of works that would be reasonably representative of the main trends in mechanics as judged by present historical knowledge, we were able to confine ourselves to writings from Italy by reason of the publication elsewhere of English translations of the works of Simon Stevin. We were then faced with the problem of securing a general introduction to the translations as a guide to their interpretation and to further research. Because the interweaving of threads in every phase of Italian renaissance culture is extraordinarily complex, we sought the assistance of an Italian specialist to provide the introduction. But nothing had been received from him when the translations were completed, and neither of us felt fully qualified or sufficiently free from other occupations to provide a suitable introduction.

After the unexpected and premature death of Professor Drabkin early in 1965, I made inquiries in the United States and England for a specialist in the history of mechanics willing to undertake the writing of a more limited introduction. Failing in that, I have determined to proceed with the publication without further delay. An introduction of restricted scope has therefore been provided, incorporating two special prefaces written by Professor Drabkin in connection with his translations. In order partly to offset deficiencies in the introduction, Professor Miriam Drabkin has compiled an extensive bibliography relating to sixteenth-century mechanics in Italy that will enable others to pursue vigorously the kind of studies that the translators were intent upon assisting.

The University of California Press has kindly permitted the use of an excerpt from Professor Edward W. Strong's *Procedures and Metaphysics* (Berkeley, 1936), an indispensable work for the study of scientific method in the sixteenth century. I am obliged to Mark Drake, Frank Soriano, and J. Boag for preparing the diagrams in the book.

I am deeply grateful to the University of Wisconsin Press for its having made possible the publication of this anthology of translations, in which I had the privilege of collaborating for the second time with that fine gentleman and distinguished scholar, Professor I. E. Drabkin.

Stillman Drake

Toronto, Canada
February, 1968

Contents

Selections from Galileo Galilei
Translated and annotated by I. E. Drabkin

Mechanics in Sixteenth-Century Italy

Introduction
Stillman Drake

The scientific revolution, in which Galileo played a leading
role and which is in many ways aptly symbolized by his many-
sided struggles, may be conveniently thought of as beginning
with the seventeenth century. It was in the year 1600 that
Johannes Kepler met Tycho Brahe, an event that had the most
profound consequences for the history of astronomy. It was
also in 1600 that William Gilbert published his *De magnete*
and thereby set a pattern for the pursuit of experimental investigation of the physical world. The same year saw Galileo making the final revisions of his treatise on mechanics, a work
widely circulated in manuscript copies even before 1600,
though not published during his lifetime. François Viete was
then expanding algebra to the treatment of geometrical magnitudes as well as arithmetical problems. Simon Stevin had
already introduced decimal fractions, greatly simplifying
calculation, and had also considerably widened and strength-
ened the sciences of statics and hydrostatics. Thus throughout
Europe there were arising new streams of astronomy, mathematics, and physics, soon to be unified in a river of science
that would break the dams of speculative philosophy.

Science since 1600 has been, and is being, actively explored
by historians of ideas. The sources of material are numerous:
printed books, the journals of learned academies, and the
plentiful correspondence of scientists after 1600. The impact
of a new world-view on traditional philosophy and literature
can also be traced in many nonscientific writings. A large
proportion of all scientific works after 1600 were written in
living languages. In addition, English translations of both Latin
and vernacular scientific works abound. The science of classical antiquity has received special attention from translators
and is no longer inaccessible to historians of science generally.
Medieval science has come under close scrutiny in the present
century, as a result of which a great wealth of material has
been discovered, analyzed, and translated, with a view to the
establishment of linkages between the scientific revolution
and the works of the remote past.

To a very large degree, the program that has been followed
in the examination of medieval science was inaugurated by

Pierre Duhem about the beginning of the present century. Duhem, extending Charles Thurot's earlier researches into medieval manuscript material in the libraries of Europe, brought to light many treatises on motion and mechanics dating principally from the thirteenth to the fifteenth century. In conjunction with printed editions of earlier works published in the late fifteenth century and throughout the sixteenth century, those manuscripts suggested a previously unsuspected continuity in scientific thought from medieval to modern times that required a thorough overhauling of conventional historical ideas. Duhem saw in the manuscripts of Leonardo da Vinci a point of convergence of medieval scientific thought and a point of its dissemination into new channels during the sixteenth century. Duhem's conclusions have been considerably modified by later scholars, but the program which he initiated has endured; it is essentially the quest for possible earlier sources in the work of each great innovator in the field of science.

Perhaps the weakest link in Duhem's hypothetical chain of succession was the assumption that Leonardo's rich treasure of manuscript notes was extensively studied and used during the sixteenth century. That bold hypothesis, which served to organize history of science as a field of inquiry and stimulated scores of able and willing workers in the field, must in-turn come under closer scrutiny — not, indeed, to call into question the validity of discoveries made under any working hypothesis, but rather to make sure that other possible trains of scientific thought, presently obscured, may not be historically more important in the actual emergence of modern science than the pattern of linkage traced or conjectured by Duhem and his followers.

There is no need to repeat here the story of the Leonardo manuscripts, their dispersion, and their ultimate repositories. It suffices to remark that, so far as their consistent theoretical content is concerned, relatively little could have been learned from them that was not to be found in earlier documents, even by a sixteenth-century student intent on creating a system of mechanics. Whether any scholar of the time envisioned such a science is questionable. Leonardo's reputation as a painter undoubtedly inspired some artists to consult, if possible, his manuscripts and to obtain drawings from them. The machines he designed failed to fire their imaginations as they do ours. A polymath like Girolamo Cardano may have taken the trouble to decipher Leonardo's mechanical speculations if those cryptic writings fell under his eyes, but the hypothesis that he did so is gratuitous in the light of his actual publications and the sources he acknowledged therein.

The transmission of old scientific ideas and the origination of new ones from the time of Leonardo to that of Galileo seems on the whole to have been conjectured rather than studied. For a proper study we need to learn more about the actual goals and methods of the men throughout Italy who wrote on mechanics from many quite different viewpoints — so many, indeed, that the whole culture of the sixteenth century in Italy promises to be a rewarding field of inquiry for historians of science.

The word "mechanics" covers several more or less distinguishable areas. The most obvious of these — that is, the actual construction and use of machines — belongs to the history of technology rather than to that of science. The question whether science has led to more advances in technology, or vice versa, is interesting in itself; but that is a question with which the present book is not directly concerned. Nor is it concerned directly with philosophical discussions of motion. Limiting ourselves to mechanics in the sense which that word has in physics — the study of statics, kinematics, and dynamics — we shall be concerned here only with attempts to describe and explain systematically the observable phenomena of equilibrium, of applications of force, and of manifestations of energy in the form of motion or resistance to motion.

Ancient and Medieval Traditions

In the science of mechanics in antiquity we may distinguish four traditions, to each of which a name may be assigned in order to facilitate our discussion of sixteenth-century writers. Two further traditions entered mechanics in medieval times.

First, in chronological order, was an Aristotelian tradition represented by the *Questions of Mechanics,* formerly ascribed to Aristotle and certainly belonging to the early literature of his school. This book dealt with a miscellany of problems drawn chiefly from common observation and having to do with motion and the application of force. Typical examples include the lever, the actions of oars and rudders, the strength of sticks and timbers, the use of wedges and pulleys, the construction of nutcrackers, and the behavior of solid particles caught in fluid vortices. All these problems were related to dynamics. Questions concerning the balance as discussed in the same work, though static in nature, were dealt with on

essentially dynamic principles. The discussions were, for
the most part, qualitative rather than mathematical.

Second in point of time was the Archimedean tradition,
in which physical postulates were introduced in order to subject mechanics to purely geometrical reasoning. In that way
Archimedes was able to deal mathematically with problems
of statics and of hydrostatics: notably with the balance, with
centers of gravity, and with the equilibrium of bodies on or in
water. The Archimedean tradition in mechanics is to be understood here as that which is represented by two specific works:
On Plane Equilibrium and *On Bodies in Water*. The Archimedean tradition differed markedly from the Aristotelian in
its emphasis on static rather than dynamic problems and on
the application to their solution of rigorously mathematical
demonstration rather than qualitative reasoning.

Two distinguishable traditions dominated the Alexandrian
period. Both are preserved in the writings of Ctesibius, Philo
of Byzantium, Hero of Alexandria, and Pappus, as well as in
treatises on the balance and on hydrostatics customarily (but
erroneously) attributed to Euclid and Archimedes by early
writers. Alexandrian works bearing on theoretical mechanics
I shall distinguish as the tradition of Hero; those bearing on
empirical rules or on the construction of actual machines
belong properly to the technological tradition, discussed
below. Hero's *Mechanics* was known only in part to European
writers before the nineteenth century, when an Arabic translation was discovered. Pappus, writing in the fourth century
of our era, could find it only in incomplete manuscripts, which
he preserved in his *Collections*. Hero's *Pneumatics*, which is
also of theoretical interest, survived intact in Greek. His
Automata and *Belopoeeca* survived also, but belong to the
technological tradition. For our purposes, the tradition of
Hero consists mainly of his *Pneumatics*, the eighth book of the
Collections of Pappus, and the pseudo-Euclidean and pseudo-
Archimedean treatises of the Alexandrian period.

The technological tradition embraces Alexandrian works
on practical mechanics and treatises such as those of Vitruvius on architecture and Oribasius on medical mechanics;
that is, writings on the construction and use of machinery
rather than on its theory. To this tradition, the sixteenth
century added a great deal. Examples are the books of Biringuccio and Agricola on machinery used in mining and metal-
lurgy; of Fontana, Agrippa, and Pigafetta on the moving of
the Roman obelisks; of Ramelli and others on machinery in

general. Unlike the other traditions with which we are concerned, the technological tradition is timeless; it is a living
and growing body of knowledge existing in all periods and
transmitted largely by oral instruction to engineers, military
men, builders, and artisans. Much of the technological tradition has always been accessible without recourse to written
sources.

One class of theoretical treatises that was to become of
central importance to the development of medieval mechanics
may have had its roots in the technological tradition in a somewhat special sense. Such are the treatises on the balance, the
steelyard, and hydrostatics that lay at the base of the medieval
science of weights. It is a striking fact that the earliest surviving manuscripts in Latin which deal with those topics belong
to the opening years of the thirteenth century. Among them
are some attributed by the scribes to writers of the Alexandrian
period, others belonging to the intervening Arabic literature,
and still others that may be European commentaries or original
treatises based on older lists of theorems. The rather sudden
and widespread interest in precision of weighing gave rise to
much activity on the part of medieval mathematicians, but
the interest itself may have been inspired by practical problems
of commerce and finance. The construction of accurate balances and the means of judging the reliability of weighing
devices constituted such problems; these had been solved
with precision in Alexandrian times and the solutions had
been preserved by traveling merchants and bankers among the
Arabs. Similarly the problems of determining the composition
of alloys of precious metals and the specific gravities of gems
by hydrostatic balances had been solved and preserved. Theoretical investigations of the science of statics and the attempt
to establish exact theorems on axioms and systematic proofs
appear to have been stimulated by the circulation of technical
manuals in the revival of European commerce and banking.

The two distinctively medieval traditions which concern us
here have been thoroughly discussed and illustrated in two
previous volumes of this series of publications in medieval
science, *The Medieval Science of Weights* (edited by E. A.
Moody and M. Clagett) and *The Science of Mechanics in the
Middle Ages* (by M. Clagett). Here it will suffice to describe
them briefly. One was the science of weights; the other was
the philosophy of motion.

The science of weights was an eclectic body of works dealing principally with problems of equilibrium but including

discussions of motion. It will be convenient to take a single name to symbolize the medieval science of weights, and the best is that of Jordanus Nemorarius, a writer of the early thirteenth century. Under the name of Jordanus we shall accordingly refer to the entire tradition embodied in a group of treatises that circulated widely in the Middle Ages, including some older works which had survived in abbreviated or mutilated form. The Jordanus tradition, however, must be regarded as medieval rather than classical.

The Jordanus tradition had one thing in common with the Aristotelian tradition of antiquity: its proofs were based on dynamic assumptions, or rather on postulates concerning motion. But the medieval science of weights was apparently independent of the ancient *Questions of Mechanics*. It is not difficult to explain the dynamic approach of the medieval writers, for they were quite familiar with the *Physics* of Aristotle, in which nature is defined in terms of motion and motion is closely linked to weight. It was therefore natural for them to draw on those ideas for the postulates of their science of weights, just as the writer of the *Questions of Mechanics* had done before them. But where he applied the ideas in discursive style as causal explanations, Jordanus and his followers molded them into postulates from which they could draw mathematical deductions in Euclidean form.

The Archimedean tradition was also reflected in the writings of Jordanus. Indeed, in one or two instances, theorems for which proofs were offered by Jordanus were genuine propositions of Archimedes, which had survived in apocryphal treatises or in fragments without proofs. But just as the Aristotelian and Jordanus traditions differed radically in form, so the Archimedean and Jordanus traditions differed radically in postulates. Paradoxical as it may sound, while the thirteenth-century writers had the genuine works of Aristotle but lacked the genuine works of Archimedes, they more nearly approached the Archimedean than the Aristotelian tradition in method. Their originality is shown by a striking difference between the key idea used in the Jordanus proofs and that which was foreshadowed in the *Questions of Mechanics*. The medieval writers assumed that the product of weight and vertical displacement must be equal at the two ends of a connected system, such as the lever. The ancient tradition linked speed and force rather than weight and distance, while it neglected the important distinction between vertical displacements and entire motions along the arcs of different circles. The Jordanus correction of this led to the ideas of work and of static moment, whereas the Aristotelian tradition suggested the ideas of force and of virtual velocities. The two approaches were not brought together effectively until the time of Galileo.

The other medieval tradition—the philosophy of motion—comprised a body of commentaries on the *Physics* of Aristotle in which logical and mathematical reasoning was applied to the classification and analysis of various kinds of actual or possible change. Those commentaries had important implications with respect to change of place, or local motion. It would be misleading to select the name of a single commentator to characterize this tradition, which contributed to sixteenth-century mechanics such important ideas as impetus, impressed force, and the mean-speed theorem. Though dominated by Aristotelian commentaries, it must be distinguished from the tradition in mechanics that I have called Aristotelian above. There are two reasons for the distinction. First, Aristotle himself did not write specifically on mechanics, but he did write on motion, and it was to those authentic writings that these commentaries were directed. Second, the commentaries were often in fundamental opposition to Aristotle's own doctrines, particularly with regard to projectile motion. Since Aristotelian orthodoxy exists in the *Questions of Mechanics,* his name is best applied to that tradition despite its later authorship. The body of doctrine on motion based on authentic works of Aristotle (but often incompatible with strict orthodoxy) may then be reasonably characterized for our purposes as the Peripatetic tradition.

In the universities of sixteenth-century Italy, it is certain that the medieval philosophy of motion continued to be discussed, but it is by no means certain that the traditions of Archimedes, Hero, or Jordanus occupied a place in university instruction. The absence of any known Latin translation of or commentary on the *Questions of Mechanics* before the sixteenth century makes it doubtful that that book ever received attention in the universities among philosophers. Both Galileo and his predecessor at the University of Padua, Giuseppe Moletto, lectured on it occasionally as professors of mathematics; consequently it is likely that their colleagues in philosophy neglected this work.

A considerable part of the development of mechanics in the sixteenth century took place outside the universities and had little or no impact on them. Of the four sixteenth-century Italians whose works are represented in the ensuing translations, only the last, Galileo, was ever a university professor. Neither Tartaglia nor Benedetti received a university education. Guido Ubaldo studied at the University of Padua, but he obtained his mathematical training elsewhere. As a general rule for the study of the history of mechanics in sixteenth-century Italy, therefore, one may reasonably expect a separation between the Peripatetic tradition (the philosophy of motion) and all the other traditions we have identified — the Aristotelian, the Archimedean, that of Hero, that of Jordanus, and, of course, the technological tradition.

The Older Traditions in the Sixteenth Century

Next it will be of interest to note the approximate times at which these various traditions were made accessible to the general reader outside the universities, and the manner in which they were disseminated. Only a fragmentary sketch can be offered here, based principally on published books. Much remains to be done, chiefly by the inspection of sixteenth-century manuscripts, to fill out our knowledge of the actual diffusion of interest in motion and mechanics. Clearly the publication of a Greek text had a much narrower field of influence outside the universities than did the printing of a Latin translation, and that in turn had less widespread influence than a vernacular translation. Every vernacular translation must have been inspired to some degree by a belief that reasonably wide interest in the text existed outside the ranks of men capable of reading Greek or Latin.

For the purpose of this sketch, publications of medieval and Renaissance commentaries on motion in the Peripatetic tradition will be neglected, as it is presumed that manuscript copies of those works were generally known and accessible to true scholars at all times, and that printed editions did little more than to make them better known and more easily accessible. Such editions had in any event begun to appear even before 1500, as had books in the technological tradition, which are likewise omitted from this section.

The chronological order of appearance of ancient and medieval traditions in mechanics during the sixteenth century was quite different from the order in which those traditions historically originated. First to appear in print was the tradition of Hero, set forth in 1501 in an encyclopedic work by Giorgio

Valla, *De expetendis et fugiendis rebus (Of things to be sought and things to be avoided)*. Hero's devices operated by water, air, and steam were portrayed and discussed in that book as a mechanical appendix to the section on geometry. Valla also gave some mathematical theorems of Archimedes, but did not describe his mechanical books. (Valla owned the oldest Greek manuscript of Archimedes then extant, from which all Renaissance copies ultimately derived.) A few of Hero's machines were again described by Girolamo Cardano in his *De subtilitate,* first published in 1550. In 1575 Hero's *Pneumatics* was published in a Latin translation by Federico Commandino, and in 1588 Commandino's Latin translation of the *Collections* of Pappus of Alexandria was printed, embodying the most complete account of Hero's *Mechanics* that was then known. In 1589 two Italian translations of Hero appeared: the *Pneumatics* as rendered by Giovanni Battista Aleotti and the *Automata* translated by Bernardino Baldi. To the latter work was appended Baldi's history of ancient mechanics and a description of some later machines. Another Italian translation of the *Pneumatics* was published in 1592 by Alessandro Giorgi.

The Aristotelian *Questions* was not printed in Latin until 1517, and in Italy not until 1525, when the *Questions of Mechanics* was published in Latin translation with brief commentaries by Niccolò Leonico. It was popularized in a paraphrase published by Alessandro Piccolomini at Rome in 1547 and reprinted at Venice in 1565. Cardano commented on several of the *Questions* in his *Opus novum . . .* of 1570. In 1582, Piccolomini's paraphrase appeared in an Italian translation by Oreste Biringuccio, who added an appendix concerning the simple machines based on Guido Ubaldo's *Mechanics.* Benedetti's commentary on the *Questions* is included among the present translations; it was published in 1585. An important commentary by Bernardino Baldi was composed about 1590, but was not published until 1621; like his translation of Hero's *Belopoeeca,* it probably was little known in the sixteenth century.

The Jordanus tradition was first represented in print in 1533, when Petrus Apianus published the *Liber de ponderibus* at Nürnberg. The work thus printed was inferior to a later version that had been corrected and expanded by an anonymous follower of Jordanus. Yet the 1533 edition sufficed to put in the public domain a text of the outstanding medieval writer on the science of weights, together with a thirteenth-century commentary. Tartaglia came into possession of a manuscript of the superior *De ratione ponderis* in 1539. On the verso of the title page of his Italian translation of Euclid, published in 1543, he mentioned the science of weights as a branch of mathematics on the same footing as optics, perspective, astronomy, music, and the other traditional "mixed sciences" using mathematics, and in the preface he remarked that the science of weights grew out of practical mechanics "as Jordanus demonstrates in *de ponderibus*." Merging the latter work with bits of the pseudo-Euclidean and pseudo-Archimedean treatises in the Hero tradition, and making some modifications of his own, Tartaglia gave additional currency to the science of weights in his *Quesiti* of 1546, this time without mentioning Jordanus. It was not until 1565 that the text of *De ratione ponderis* (together with the previously mentioned text attributed to Archimedes) was printed from papers left by Tartaglia to his publisher, Curtius Troianus.

The Archimedean tradition in mechanics, comprising the books *On Plane Equilibrium* and *On Bodies in Water,* was first printed in Latin at Venice in 1543. Tartaglia published the work; the translation was that of William of Moerbeke, completed in 1269, and the printing was done from a copy containing many errors. The first Greek text of Archimedes (lacking *On Bodies in Water*) was published at Basel in 1544 with the Renaissance Latin translation of Jacobus Cremonensis. In 1551, Tartaglia published an Italian translation of and commentary on the first book of Archimedes' *On Bodies in Water.* The first critical Latin translation of several mathematical works by Archimedes was that of Federico Commandino, published in 1558 with scholarly commentaries. Another translation, by Francesco Maurolico, was in preparation about the same time at Messina, but it remained unpublished until late in the seventeenth century. In 1560 Tartaglia's Italian translation of Archimedes' *On the Sphere and Cylinder* was published in his encyclopedic work on mathematics. Two editions of the Moerbeke translation of both books of *On Bodies in Water* (of which the second book had not appeared in earlier editions) were published in 1565, one at Bologna by Commandino, carefully edited and reconstructed, and one at Venice from the posthumous papers of Tartaglia, who had not succeeded in editing it properly. In 1588, Guido Ubaldo published his paraphrase of and commentary on the work of Archimedes, *On Plane Equilibrium.*

The Two Italian Schools of Mechanics

Several of the men who revived, translated, or commented on older works belonging to the four main traditions discussed above were also active in producing works of their own on mechanics and in instructing others in that subject and in mathematics. These men may be divided into two clearly recognizable groups, distinguished on the one hand by geographical location and on the other by differing concentrations of interest. The significance of and the reasons for that separation of groups must remain for further research. Both groups are of great importance to the history of mechanics in Italy.

The first group flourished in northern Italy and included Tartaglia, Cardano, and Benedetti, all of whom were conspicuously interested in practical aspects of mechanics. The second group developed in central Italy and included Commandino, Guido Ubaldo, and Baldi. That group concentrated its interest on works of classical antiquity and on the rigorous application of mathematics to mechanics.

Certain facts concerning the two separate groups are striking. In both, the Archimedean revival was of central importance. Both groups developed outside the main stream of conventional university education. Both converged in direct lines on Galileo, though perhaps not in quite the way that is ordinarily believed. Yet no member of either group seems to have had a good word to say about any member of the other. Indeed, the two groups appear to have been scarcely aware of one another's existence, though all the men concerned were working actively and publishing in a new and exciting field of study. On the surface it might appear that they were held apart by the seemingly opposed objectives of applied mechanics in the north and theoretical mechanics in central Italy, while they had in common only the Archimedean tradition. Since Archimedes himself had been no less skilled in applied mechanics than in pure mathematics, there would be nothing startling in such a division. But as we shall see, they had much more than Archimedes in common, and their objectives were not in as clear-cut opposition as it may seem.

Again, it might appear that while the northern group was trying to create a science of dynamics, the central Italians were striving to perfect the theory of statics. If that were true, it might account for Tartaglia's acceptance of the Jordanus tradition and Guido Ubaldo's rejection of it. But the dynamic

tradition embodied in the *Questions of Mechanics* found a
patient expositor in Bernardino Baldi, whereas Tartaglia (and
Benedetti) considered that work to be seriously defective.
Moreover, the tradition of Hero, that ingenious constructor of
machinery, found its main welcome in central Italy rather
than among the more practically oriented writers in the north.

A key to some of these puzzles is to be found in the underlying attitude of each group toward Aristotle himself. In the
northern group, Tartaglia and Benedetti were outspoken critics
of Aristotle, a pattern in which Cardano did not follow them.
For Tartaglia, if not for Benedetti, the Jordanus tradition had
replaced the Aristotelian. Now the central group paid so little
overt attention to Aristotle that one might take it to have been
hostile to him in favor of Archimedes, or at the most to have
been indifferent to the Aristotelian tradition. But that was not
the case at all, as may be seen from the following passage in
Baldi's "Life of Archimedes," composed about 1590 but first
published in 1887.

Shortly after this [the time of Eudoxus] came Aristotle, whose divine intellect would omit nothing. He also labored on the causes
behind the marvellous effects of machines, in that book that is accordingly called the [*Questions of*] *Mechanics*. He revealed in that
work many important things necessary to the understanding of the
effects of which we are speaking. But mechanics being a physical subject, and nevertheless demonstrable by mathematical reasoning, it
seems that Aristotle, leaving aside the mathematical aspects, preferred to draw his demonstrations from [his] physical principles, which
are of such force that, [if] accompanied by an understanding of
mathematics, they are capable of bringing a complete and whole
theory [*dottrina*] into the actuality of machines. Archimedes thus
seeing—as is likely, and as Guido Ubaldo also seems to believe in
the preface to the first book of his [paraphrase of Archimedes on]
Plane Equilibrium—that this book of Aristotle's was quite sound in
its principles, though only implicitly so and not quite clearly, the
adding of mathematical proofs to the physical principles [would]
render it plainer and smoother and bring it to more specific conclusions [*cose*]. Thus when Aristotle discloses the reason for which the
lever moves a weight more easily, he says that this happens because
of the greater length on the side of the power that moves; and this
[accords] quite well with his first principle, in which he assumes that
things at the greater distance from the center are moved more easily
and with greater force, from which he finds the principal cause in the
velocity with which the greater circle overpowers the lesser. So the
cause is correct, but it is indeterminate; for I still do not know, given
a weight and a lever and a force, how I must divide the lever at the
fulcrum so that the given force may balance the given weight. Therefore Archimedes, assuming the principle of Aristotle, went on beyond him; nor was he content that the force be on the longer side of
the lever, but he determined how much [longer] it must be, that is,

with what proportion it must answer the shorter side so that the given force should balance the given weight; and thus he discovered and proved that which is the basis and foundation of every necessary power [*facoltà*]; that is, that the arms of the lever must be inversely as the force to the weight. . . . These things he found and proved most acutely in the first book *On Plane Equilibrium*, which, as Guido Ubaldo notes, is the book of elements of the entire mechanical domain [*genere*]. He thus shows in the preface of this book that Archimedes followed entirely in the footsteps of Aristotle as to the principles, but added to them the exquisite beauty of his proofs.

Thus Guido Ubaldo's rejection of Jordanus (discussed further, below) arose not from a difference in program but from a conviction that the program could be properly carried out in only one way. He would accept the Aristotelian tradition, but only as an original source for the physical postulates of the Archimedean tradition. This was a curious theory of the history of mechanics, but it did manage to relegate everything good to classical antiquity. The Jordanus tradition had also sought a reconciliation between Aristotle and Archimedes, but, lacking the authentic works of either in mechanics, Jordanus attempted to establish postulates from Aristotle's *Physics* by means of which various theorems on equilibrium, some of which were propositions of Archimedes that had survived without proofs, might be demonstrated. The central Italian mathematicians, in possession at last of the authentic works of Archimedes, turned their backs on the less rigorous attempts of the medieval writers to link dynamics with statics. In so doing, they made it very difficult for themselves to go beyond the ancients in mechanics.

It is easy to see the grand continuity of a landscape from a great distance, or that of history taken several centuries at a stride. Closer inspection causes us to lose sight of the continuity, but often provides further information by means of which we may ultimately understand both the long-range continuity and the small-scale interruptions. Viewed in a perspective of twenty centuries, the passage from Aristotle, Archimedes, and Hero to seventeenth-century mechanics may appear inevitable. In a four-century perspective, the passage from impetus theory in Buridan to inertia in Galileo presents no problems. Viewed more closely, however, the simple notion of an orderly (if interrupted) transmission breaks down. Much remains to be filled in, and nowhere more than in the sixteenth century, decade by decade.

For a proper understanding of the actual evolution of mechanics in the sixteenth century, we must go far beyond the

program of Duhem. A simple search for possible sources and
a general assumption that liberal borrowing (or outright theft)
took place wherever possible is not enough. For one thing,
too many men were writing on mechanics from quite different
viewpoints. For another, it is the nature of mechanics that
results will often be independently duplicated; the phenomenon of simultaneous scientific discovery in more recent times
abundantly illustrates that point.

We shall need explanations for facts like these: Tartaglia
accepted from Jordanus the correct inclined-plane rule; Guido
Ubaldo rejected it for an erroneous theorem of Pappus;
Cardano rejected both Jordanus and Pappus in favor of an
incorrect rule of his own. Why did two men support erroneous
theorems after the correct one had been widely published?
Why did one of them favor an ancient error and the other a
novel fallacious approach? Why did neither of them resort to
a simple actual test, and why did the other writer not adduce
experience in support of his correct rule? For answers to such
questions we shall have to search into traditions, methods,
goals, and even personalities. But in return for our trouble we
shall learn more about the history of ideas in mechanics than
the mere dates and forms of their earliest (surviving) appearances in manuscript or print and the names of men to be
credited.

Correlation of scientific traditions with actual methods and
goals may in turn help us to understand the interests and
activities of the men who wrote on mechanical topics in the
sixteenth century. Different methods in achieving similar
goals may explain the changing emphasis on older traditions.
Prevailing traditions inside and outside the universities need
examination and explanation. Actual interests and pursuits
of individuals, as well as groups, throw light on changes in
pace of scientific progress. Keeping such possibilities in mind,
and with a tentative list of traditions before us, let us now turn
to the biographical examination of our central figures and their
associates in the two chief Italian schools of mechanics in the
sixteenth century.

Niccolò Tartaglia

Niccolò Tartaglia was born at Brescia in 1499 or 1500. His
father was named Michel; he was a post-rider and was called
Micheletto from his diminutive stature. No more is known of
him except that he died when Niccolò was still a small child.

It is usually said that the family name was Fontana, because Tartaglia in his will mentioned a "brother" of that name. Even that is not decisive, however, for the usage of the time makes it possible that this was a half-brother. The early death of Niccolò's father and the evidence that Niccolò had no surviving relatives at Brescia when he returned to that city for a time in 1548–49 make it probable that even his true family name has been lost, despite the clue of the will.

The nickname *tartaglia* ("stutter") was given to Niccolò because of a permanent speech defect resulting from sword cuts he received from French soldiers during the sack of Brescia in 1512. His mother had taken him into the basilica for sanctuary, but he was nevertheless mutilated and left for dead, one cut splitting his jaw. His mother preserved his life, and he took the nickname as a badge of honor and a tribute to her loving care, signing himself Tartaglia (or Tartalea) in all his published works and letters. Niccolò received no formal education, according to his own account, except from a writing-master who taught him the first half of the alphabet. The balance he learned by himself when poverty prevented the payment of further fees for his education. He learned to read Latin sufficiently for his needs, but (except for one brief and unfortunate preface) he wrote only in Italian, and in such a rough and idiomatic style that it often moved the classics scholar Bernardino Baldi to mirth. Whether he knew any Greek is problematical.

Most biographical material concerning Tartaglia must be gleaned from his books. From Brescia he moved to Verona about 1521 and remained there until 1534, giving lessons in mathematics to private pupils. During this period he learned to solve certain forms of cubic equations. In 1534 he moved to Venice, and during the next year he mastered two more difficult cases of the cubic equation. In 1535 he entered into a celebrated contest with a rival mathematician and won decisively by his superior algebraic skill.

In 1537, Tartaglia published his first book, the *Nova Scientia* or *New Science*, which dealt principally with the trajectories of cannonballs. The engraved title page to the book presents us with a most interesting allegory from which we can clearly deduce the author's conception of the relations among mathematics, mechanics, and philosophy, as well as the positions of other sciences. I cannot improve on the description of it written by Professor Edward W. Strong three decades ago in *Procedures and Metaphysics:*

Disciplinæ Mathematicæ loquuntur
Qui cupitis Rerum varias cognoscere causas.
Discite nos: Cunctis hac patet una via.

The relation of mathematical disciplines to the sciences and the standing of Aristotle and Plato in respect to them is delightfully pictured in the frontispiece to the *Nova Scientia Inventa da Nicolo Tartalea. . . .* We are shown an enclosure encircled by a high wall. Steps are depicted mounting up to a single gate guarded by a venerable man entitled *Euclide.* At the foot of the ladder on the terrain outside the wall stand two gentlemen. One gestures toward the discipline of the stairs with its geometrical guardian, but his companion appears to be reluctant. Only by this discipline is one admitted to the sciences within the enclosure. At some distance down the wall is an individual trying to gain admittance by his own ladder, but the wall is drawn much too high or the ladder much too short to admit anyone who fails to pass *Euclide.* Perhaps the gentleman who hesitates to learn geometry and the one who tries to gain admittance by other means fail to recognize the nature of the sciences represented within. These are portrayed as a chorus of women each bearing in one hand the label of a science. In the fore and flanked by the array stands Niccolò Tartaglia. On his right hand are *Aritmetica* and *Musica*; on his left are *Geometria* and *Astronomia*; in the rear are *Hydromantia, Geomantia, Architectura, Astrologia, Cosmografia, Necromantia, Presti[di]g[itat]io, Sortilegio,* and others. Passing from the right side of the enclosure, where these symbolic figures stand facing the gate, we see on the left side two fieldpieces firing, the trajectory of the balls being plainly indicated. This mathematical diagram relating to the new science is just behind the backs of two neophytes in this scientific world. They gaze timidly or perhaps respectfully upon Niccolò and his vanguard.

Diametrically opposite the entrance admitting to the sciences by way of *Euclide* is another set of steps leading up into a smaller enclosure, at the far end of which enthroned in solitary state sits *Philosofia* herself. Near the foot of the flight of stairs stands Aristotle, who is thus nearer to the realm of science than is Plato. Plato stands at the head of the steps by the entrance to the inner sanctum and is consequently farther removed from the affairs on the plain of science. Plato bears in his hand a banner inscribed NEMO HUC GEOMETRIE EXPERS INGREDIATUR. Yet although this banner announces that no one enters without a knowledge of geometry, Tartaglia does not treat geometry nor mathematics in general as a preparation for philosophy. The "divine" Plato is beyond science, but only those enter into Philosophy who can salute that banner.

Tartaglia's "new science" was inspired by a practical problem of gunnery presented to him by a professional artilleryman at Verona in 1531. In this earliest attempt to apply mathematics to the motion of projectiles, the pioneer scientist satisfied himself that he had solved the problem, but was then moved by moral scruples to suppress its publication. When a religious element entered a few years later, however, his moral views changed entirely and he printed his theory of gunnery. The story, curiously prophetic of modern events, is set forth in the letter of dedication to the *Nova Scientia.*

It is noteworthy that in Tartaglia's first book there is no hint of the theory of impetus, which developed in the Middle Ages, but there is a strong element of orthodox Aristotelian physics in the description of natural and violent motions, in their supposed incapability of mixture, and in Tartaglia's animistic account of acceleration. All the medieval discussions of the proportionality of speed to time elapsed or to space traversed are lost on Tartaglia, the self-taught theorist; he believed that a falling body hastens as it approaches its natural resting place, just as a returning traveler hastens his homeward way. But at the same time Tartaglia, the observer, was not really convinced that natural and violent motions cannot mix; his illustrations and his text both show that he believed the actual trajectory to be curved at all points. In this first stage of Tartaglia's thought, his approach to dynamics is a mixture of ancient and novel ideas uninfluenced by medieval speculations on motion, impetus, or the science of weights.

In 1538 the Duke of Urbino, who had been retained by the Venetian government to defend its territory against a threatened Turkish attack (which never materialized), talked to Tartaglia about various matters discussed in the *Nova Scientia,* which had been dedicated to him. In the light of those discussions, as reported in the *Quesiti* of 1546, the duke promised to carry out experiments on his return to Pesaro. Pesaro was later the home of Guido Ubaldo, and with Urbino it became a principal seat of the central Italian mathematical school founded by Federico Commandino. Thus a genetic relationship may have existed between the two rival centers of sixteenth-century thought in mathematics and mechanics, the practical school being prior to but perhaps to some extent responsible for the purist school. Commandino, founder of the latter, was closely associated with the son of the Duke of Urbino by whom Tartaglia was consulted; the son succeeded the father in military service to the Venetian Republic. Commandino was also teacher of mathematics to the succeeding Duke of Urbino, to Guido Ubaldo, and to Bernardino Baldi.

In 1539 the physician, astrologer, and mathematician Girolamo Cardano (1501-76) wrote to Tartaglia inviting him to visit Milan. Tartaglia did so, and a correspondence between the two men ensued in 1539-40. Cardano, having heard of Tartaglia's victory in the mathematical contest of 1535, induced him under vows of secrecy to impart his method of solving certain algebraic equations of the third degree. This knowledge was Tartaglia's most carefully guarded secret, which he intended to publish in time as the jewel of an original work on mathematics. In 1541, when his English pupil, Richard Wentworth, asked him for the formula, Tartaglia withheld it, saying (as reported in the *Quesiti*): "My friend, certainly there is nothing that would gratify you which I ever could or should deny you, because of my deep obligations to you. But ... I have decided, as soon as I have finished translating Euclid and correcting the diagrams and other errors made by the scribes and translators of Archimedes of Syracuse, to compose a work on the practice of arithmetic and geometry, and together with it a new algebra, in which I shall put not only the rules found by me for those equations with all their reasons and foundations, but many other things as well. . . ."

Tartaglia's editions of Euclid and Archimedes were published in 1543, but in the same year he suffered a severe personal blow in the loss of nearly his entire family. And before Tartaglia composed his next book, Cardano broke his vow of secrecy, publishing the *Ars magna . . .* in 1545 and including there the information he had obtained from Tartaglia concerning cubic equations.

Tartaglia's Italian translation of Euclid—the first published translation of the *Elements* into any living language of Europe—was an event of great importance to the progress of mechanics, and indeed of all applied sciences. For the first time the principal treasury of rigorous mathematical reasoning was opened to men who knew neither Greek nor Latin. The implications of that event for the science of mechanics were great because literacy in Italy was very high, especially among engineers and artisans. Those men were by no means always university trained, as evidenced by the fact that as late as 1606 Galileo wrote his book on the proportional compass in Italian, precisely in order to make it useful to engineers and military men. The few propositions of geometry in daily use among surveyors, architects, and engineers were, of course, widely known, but the opportunity for such men to learn general propositions based on rigorous mathematical deduction (rather than on use alone) was a new thing. Tartaglia made it the aim of his translation to provide explanatory comments throughout, and on the title page he declared that the book would render geometry easily comprehensible to anyone without undue effort. Tartaglia's translation was made not from Greek but from the two existing Latin versions (those of

Campanus and Zamberti), and he supplied in the margins the numbering of propositions used in those editions, which were not always in agreement with each other or with his own.

Publication in Latin of Archimedes' *On Bodies in Water* even in part, especially with the addition of the work *On Plane Equilibrium,* was a still greater event in the history of mechanics. Previously the only works of Archimedes to have been published were mathematical treatises on the circle and the parabola, also reprinted by Tartaglia in 1543. Now the two Archimedean works relevant to mechanics were available for the first time, and even after the Basel edition of Archimedes was published in 1544, the work *On Bodies in Water* was accessible only in Tartaglia's edition.

But here a cloud hangs over Tartaglia. Instead of being praised by historians of science for opening the Archimedean revival of the sixteenth century, he is charged with having passed himself off as the translator, when in reality he merely published, with a few comments and new diagrams, the medieval translation of William of Moerbeke. Now, that charge cannot be entirely answered, but it can be reduced to a large extent. Tartaglia frequently mentioned this work elsewhere without claiming the translation as his own. One part of it had, in fact, been previously printed by Luca Guarico. In 1541 Tartaglia spoke to Wentworth not of translating, but only of amending some diagrams and some scribal and translator's errors in Archimedes. In 1546 he published that statement in the *Quesiti,* repeating it in the second edition of 1554. The title page of the 1543 Archimedes, moreover, makes no explicit claim that Tartaglia is the translator, though a Greek copy is mentioned there as defective in its diagrams. Tartaglia's part in the work is described only as that of publishing it with corrections, commentaries, and revised diagrams. Later, in 1551, when he published an Italian translation of the first book of Archimedes' *On Bodies in Water,* Tartaglia referred to the 1543 translation not as his, but merely as published by him; he acknowledged its lack of clarity and even accuracy; he spoke of it in disparagement, and he called it "the said translation" when the normal phrase would have been "my translation" had he intended to claim it. Yet despite all this, the 1543 title page mentions a Greek copy, and the letter of dedication (to Richard Wentworth) begins by saying that Tartaglia had run by chance into a text of Archimedes *manu Graeca scripta,* in poor condition and hardly to be read. He goes on to say that he much desired a translation *nostram*

in linguam, which ought to mean "in Italian," the language
that he used with Wentworth as a pupil. I should think that
here he was speaking of an Italian translation he projected,
similar to his translation of Euclid. But he then talks of the
1543 Latin translation as having been carefully amended
through his diligent studies, in terms which would leave any
reader with the impression (though a flat assertion is always
avoided, and the word "Latin" never occurs) that the entire
work being dedicated was his own production. The fact that
Tartaglia does not identify it as Moerbeke's is, of course, irrelevant; he could not have known whose it was. But he could,
and should, have said clearly in the dedication that it was not
his own.

Tartaglia's manuscript of the Moerbeke Archimedes may
have been taken to him in the autumn of 1539 by Don Diego
Hurtado de Mendoza, when Mendoza went to Venice as ambassador of Charles V of Spain, remaining until 1546. The
manuscript from which Tartaglia published the 1543 edition
and from which he copied the Jordanus *De ratione ponderis*
and the pseudo-Archimedean *De ponderibus* is preserved
at the National Library in Madrid. Mendoza was a pupil of
Tartaglia's, and it was to him that Tartaglia expounded the
Aristotelian theory of the balance and the medieval science
of weights. From Tartaglia's letters to Cardano it appears
that he was at work on his translation of Euclid in August,
1539, and that he then knew Mendoza. It was early in 1540
that he first mentioned the work of Archimedes, which in
1541 he again mentioned to Wentworth, complaining of the
errors of scribes and translators. Since it was to Mendoza
that he expounded the science of weights, using for that
purpose treatises that are bound together with the Madrid
manuscript, it is possible that Mendoza brought the manuscript
to Venice, or acquired it there; that he took it to Tartaglia
for discussion and explanation; and that he returned to Spain
with it after allowing Tartaglia to use it and perhaps to copy
from it. But there is also internal evidence that the same manuscript, or an extraordinarily faithful copy of it (errors and all),
was used at Venice in 1565 for the printing of Archimedes'
On Bodies in Water (of which the second book had been omitted in the 1543 edition).

Cardano's publication of the solution of cubic equations in
1545 caused Tartaglia great distress of mind. In 1546 he published his *Quesiti,* drawn from a variety of discussions he had
held over a period of many years with pupils and with visitors.

In the *Quesiti* he published his side of the Cardano affair,
which in essence was never disputed by Cardano. Cardano
had acknowledged in his *Ars magna* that Tartaglia was his
source of certain information concerning cubic equations.
Tartaglia revealed the confidential nature of that communica-
tion. Instead of receiving an apology, however, he found himself the subject of attack.

Cardano's pupil, Ludovico Ferrari, in a series of published
letters, defended his teacher against Tartaglia's exposures
in the *Quesiti*. One of Ferrari's retorts against Tartaglia's
documentation of Cardano's plagiarism was to say that Tartaglia himself had plagiarized the entire treatise of Jordanus
without so much as mentioning his name. To this Tartaglia
replied that, among mathematicians, little credit is given for
propositions alone, as it is the demonstrations that count. He
said that he had supplied proper proofs for the science of
weights, which Jordanus had failed to do. This is generally
regarded as a lame excuse. But it is seldom pointed out that
the science of weights was already in the public domain.
Apianus had published *De ponderibus* in 1533. Tartaglia, in
his 1543 Euclid, had spoken of the science of weights in the
same terms as optics or perspective, and he had named Jordanus as its founder. Now, the science of weights as published
by Tartaglia in 1546 was, in fact, drawn from at least three
medieval treatises, not just that of Jordanus, and it differed
from all of them in the presentation of assumptions and definitions. The order of propositions is different, as is the wording
of several proofs. In view of all this, the omission of the name
of Jordanus does not seem reprehensible or even unusual.
Tartaglia cannot have meant to claim the science of weights
as his own invention after describing it as a standard branch
of mathematics in his translation of Euclid. Nor is any claim
made in the *Quesiti* that the science of weights or any part
of it presented there is Tartaglia's own. Finally, Tartaglia
willed to his printer a copy of the Jordanus work in which the
few corrected or added diagrams were marked with Tartaglia's
name, and this was published posthumously in 1565. To will
his edited copy of Jordanus to a publisher would certainly
have been a curious way to perpetuate the false claim to
authorship with which Tartaglia is often charged.

Tartaglia had now given the world its first vernacular
Euclid, the first published Latin text of the two mechanical
treatises of Archimedes, and the first printed text of many
important propositions in the medieval science of weights.

But he was by no means through with his contributions to the development of mechanics. In 1551 he published a book on the raising of sunken vessels, a practical mechanical treatise of particular interest to the Venetian state as a maritime nation. To this he appended his previously mentioned Italian translation of the first book of Archimedes' *On Bodies in Water*. As will presently be seen, that publication may well have been the stimulus of an extremely important event in the history of mechanics, namely, the first open challenge to Aristotle's law of falling bodies. Finally, Tartaglia published an Italian translation of the first book of Archimedes' *On the Sphere and the Cylinder;* this appeared in his last book, an immense treatise on mathematics. Tartaglia died at Venice in December, 1557. Among his posthumously printed works was the first experimental table of specific gravities of materials.

The principal ideas of lasting value that were either introduced by Tartaglia or given a clear place in mechanics through his treatment include the following. In ballistics, he declared that the trajectory of a projectile is curved at every point and does not consist simply of straight lines and circular arcs. He stated that the maximum horizontal distance covered by a projectile is achieved by firing the artillery piece at an elevation of 45 degrees, and that any intermediate distance can be reached by firing at either of two complementary angles. He made it clear that resistance of the air is a real practical problem, against Aristotle's theory that the medium moves the projectile, and accordingly he introduced the principle of limiting discussion to bodies of such density and shape as to minimize the factor of air resistance. In statics, he attempted to give mathematical rigor to the demonstrations of his predecessors, chiefly by the addition of postulates of a physical character concerning motion to supplement the definitions and axioms found in medieval writings. The results are generally far from rigorous, but in every instance they afford some improvement. The most interesting case is that of the celebrated theorem of Jordanus concerning equilibrium on inclined planes. That theorem is quite correct, and its statement contains a principle of the highest importance for the development of mechanics. The proof offered by Jordanus is, however, liable to serious objections. Tartaglia attempted to obviate those objections without altering the proof more than he believed necessary. In our time it is difficult to discuss this proof objectively because so much has been made to hinge on

it, especially by followers of Duhem's program. But before
that program was inaugurated, the proof as given by Jordanus
had been criticized by Rafaello Caverni, who considered
Tartaglia's revisions of it in the *Quesiti* to have created the
first valid demonstration. Amusingly enough, Caverni (writing in 1895) mistakenly ascribed both the original proof and
its correction to Tartaglia, so that his judgment was not
affected by a knowledge that the former was of medieval
origin.

Tartaglia's influence on mathematics and mechanics was
very widespread. Many of his pupils were foreigners. Some
of his books were translated into English, French, and German
during the sixteenth century, and others in the seventeenth
century. His theory of trajectories was preserved in military
books long after the theoretically correct (but practically
inapplicable) parabolic law of trajectories had been set forth
by Bonaventura Cavalieri and Galileo.

Within Italy, Tartaglia's most important pupil was Giovanni
Battista Benedetti, though that relationship was brief and
Benedetti was later critical of Tartaglia's work. It is also
believed that Ostilio Ricci, Galileo's teacher of mathematics,
had been a pupil of Tartaglia's.

Girolamo Cardano

Girolamo Cardano has already been mentioned as a rival
of Tartaglia's in mathematics. His place in the history of
mechanics belongs with the north Italian school, with its
orientation toward practical applications and unsystematic
but original speculations. Cardano's writings on mechanics
form but a small part of his works, which were chiefly medical
and astrological. Of Cardano's varied and often conflicting
views on motion and mechanics, I shall mention only those
which appear to me likely to have been influential on Galileo
in his early writings, or to have been novel and potentially
useful to later writers.

The sources of Cardano's ideas on mechanics are contro-
versial. It was Duhem's conviction that he borrowed liberally
from the manuscripts of Leonardo da Vinci. His father was a
friend of Leonardo's, and the manuscripts reposed for a long
time near Milan, where Cardano spent much of his life. It is
possible that he examined them, though the evidence offered
appears very dubious to me. Cardano specifically cited Archimedes, Ctesibius, Hero, Vitruvius, and some of his own con-

temporaries, but his works also show him to be basically an orthodox Aristotelian. His brief and rare references to Leonardo are inconclusive; the most that I can infer from them is that he had probably read the *Treatise on Painting,* which circulated in manuscript. I have not noted in his works any references to Jordanus, but that tradition was certainly known to Cardano, at least through Tartaglia's *Quesiti* and his pupil Ferrari's controversy with Tartaglia over it. He commented on many of the *Questions of Mechanics,* and questioned its authenticity as a work of Aristotle. To Cardano alone of the principal writers discussed here, the Archimedean revival seems not to have been of central importance. He discusses the Archimedean screw and the principle of buoyancy only in a practical context. In his theoretical discussions, the concept of centers of gravity plays but a minor part. Cardano was doubtless widely read in the Peripatetic tradition, but his only notable departure from Aristotelian orthodoxy is the concept of impetus, of which he makes extensive use. Of all writers on mechanics before Galileo, Cardano appears to have had the widest knowledge of and interest in the various traditions of the past; had he concentrated his energy on mechanics systematically, a fruitful synthesis of past traditions might well have emerged.

Problems of statics, chiefly concerning the balance, occupy the final quarter of the first book of Cardano's *De subtilitate,* first published at Nürnberg in 1550 and often reprinted in Italy and abroad. His discussion there of the balance shows little originality; like the other northern Italians, he was more interested in machines. Among those described in *De subtilitate* are the siphon, furnaces, the Archimedean screw, and devices suggested by Hero. Though this book was widely read and may have popularized the study of mechanics, there is little in it that could have advanced the science. The same may be said of his *De rerum varietate.*

In the *Opus novum de proportionibus* . . . of 1570, however, there is much of interest and ingenuity. Some questions of statics are taken up with great insight, but the novelty of the work lies in its discussions of problems of motion. The possibility of unifying statics with dynamics, or at least of mathematically connecting the two disciplines, seems to have captured Cardano's imagination. The influence of this book was probably limited, since it was printed but once, at Basel, and may not have circulated widely in Italy. Unlike *De subtilitate,* it is certainly a very rare book now; I have had to rely on the text of the collected edition of Cardano's works published at Lyons in 1663, which is defective both as to language and as to the important diagrams.

Cardano's account of acceleration is wholly Aristotelian; he remarks that in natural motion the body has an appetite to approach some end, whence the end must be good, and therefore the body hastens as it approaches the end. He holds that since the medium is divided and driven aside beneath a falling body, it must force upward with it the neighboring parts of the medium. Those parts then press in above the body to prevent the formation of a vacuum, and in so doing they press down on the body and speed its motion. To this concept of antiperistasis he then adds that in both violent and natural motions there is an increase in speed at least up to some point, by which he explains the need in war machines for space through which to act in order to increase the violence of their projectiles. In one proposition, acceleration is linked to time, but Cardano's reasoning for this depends again on antiperistasis.

In discussing the motion of projectiles, Cardano asserts that motion in some part of the horizontal (initial) path is uniform, and he says that, as the path turns downward at the end of that part, the projectile is slowed; hence he believes that it will reach the ground later than it would have reached the corresponding point on the initial horizontal line. This idea is consistent with his argument elsewhere that there is always a conflict between motions of different kinds, rather than a simple composition. But despite its overall Aristotelian orthodoxy, Cardano's discussion of projectile motions is of interest because of his clearly expressed view of speed as a ratio of space to time. This concept, inspired by Cardano's algebraic approach to mathematics, was never grasped by Galileo or his contemporaries.

Still more striking is Cardano's classification of motion into three kinds rather than two: natural, violent, and "voluntary." Voluntary motion is exemplified by circulation of the celestial spheres around the center of the universe; other circular motions, for Cardano, are either violent or mixed motions. In voluntary motions, the body as a whole remains in one place. Cardano considers such motions to be uniform and to be simpler than other motions. This discussion by Cardano is a probable source of Galileo's reflection, added as a note to his *De motu,* concerning neutral motions. That reflection, discussed in the final section of this introduction, led ultimately

to the inertial concept. But Cardano remained faithful throughout his works to impetus theory — possibly an example of the
well-known importance of terminology in science, for the word
"voluntary" has animistic implications, whereas "neutral"
suggests indifference to motion.

Cardano's impetus discussions led him to some potentially
fruitful reflections concerning weight and speed, and, though
these turned out to be tautologous rather than physical, they
may have been turned to good account by later writers who
were able to discard the Euclidean theory of proportions and
apply algebra to physical concepts.

Another valuable and probably original concept of Cardano's is that of concealed motion in a resting weight (*occultus
motus quiescendo*). That concept he repeatedly applied, not
only in a sense in which it adumbrates a sort of potential
energy, but also in an action-reaction sense similar to that of
Galileo's *De motu*. Cardano was aware that a sphere on a
horizontal plane could be moved by any force sufficient to
divide the surrounding air. Like Nicholas of Cusa, he limited
this argument to the sphere, whereas Galileo later extended it
to horizontal motions of any body. Yet Cardano's proposition
is linked in another way to Galileo's, for both men use the
idea of constant distance of the moving body from the center
of the world. Cardano also speculated on the possibility of
finding, between the tangential plane to the earth and the
earth's spherical surface, a plane which would appear to decline and yet on which a heavy body would not spontaneously
move. But he remarked that this would be hard to accomplish,
especially for short distances.

Cardano's treatment of the inclined plane is curious. Accepting neither the correct theorem of the Jordanus tradition nor
the incorrect theorem of Pappus, he offers a proof that the
effective weight of a body on any inclined plane is proportional to the ratio of the angle of the plane to a right angle. It
is interesting that in the course of this proof he remarks that
it is a matter of common knowledge that no (appreciable)
force is required to move a body horizontally. Yet in none of
these views is he truly consistent, for elsewhere (using an
argument that seems to assume speed on an inclined plane
proportional to effective weight on that plane) he declares that
the speed is not proportional to the angle, but increases more
rapidly than the angle. And in several propositions he discusses the force needed to draw or push a body along the
horizontal, relating this to the shape of the body and the posi-

tion of the applied force. Still more curious is the proposition
immediately preceding that of the inclined plane, a proposition
that Duhem takes to refer to a screw-jack, though the diagram
is hard to reconcile with that. But whatever its application,
the proposition certainly relates the power used in raising a
weight to the length of the path over which it is moved, a correct relation which is ignored in the inclined-plane proposition
that follows.

Cardano offers many propositions concerning the speed of
fall. I shall mention the last of these first, as it is the best
known, most interesting, and most puzzling of all. In it he
seems to assert that all spheres of the same material falling
from the same place through air will reach the horizontal plane
at the same time. His argument is difficult to follow, partly
because of the typography, for in two places *inaequali tempore*
must be read as *in aequali tempore;* but the gist of it seems to
be this. If one sphere is triple the other, then the weights are
as 27 to 1, the volumes of the cylinders of air beneath them are
as 9 to 1, and the density of that air is as 3 to 1. The greater
impetus of the larger sphere is thus able to drive away nine
times as much air three times as dense as the smaller sphere
needs to do, for which tasks the respective weights are exactly
sufficient. The only difference in the time of their reaching the
plane, he concludes, results from the difference of their diameters; and the same is said to hold true for fall through water.

This curious argument may have been an isolated reflection
of Cardano's, based on the fourth part of *De ratione ponderis,*
or an attempt on his part to find an Aristotelian explanation,
in terms of density of medium, for the observed fact of equal
speed of fall. As we shall see, that fact became widely known
and discussed in 1553–54 as a result of Benedetti's first published book. It was probably discussed again when his argument was republished in 1562. The idea that the fact of equal
speed was known and needed explanation is supported by an
unpublished dialogue composed in the 1570's by Giuseppe
Moletto, Galileo's predecessor at Padua. Moletto repeated
Cardano's fantastic argument with guarded approval, mentioning him but not Benedetti, whose Archimedean argument
appears not to have been known in university circles at that
time.

With regard to fall in different media, Cardano asserted that
the weights of two bodies falling in the same time through the
same interval will be as the squares of the rarities (that is,
inversely as the squares of the densities) of the two bodies. As

a scholium he adds that the argument does not apply to media as widely different as air and water, "for a ball of wood weighing 100 pounds no more descends in water than a wooden ball of one pound." Cardano attempted to determine the relative densities of air and water by the speeds of descent of the same body in both, obtaining an estimate of 50 to 1. It is noteworthy that Cardano did not use the Archimedean method of weighing alloys in air and water; for that he substituted the clumsier method of weighing water, metal, and container together in the required combinations.

Cardano may have been the first writer on mechanics to attempt a discussion of impact. In some propositions on percussion he multiplied weight by impetus, which he did not associate directly with velocity. Impetus, speed, force, and motion were left undefined by Cardano, and he often substituted one for another without apparent system. He gave several propositions on impacts between boats of different sizes and loads, and others regarding sails and wind directions, all suggested by the *Questions of Mechanics.*

The foregoing summary may give a greater appearance of order and coherence to Cardano's *Opus novum* than is actually found in it. Cardano appears not to have sought a system of mechanics, but rather to have recorded his reflections at various times and then to have published them without further editing. Those selected above are interspersed among a wide variety of mathematical and philosophical speculations. The central Italian mathematicians appear not to have known of Cardano's *Opus novum*, and they paid scant attention to his mechanical observations in *De subtilitate*.

Giovanni Battista Benedetti

Giovanni Battista Benedetti is not mentioned by Tartaglia in any of his books. He was nevertheless a pupil of Tartaglia's at Venice for a short time, probably about 1550. In the preface to his *Resolutio* of 1553, Benedetti acknowledged (in language that amounted to saying "to give the devil his due") that he had studied the first four books of Euclid under Tartaglia. According to his own account, this was his only formal schooling after his seventh year, his education having been received principally from his own father.

Benedetti was born at Venice on 14 August 1530. Although there were several ancient families of that name in Venice, he has not been identified with any of them by documentary record. There is no question of his patrician status at Venice, which is attested to not only by his own title pages but by documents connected with his later acquisition of similar status in Savoy. Yet Benedetti's father was described by a contemporary, Luca Guarico, as a Spaniard. Since it is evident that Benedetti was well acquainted with distinguished Spaniards who visited Venice, it appears likely that his father moved to Venice as a young man and had acquired distinction there. Guarico calls him a philosopher and *physicus,* which I believe is more likely to have meant "student of nature" than "doctor of medicine."

Benedetti's first book, the *Resolutio* of 1553, is of outstanding importance in the history of the laws of falling bodies. The text of this book dealt with the resolution of all Euclidean geometrical problems using a fixed setting of the compass. That topic was one of the challenge problems that arose in the controversy between Tartaglia and Ferrari in 1547–48, and, though Benedetti said nothing of the feud, that was probably the source of his interest in the geometrical problem. His passing reference to the study of Euclid under Tartaglia seems designed to disclaim any indebtedness to the latter for the results published in the book; and indeed in this matter he far surpassed his former teacher, though he was but 22 years of age when it was published. Tartaglia's own solution of the same problem, published posthumously in his great treatise on mathematics, is less systematic and less complete than Benedetti's. Significantly Tartaglia makes no mention there of his former pupil's work.

Now, the importance of the *Resolutio* to mechanics lies not in its text, but in Benedetti's letter of dedication to Gabriel de Guzman, a distinguished Spanish Catholic priest who was entrusted with many delicate diplomatic missions. Guzman had talked with Benedetti in Venice in the summer of 1552, when it appears that Benedetti first disputed Aristotle's law of falling bodies. He asserted that bodies of the same density would fall with equal speed regardless of weight, and his argument depended on the application of the principle of Archimedes to the analysis of free fall. Clearly his novel idea was of recent origin, since he stated that this publication in an otherwise inappropriate place was intended to prevent its theft by others. Benedetti's argument could hardly have been conceived except for the Archimedean revival, and it is a striking fact that Tartaglia had published in 1551 an Italian translation of and commentary on the relevant work of Archimedes. It is therefore probable that Benedetti's inspiration for the law of equal speed of fall came from his former teacher's *Travagliata inventione* of 1551.

Here I cannot refrain from commenting on the sole possible anticipation of Benedetti's argument that has been suggested. Benedetto Varchi is said to have composed in 1544 a work on alchemy which was not published until the nineteenth century. That work mentioned, in passing, the equal speed of fall of bodies differing in weight. The date 1544 cannot, of course, be absolutely substantiated, and the Archimedean principle is not directly involved in the statement made by Varchi. But it is possible that Varchi was first led to his remark by similar reasoning, and had later found it to be approximately true by test. At any rate, Tartaglia had first published the Archimedean work in Latin in 1543, the year before Varchi's work is said to have been composed. Thus all sixteenth-century attacks against Aristotle's laws of falling bodies came after the publication of the principle of Archimedes.

The *Resolutio* of 1553 occasioned a great deal of discussion, as evidenced by Benedetti's new letter of dedication to the *Demonstratio* of 1554. That discussion probably accounts for a curious allusion to a similar idea (equal speed of fall of bodies of equal weight but different material) in a book on cryptography published at Venice in the same year (1553) by G. B. Bellaso. Some adversaries at Rome declared that Benedetti's rule did not refute Aristotle's. Thus Benedetti felt impelled to amplify his argument and to give specific references to the contradictory passages in Aristotle. This he did early in 1554, in a pamphlet of which the title page states it to be an overt attack on Aristotle. No sooner was this booklet printed than Benedetti re-issued it with an important correction. Professor Drabkin, who first discovered the significant difference between two surviving copies, describes it hereunder.

In 1558 Benedetti was invited by Duke Ottavio Farnese to serve as court mathematician at Parma, where he went with his only daughter, born in 1554. Nothing is known of his wife, who perhaps died in childbirth. Benedetti remained in the service of Farnese for about eight years. He gave instruction at the court, advised on the construction of public works, designed and constructed elaborate sundials, and served as court astrologer. During this period he also carried out astronomical observations, which appear to have been for astrological rather than for theoretical purposes of astronomy.

While still in the service of Farnese, Benedetti gave a course of lectures at Rome on the science of Aristotle at least once, during the winter of 1559–60. There he was heard by Girolamo Mei, who praised highly the young man's intelligence, memory, eloquence, and independence of thought. Mei was the greatest living authority on the music of the ancients, and in later years supplied Galileo's father, Vincenzio Galilei, with most of the information on that subject that appeared in his published books, to which Galilei added his own experimental and musical theories of correct intonation. There is no evidence, however, that Mei knew about Benedetti's own contributions to musical theory from a physical standpoint, or that they would have interested him.

It was probably by hearing Benedetti's public lectures at Rome about 1560 that Johannes Taisnier learned of Benedetti's theory of free fall. At any rate, Taisnier took from Italy to Cologne a copy of the 1554 *Demonstratio* in its first (uncorrected) form and published it there in 1562 as his own, word for word, in the *Opusculum . . . de natura magnetis . . . item de motu continuo, demonstratio proportionum motuum localium contra Aristotelem* etc. Taisnier's publication helped keep alive the knowledge of the fact of equal speed of fall, separately from Benedetti's name.

Benedetti's buoyancy theory of fall thus gained a European circulation, and was even translated into English about 1578 by Richard Eden (*A very necessarie and profitable Booke concerning Navigation . . .* by Joannes Taisnerius). This lone sixteenth-century English translation of Benedetti's work is still listed in the British Museum catalogue under the authorship of Taisnier. It was Taisnier's plagiarism from Benedetti that inspired Simon Stevin to carry out (about 1586) the experiment of dropping balls of different weights from a high place and thus to confirm the equal speed of fall. But ironically, because Taisnier had stolen the uncorrected version of Benedetti's argument, he (and not Benedetti) was criticized by Stevin for neglect of the frictional resistance of the medium, that being proportional to the area and not to the volume of the falling body.

It was probably in 1563 that Benedetti wrote to Cipriano da Rore two letters concerning the physics of sound and in particular the mechanics of production of musical consonances. Those letters, first published in 1585, contain a description of the inverse relationship of frequency of vibration and length of string for like strings under equal tension. The resultant air waves, agreeing with or disturbing one another, are then related to musical consonance, for which Benedetti gave a numerical scale of measure. Vincenzio Galilei studied at Venice with Gioseffo Zarlino, who in 1565 succeeded da Rore as chapelmaster at St. Mark's in Venice, but who held a sharply opposing theory of musical consonance. Galilei was his pupil at Venice about 1560, too early to have learned there of Benedetti's views. Galilei's later theories of musical intonation, which brought him into sharp controversy with his former teacher, were based on experimental researches similar to those of Benedetti.

In 1567 Benedetti left Parma at the invitation of Emanuele Filiberto, Duke of Savoy, to serve as mathematician and engineer at Turin. The duke had a grand plan for the rehabilitation of Piedmont in which extensive public works and the revival of higher education played a principal role. The University of Mondovì was no longer functioning regularly, and, though it was ultimately supplanted by the University of Turin, that institution also was then of little repute. Traditions ascribe a teaching role to Benedetti at both universities, but they are unsupported by contemporary records. Benedetti served as adviser to the duke with respect to the University of Turin, as he secured the appointment of Antonio Berga to the chair of philosophy in 1569. But that he himself held no position there is evident from the title page of his book embodying a controversy with Berga, in which he gives Berga's title as professor and his own as philosopher to the duke.

Benedetti gave instruction in mathematics and in science to the duke's heir and to other persons of the court, but his principal activities appear to have been the design and construction of public works, such as fountains and sundials, and the inspection and improvement of military installations. In 1570 he was granted a patent of nobility in Savoy in recognition of the excellence of his services.

Although Benedetti published several minor works at Turin, he did not deal with mechanical topics again until the publication in 1585 of his chief work, the *Diversarum speculationum mathematicarum, et physicarum,* of which parts are translated here. On the ideas in that work, Professor Drabkin comments below. Benedetti died on 20 January 1590 (N.S.), two years earlier than he had predicted from his own horoscope. It is interesting that on his deathbed he made a recomputation and concluded that there must have been an error of four minutes in the original data—presumably the exact hour of his birth.

The question of Benedetti's influence, particularly on the young Galileo, is one of great interest and importance in the history of mechanics in the sixteenth century. It is usually said that the many close parallels between Benedetti's last work and Galileo's first (unpublished) essay and dialogue on motion cannot be explained except by direct line; that is, by an assumption that Galileo had read Benedetti's book, published in 1585, before he wrote his own treatise on motion about 1590. No historian can reasonably question the possibility; yet I think that no prudent historian can accept the hypothesis as an established fact. Greater difficulties are entailed by the generally accepted hypothesis, in my opinion, than by a different view. Without taking a final stand on the matter, I shall outline the nature of those difficulties and the character of a different approach.

To begin with, there are several very illuminating ideas in Benedetti's final work that would have been useful to Galileo in solving some of the problems that remained in his youthful work, problems that probably discouraged him from publishing it. Conspicuous among these is Benedetti's idea that acceleration would result from the successive accumulations of impressed force on a body already in receipt of such force, an idea with roots in the Peripatetic tradition. Had that notion come clearly to Galileo's attention, it appears likely that he would have preferred it to his own Hipparchian theory of residual impetus at the end of forced motion or the beginning of natural motion. Or, if he knew and did not adopt the better physical hypothesis, he would in all probability have attempted to refute it as a rival theory to his own.

Likewise, Benedetti's idea of a resistance of the medium proportional to the surface of the moving body, correcting the simple buoyancy theory of fall, would very likely have been accepted by Galileo (because of its mathematical form and its anti-Aristotelian character) if it had come to his attention. The development of Benedetti's thought in 1554, Stevin's prompt criticism of Taisnier's neglect of the same factor, and the later evolution of Galileo's own views on it, all point in this direction. As a final example, Benedetti's notion of tangential rather than radial projection in circular motion would have appealed to Galileo as a young professor intent on refuting Aristotle at every turn. That idea was not taken up by Galileo until much later, in the *Dialogue* of 1632. Had he encountered it before 1590, even if he did not adopt it, one would expect him to have attempted its refutation in *De motu*.

These are some of the difficulties in the path of any rational explanation of selections and rejections by Galileo from Benedetti, on the assumption that he had in fact read the latter's book before he composed *De motu*.

Equally troublesome are some psychological and biographical difficulties entailed by the plagiarism hypothesis. Galileo's situation at the time of composing *De motu* was that of a young professor anxious to publish novelties and gain favorable notice. If Benedetti's book was known to him, it would presumably also be in the hands of his colleagues; indeed, the book is often supposed to have been put in his hands by another professor. But for Galileo to claim as novelties of his own any material freshly off the press elsewhere in Italy would achieve the very reverse of his intention. For a young and unknown writer who was composing a direct attack on Aristotle at that period it would have been customary and prudent, and moreover would have heightened the probable effectiveness of the work, to cite (not to conceal) another writer's authority in favor of the views presented.

It is widely supposed that Galileo's failure to mention Benedetti's name is to be explained by his wish to conceal a main source of his ideas. But of course it may be simply that Galileo had not yet heard of Benedetti. That is a simpler hypothesis, and no less in accord with the present evidence. It was only in 1596, four years after Galileo had moved to Padua, that Benedetti was cited in a book by Galileo's former colleague (not teacher, as is often said), Jacopo Mazzoni. None of Benedetti's known correspondents was connected with Pisa or with Galileo. No mention was made of Benedetti in the thousand-page book on motion published in 1591 by Galileo's teacher at Pisa, Francesco Buonamico. Benedetti's chief book was not mentioned in print by anyone before Mazzoni, to the best of my knowledge. The only books of Benedetti cited by other authors up to 1596 are unrelated to mechanics; Christopher Clavius mentioned his work on dialing, and Agostino Michele referred to his book on the relative sizes of earth and water. Benedetti's books, printed at Turin, probably did not circulate rapidly; the Venice edition certainly did not, for that was reissued in the same sheets with a new title page fourteen years later.

But without Benedetti's book, Galileo may very well have been influenced by the buoyancy theory of fall, and even without knowing it to be Benedetti's. For that theory caused enough of a stir in 1553–54 to bring reports to Venice of dis-

cussions about it in Rome. The theory was carried abroad by
Taisnier, published again in 1562, read by Stevin, and translated into English by Richard Eden. Galileo's attention may
first have been attracted to this anti-Aristotelian view by
others who had heard of it, and once having the idea in hand,
he could scarcely help developing it along the same lines as
those used by Benedetti in his earliest version. Such a course
of events seems to me more likely than that Galileo actually
read the corrected version of 1585 and then reverted to the incorrect one before writing *De motu*. But I cannot reject that
possibility or the possibility that Galileo independently hit on
the same application of the principle of Archimedes against
Aristotle that Benedetti developed nearly forty years before,
just as Varchi may have hit on it still earlier. The evidence thus
far adduced simply appears to me to be inconclusive. In the
circumstances, it is perilous to erect a theory of the transmission of ideas and of the character of Galileo on the basis of
parallels between works of nearly even date. And apart from
the buoyancy theory of fall, the parallels are not striking; they
relate mainly to impetus theory, which was part of the Peripatetic tradition and widely accepted.

In connection with his translations of the excerpts from
Benedetti's works included here, Professor Drabkin has left
us the following remarks:

I mention some of the ideas that are found in these selections
from Benedetti.

In the preface to the 1553 *Resolutio* it is held that the distance traversed in a given time by a body falling freely in a
medium is proportional to the difference between the specific
weight of the body and that of the medium. Thus, in a given
medium the ratio of the speeds of natural motion of two bodies
of the same shape and size but of different material would be
equal to the ratio of their respective specific weights (each
diminished by the specific weight of the medium). A corollary
of this proposition is that bodies of the same shape and material but of different size, falling freely in a given medium,
fall with the same speed, i.e., traverse the same distance in a
given time.

This approach characterizes the anti-Aristotelian tradition
in emphasizing "specific" rather than "total" weight and in
treating the density of the medium as a subtrahend rather than
as a divisor. It is the same approach as that taken by Galileo

in the early writings *De motu,* and it reflects the influence of Archimedean ideas.

On the other hand, the underlying principle that the speed of fall is dependent on weight is basically Aristotelian. There is no concept here of mass as differentiated from weight, nor of the offsetting of gravitational by inertial considerations.

In the first version of the *Demonstratio,* as represented by the Vatican copy, the argument is essentially identical with that in the dedication to the *Resolutio.* In a later version, of which a copy exists today in the library of the University of Padua, a modification is introduced. The medium, in addition to affecting the weight of the body immersed in it, offers resistance to penetration as the body moves through it; this resistance is held to be proportional, not to the weight of the body, but to its surface. Thus, while the 1553 formulation (repeated in the Vatican version of the *Demonstratio*) for bodies of the same material but different size still holds for motion in a void, it no longer holds for motion in a plenum. In all problems of motion in a plenum, allowance must be made for this additional factor of resistance.

This was the view that Benedetti ultimately maintained. It is part of his systematic treatment of the problem of falling bodies in the section entitled "Disputations" *(Disputationes)* of the 1585 *Speculationum.*

Both versions of the *Demonstratio* also assert (inconsistently with the foregoing) that all bodies move with the same (finite) speed in natural motion in the void. And the same doctrine seems, at one point, to be implied in the "Disputations," through a confusion between gross and specific weight.

The "Disputations," a critique of Aristotelian physics, deals also with other aspects of motion and such related problems as the existence of the void and of absolute (as opposed to relative) heaviness and lightness. On these matters, too, the views of Galileo in his early writings are strikingly similar to those of Benedetti; both were heavily indebted to the dynamical theories of the scholastic authors.

In the *Speculationum,* both the "Disputations" and the "Letters" *(Epistolae),* from which selections are given here, deal largely with the traditional questions of dynamics. An effort is made, though it is hardly successful, to unify the treatment of natural and violent motion. Benedetti explains the persistence of the motion of projectiles after contact between projector and projectile is broken (again in the anti-Aristotelian

tradition) on the basis of a theory of impressed force (*impetus*)
like that taught by Buridan and widely held in the later medieval period. The acceleration of freely falling bodies is ascribed
to a continuous increase in impetus occasioned by the continuous action of the body's weight. The denial of an interval of
rest at the top of the trajectory of a missile thrown vertically
upward is based on the notion of a continuously diminishing
impetus during the upward path.

On these last points Galileo in his *De motu* holds views
similar to those of Benedetti. On questions of acceleration,
however, there are interesting differences between their views.
Galileo (*De motu, Opere di Galileo Galilei* [Ed. Naz., 1890],
I, 329) does not admit that the acceleration of a freely falling
body persists indefinitely (Benedetti, *Speculationum*, p. 184).
And even in the case of a body projected vertically downward,
Galileo (*De motu*, p. 408) holds that the impressed force of
projection is ultimately dissipated. This is at variance with the
view of Benedetti in the works of 1553 and 1554.

The notion that a given impetus impressed upon a body
tends to produce rectilinear, not curvilinear, motion is central
to Benedetti's discussion of the traditional problems of the
frictionless rotation of a millstone, the motion of a spinning
top, and the dynamics of the sling shot (*Speculationum*, pp.
160, 285). Benedetti also applies the theory of impetus to such
problems as the relation of the angle of elevation to the trajectory of the missile (*Speculationum*, p. 258). In approaching
the problems of exterior ballistics, he adopts the then-current
conception that the trajectory consists of two rectilinear (or
virtually rectilinear) segments joined by a circular segment. He
considers the curved part to be the resultant of two forces —
the force of projection and the natural tendency of weight
(*Speculationum*, pp. 160f.). Young Galileo's analysis of the
factors determining the length of the initial segment (*De motu*,
pp. 337ff.) is very similar to that of Benedetti. But in his early
essay, Galileo still uses the traditional analogies of heat and
sound in developing the general theory of impetus; these
analogies are not present in Benedetti's work.

Benedetti discusses the perpetual motion, in response to a
minimal force, of a frictionless ball upon a spherical surface
concentric with the earth (*Speculationum*, p. 156; cf. Galileo,
De motu, p. 301). He describes the oscillatory motion of a
body freely falling through a tube extending the whole length
of the diameter of the earth (*Speculationum*, p. 369). His

dynamical principles (and their limitations) may be studied in these chapters.

The section "On Mechanics" (*De mechanicis*) of the *Speculationum* commences with an "explanation" of the law of the lever based on a concept of the degree to which a weight pushes or pulls upon the center (Chs. 1–5). In the discussion of the bent lever (Ch. 3) the basic notion of statical moment is developed. The rest of the section is taken up largely with a running commentary and critique (Chs. 10–25) of the Aristotelian *Questions of Mechanics* (*Mechanica*). There are also some strictures (Chs. 7–8) on Tartaglia's edition of the *Liber Jordani de ratione ponderis,* which had appeared in 1565. The subject of mechanics figures prominently in other sections as well, e.g., in the anticipation of Pascal's hydrostatic principle (*Speculationum,* p. 287). On the other hand, there are serious gaps; a discussion of the inclined plane would have been expected. — *I. E. Drabkin*

The traditions strongly discernible in Benedetti's work are the Archimedean, particularly in his use of the buoyancy principle, the Aristotelian, leading to his commentary on the *Questions of Mechanics,* and the Peripatetic tradition. In the last-named he shares with many sixteenth-century writers on mechanics a knowledge and acceptance of the concept of impetus, but appears not to know, or at least does not use or comment on, the mean-speed theorem. The Jordanus tradition, though known to him, is rather conspicuously lacking in influence on Benedetti; as Professor Drabkin has remarked, one would have expected some analysis of the inclined plane, either statically or in connection with his discussions of motion. The tradition of Hero is not apparent in Benedetti's works, but one of his letters shows that he was familiar with the *Pneumatics* before the translations were published.

Federico Commandino

We pass next to the central Italian group of writers on mechanics, which began with Federico Commandino. Though Commandino is noted particularly for his scholarly translations of and critical commentaries on mathematical works of classical antiquity, rather than for his own writings on mechanics, his career is of interest to us by reason of the work of his pupil, Guido Ubaldo, as well as for his own pioneer treatise on centers of gravity in the Archimedean tradition.

Born at Urbino in 1509, Commandino was first educated by his father, Battista, an architect who served the Duke of Urbino (uncle of the duke to whom Tartaglia dedicated his *Nova Scientia*). About 1524–26 he was put under the tutelage of Giacopo Torelli at Fano, where he learned Latin and Greek. In 1527, after the sack of Rome by the Spanish, the Orsini family went to Urbino and took with them as tutor Giovanni Pietro di Grassi, from whom Commandino learned mathematics. About 1530, Grassi returned to Rome in the service of Cardinal Ridolfi, taking Commandino with him. Shortly thereafter Grassi was made Bishop of Viterbo, but, before leaving Rome, he secured for Commandino the post of private secretary to Pope Clement VII. Commandino remained at Rome until the pope's death in 1534, when he went to the University of Padua to study philosophy and medicine.

Commandino appears to have remained at Padua for ten years. It is sometimes said that he studied mathematics under Tartaglia, who was then lecturing publicly at Venice. That story is very doubtful, as nothing is said of it by Bernardino Baldi in his account of either Commandino or Tartaglia, and Baldi would certainly have known of it if it were true. About 1544 Commandino left Padua for the University of Ferrara, taking his doctorate in medicine there about 1546. He then returned to Urbino, where he married and had two daughters and one son. It appears that his wife and son died before long, and he never remarried.

About 1552 Commandino entered the service of Duke Guido Ubaldo della Rovere, son of Duke Francesco Maria I (who had consulted with Tartaglia) and likewise employed as captain by the Republic of Venice. They spent some time together in Verona, where Commandino learned from the duke the principles of military mapping, land surveying, and similar matters. Commandino had at this time abandoned the practice of medicine "by reason of its uncertainty," and devoted himself to mathematics; but the duke falling ill, he treated him successfully and they returned to Urbino. There Commandino met Ranuccio Cardinal Farnese, became his physician, and in 1553 apparently went to Rome in the service of Farnese. In the following year they were at Venice. It was probably about this time that Commandino met Marcello Cardinal Cervini, who had become Vatican Librarian in 1548. Cervini desired a reconstructed text, from Moerbeke's poor Latin translation, of Ptolemy's *De analemmate*, the Greek original being lost. Thus Commandino began on the long series of reconstructions and translations of ancient texts for which he is noted. The same manuscript contained Moerbeke's translation of Archimedes' *On Bodies in Water*, but not the Greek text (as Baldi asserted), for no Greek copy of the second book of that work was then known. In 1555, when Cervini became pope, he appointed Commandino to his service, but he died a few weeks after his election. Commandino thereupon re-entered the service of Ranuccio Cardinal Farnese, in which he remained until Farnese's death in 1565.

The year 1558 saw Commandino's publication at Venice of a corrected Latin version of Ptolemy's *Planisphere* together with the *Planisphere* of Jordanus, both with commentaries. In the same year he published a Latin translation of several mathematical works of Archimedes, by far the best that had been made, with his own commentaries. From this edition, he omitted the mechanical works. The work on centers of gravity was held back in expectation of Francesco Maurolico's translation, while the hydrostatics was probably left out for want of a Greek text. Meanwhile he continued to seek (in vain) an Archimedean or other ancient work on the centers of gravity of solid bodies. Failing to find one, he composed such a work of his own on Archimedean principles.

Commandino was probably at Rome in 1562, when he published the corrected translation of Ptolemy's *De analemmate* begun for Cervini. From there he went to Bologna, where in 1565 he published a carefully edited text of the Moerbeke Latin translation of the two books of Archimedes' *On Bodies in Water* and his own work on centers of gravity of solid bodies, despairing of Maurolico's promised publication. In 1566 he published a translation of four books of Apollonius on conic sections, having become interested in that subject through his studies of Archimedes.

The remainder of Commandino's life was spent at Urbino. In 1569 or 1570 he was visited there by John Dee, the English mathematician and mystic, with whom he revised and published an Arabic work on the division of surfaces. Guido Ubaldo del Monte began his studies of mathematics with Commandino about this time. The young Bernardino Baldi probably studied with them for a time before he went to the University of Padua in 1573, rejoining them a year later when the university was closed during a plague epidemic. It was

also about 1570 that Commandino gave instruction to Francesco Maria II, Duke of Urbino, who induced him to make a
new Latin translation of Euclid's *Elements*. The Latin Euclid
and a translation of Aristarchus were published by Commandino at Pesaro in 1572.

Commandino died at Urbino in 1575, the year of publication
of his Italian Euclid and of his Latin translation of Hero's
Pneumatics. Bernardino Baldi was in constant attendance on
Commandino at the end of his life, and collected in that way
most of the biographical information included in the present
extract. An attempt has been made here to arrange the events
and to supply dates, omitted from Baldi's account.

Commandino's works show him to have been the representative of two traditions in mechanics to the virtual exclusion
of all others—the traditions of Archimedes and Hero. Although he was probably familiar with the Jordanus and Peripatetic traditions, he seems not to have concerned himself at
all with dynamic problems. The Aristotelian tradition was likewise passed over in silence by Commandino. Even in his long
introduction to the vernacular translation of Euclid, which he
says was made in order to bring the work to practical men, he
mentions neither Tartaglia as his predecessor in that regard,
nor the science of weights as a branch of mathematics, nor the
name of Jordanus. His only apparent concern with medieval
mathematicians was as translator of a nonmechanical treatise
of Jordanus, on the planisphere. It is interesting that Commandino, an expert judge in such matters, considered that
treatise to be of much greater antiquity than the medieval
period to which it was (and is) ascribed.

Guido Ubaldo del Monte

Guido Ubaldo, Marquis del Monte, was born at Pesaro on
11 January 1545. He entered the University of Padua in 1564,
where one of his companions in study was the poet Torquato
Tasso. On his return from the university, he continued his
studies in mathematics under Federico Commandino at Urbino;
it was probably at this time (about 1570) that he became acquainted with Bernardino Baldi. After Commandino's death
in 1575, Guido Ubaldo became Baldi's teacher of mathematics.
In 1577 he published his *Mechanics*, followed in 1579 by a
work on the planisphere and in 1581 by a book on the ec-
clesiastical calendar.

Guido Ubaldo is said by one biographer to have served in military campaigns in Hungary and against the Turks, and on his return to have settled at a family villa in Monte Baroccio, near Urbino. In 1588 he received from Galileo some theorems on centers of gravity with a request for his opinion. In this way a correspondence was opened which continued until his death in 1607. Guido Ubaldo was favorably impressed with Galileo's talents, and sent to him a copy of his second important contribution to mechanics, a paraphrase of and commentary on the work of Archimedes on plane equilibrium. The same year, 1588, saw Guido Ubaldo named visitor-general (inspector and adviser) of the cities and fortifications of Tuscany. He continued to reside at Monte Baroccio, but it was probably through his connections at the Tuscan court that he was able in 1589 to secure for Galileo an appointment to the chair of mathematics at the University of Pisa. Three years later he was instrumental in obtaining a better post at the University of Padua for Galileo.

Commandino had left among his papers a Latin translation and commentary on the *Collections* of Pappus. This was published in 1588, with diagrams drawn by Baldi. The final editing was done by Guido Ubaldo, though with customary modesty he did not affix his name to the book. In 1600 Guido Ubaldo published an important book on perspective, which includes the first discussion of scenography. He designed the ducal theater at Urbino. After his death in 1607 two further works were printed, one on problems of astronomy and one on the Archimedean water-screw.

Guido Ubaldo's two chief works on theoretical mechanics make clear his devotion to the idea of mathematical rigor of treatment and his repugnance for medieval writings on the science of weights and for Tartaglia's adherence to that tradition. The *Mechanics* contains numerous criticisms of those writers, and in the *Paraphrase* of 1588 Guido Ubaldo wrote: "And however much Jordanus Nemorarius (whose followers include Niccolò Tartaglia and others) struggled in his book *De ponderibus* to prove this same proposition of the general lever by many means, yet not any of the proofs were worthy to be called demonstrations, and were scarcely to be credited. For he put things together which in no way command conviction and perhaps do not even persuade anyone by probability, when in mathematical demonstrations the most precise reasons are required. And on that account it never seemed to me that this Jordanus should even be reckoned among writers on mechanics." In appraising probable influences on Galileo, one should remember that, before Galileo wrote anything on motion, he had received this book from his most valued patron, that it was a book on Archimedes (whom Galileo admired above all other writers), and that the opinion of Guido Ubaldo was likely to have been shared by other strict Archimedeans of the time.

In any attempt to trace the history of mechanics in the sixteenth century, we should recall that Guido Ubaldo's severe censure of Jordanus came after the revival of Archimedes. Commandino's emendations and critical commentaries of 1558 and 1565 had effectively restored the authentic texts. Thus Guido Ubaldo had before him a pattern of the most precise and the most fruitful method conceivable of reasoning mathematically about physical matters. For that reason the influence of the medieval science of weights in the first half of the sixteenth century was very much greater than it was after the Archimedean revival.

The reaction of Guido Ubaldo against the medieval pattern, after he had studied the ancient mathematicians under Commandino, was so great that he actually rejected the correct theorem of Jordanus on inclined plane equilibrium and adopted the incorrect theorem of Pappus in its place. This misplaced homage to the ancients and to the idea of absolute mathematical rigor in questions of mechanics blinded Guido Ubaldo to the possibility of important advances in the science that he would otherwise have been quite capable of making. Galileo, with a more objectively critical attitude toward his predecessors, was able to make them.

Guido Ubaldo's interminable discussion of the balance has been preserved here in translation virtually intact, in order to portray the new (if essentially illusory) quest for mathematical precision in mechanics on the part of this central Italian classicist. Guido's exaggerated concern for rigor appears in his insistence that, strictly speaking, the lines of descent of suspended weights are never mathematically parallel, but converge toward the center of the earth. Galileo's curt dismissal of this theoretical consideration in his own treatise on mechanics, composed about 1593 and elaborated in successive versions, illustrates the great differences in temperament between the two writers. Galileo was concerned not with geometrical minutiae but with the fruitful application of mathematics to mechanics, in which dynamic concepts made their appearance in the treatment of statics. But the struggle against the purists was only gradually won. The kind of problem that concerned Guido with respect to the theory of the balance re-appeared in the seventeenth century, after the foundations of mechanics had been laid, in the dispute over "geostatics" which absorbed several prominent French mathematicians for a time, prior to Newton's formulation of the law of gravitation.

The remainder of the translation included here is fundamentally a list of propositions demonstrated by Guido, most of the proofs and some of the subsidiary theorems being omitted after the pattern of Guido's approach has been clearly illustrated. The most valuable and original section of the work deals with the analysis of pulley systems. It not only is correct in method and (to the best of my knowledge) original with Guido, but is inherently productive in that it shows how one simple machine (the pulley) may be reduced to another (the lever) which superficially does not resemble it. Galileo adopted this analysis in his treatise on mechanics, but there is an important difference in conclusions. Guido missed the one crucial point of which Galileo made a general principle—that the products of force (or weight) and virtual displacement at the ends of any system in equilibrium are equal. Perhaps Guido's failure to accept that principle is related to his antagonism toward Jordanus. Even in rigidly connected systems, Guido denied on logical grounds that this principle could hold. Thus, in his fourth proposition on the lever, as a corollary, he states that, "the space of the power has the same ratio to the space of the weight as that of the weight to the power which sustains the same weight. But the power that sustains is less than the power that moves; therefore the weight will have a lesser ratio to the power that moves it than to the power that sustains it. Therefore the ratio of the space of the power that moves to the space of the weight will be greater than that of the weight to the power." Likewise, in discussing pulley systems, he says in the corollary to the 26th proposition that, "the space of the power that moves has always a greater ratio to the space of the weight moved than that of the weight to the same power," referring to the corollary on the lever cited above. Galileo, by introducing the idea of a minimal power and a virtual displacement in terms analogous to the mathematical concept of limit, bridged the static-dynamic gap for all simple machines.

In addition to his published works, Guido Ubaldo left three manuscripts which I have not had an opportunity to inspect. Two of these, said to be in the Biblioteca Oliveriana at Pesaro, concern the fifth book of Euclid's *Elements* and the theory of compound proportion. It is worth noting that Galileo lectured at Padua at least one term on the same matter, and that he intended to add to the *Discorsi* of 1638 a separate discussion of the theory of proportion. The other manuscript of Guido Ubaldo was found at the Bibliothèque Nationale in Paris by Guglielmo Libri, who quoted several passages from it in his *Histoire des Sciences Mathématiques en Italie* (Paris, 1840). The close relationship of some ideas given there to ideas used by Galileo suggests that Guido Ubaldo was a much more important stimulator of Galileo's thought than has yet been recognized. On a single page quoted by Libri, Guido Ubaldo stated that the path of a projectile was symmetrical, of mixed natural and violent motion, resembling a parabola or hyperbola; that the curve could be represented by a loosely hanging cord, inverted; and that it could be obtained experimentally by throwing an inked ball upward and forward against a steeply tilted board.

In the work of Guido Ubaldo, the traditions of Archimedes and of Hero (through Pappus) predominate. As we have seen, an orthodox Aristotelian tradition underlay his thinking, but it was masked in his writings. His belief that Archimedes had extracted physical postulates from the *Questions of Mechanics* and had then applied mathematics only to statics appears to have closed the mind of Guido Ubaldo against the possibility of any further useful application of dynamic concepts such as the principle of virtual velocities, or of work. His utter rejection of the Jordanus tradition is consistent with that view, as it is with his predilection for antiquity and for formal mathematical proof. The Peripatetic tradition is likewise absent from Guido Ubaldo's works, even with regard to impetus theory. But if, in his treatment of simple machines, he fell short of a principle that would unify statics and dynamics, his *Mechanics* was nevertheless the first truly systematic attempt at a rigorous treatment of the field, and it paved the way for Galileo's synthesis of all the traditions that had influenced mechanics in the sixteenth century.

Bernardino Baldi and Some Minor Mechanicians

The final representative of the central Italian group was Bernardino Baldi, who was born at Urbino in 1553. His early education included instruction by Paolo Manuzio in Greek and Latin, which he applied in his early youth to a translation of the astronomical poem of Aratus into Italian. He began the study of mathematics under Commandino in the company of Guido Ubaldo about 1570. In 1573 he went to the University of Padua, but his studies there were interrupted by an outbreak of plague. Returning to Urbino before the death of Commandino in 1575, he obtained and recorded the detailed story of Commandino's life. He distinguished himself at the university in classical scholarship and languages, including Arabic, Hebrew, and Slavonic tongues. He was a man of vast erudition, keen intelligence, and scholarly probity who, in addition to his many published works, left manuscripts on theological and classical subjects as well as on war machines, the paradoxes of mathematics, dialing, and geography.

Intermittently over a period of twelve years after the death of Commandino, Baldi collected information concerning every mathematician known. After he became abbot of Guastalla in 1586, he devoted much time to the writing of some 200 lives based on that material, completing the manuscript in 1588 or 1589. His biography of Commandino was eventually published in 1714, preceded in 1707 by a highly abridged version of the entire compilation. Since that time about forty of the full lives have been published, but the bulk of the work is still in manuscript. From the 1707 summary we have further evidence of the isolation of the central Italian group from the north Italian writers on mechanics. Tartaglia is there mentioned only as a writer on mathematics who had genius, but who was oblivious to literary style. Cardano is referred to only as a rival of Tartaglia. Benedetti is credited only with a book on dialing that was reproached by more expert writers (i. e., Christopher Clavius) for its lack of rigor. Baldi included in his list of mathematicians the architect Leon Battista Alberti and the musician Gioseffo Zarlino and, in general, all persons who wrote on subjects related to mathematics whether their works were published or not. The fact that he did not include Leonardo da Vinci makes it most improbable that he knew of any works left by that genius bearing on mathematics. Duhem's opinion that Baldi obtained many of his ideas on mechanics from the manuscripts of Leonardo, directly or indirectly, is, in my opinion, entirely without foundation.

Baldi's translation of Hero's *Automata*, though completed before 1576, was not published until 1589, about the time that he appears to have written his commentaries on the *Questions of Mechanics,* published posthumously in 1621. In the preface to the latter work he mentioned that he had heard of, but had not seen, a work on mechanics by Simon Sticin (Stevin). Duhem, who accepted the date 1582 for the composition of Baldi's commentaries on the *Questions,* supposed that word of Stevin's work on mechanics had reached Italy four years before its publication. But in the same preface, Baldi referred to Guido Ubaldo's paraphrase of Archimedes' *On Plane Equilibrium,* which was not printed until 1588. It is therefore likely that the date of 1582, ascribed to Baldi's commentary by his biographer, is in error.

Baldi's preface mentions, in addition to Guido Ubaldo, writings on mechanics by Archimedes, Hero, and Commandino, as well as commentaries on the *Questions* by his predecessors Leonico and Piccolomini. Tartaglia and Cardano are mentioned in the text, in connection with the discussion of the balance, as "followers of Jordanus," the precise words employed by Guido Ubaldo in the corresponding section of his *Mechanics.* Baldi's comment on Jordanus (in the brief paragraph devoted to him in the 1707 summary of the *Lives of the Mathematicians*) is that the teachings of Jordanus were barbarous and involved false assumptions. Baldi assigns the date 1250 for the death of Jordanus.

Among the interesting passages in Baldi's commentary on the *Questions of Mechanics* is a discussion of the problem why a spinning top remains erect. Baldi remarks that the inclination to fall to one side must depend on the position of the center of gravity of the top, and so long as this spins rapidly, the top can have no more inclination to one side than to another, even if defects in the material cause the center of gravity to lie elsewhere than on the geometrical axis of the top. He also describes another toy, a clown that balances on a wire by holding a pair of counterweights placed lower than the wire; Baldi's explanation is in terms of the center of gravity of the system and adumbrates the concept of stable equilibrium.

Another section of particular interest discusses the power of oars and rudders. This is placed in the context of Baldi's analysis of the steelyard, where the philosophical vocabulary of potency and act is ingeniously applied to the crude macroscopic treatment of relative arcs given in the *Questions.* In discussing the rudder, Baldi makes it clear that the relative motion of ship and river governs the position of the ship. He mentions in this connection many observations he had made of boats moving, being towed, or anchored in Italian rivers at Padua, Mantua, and elsewhere. Those observations account quite plausibly, in my opinion, for Baldi's drawings and discussion of eddy currents and vortices in water at the end of the commentary, though Duhem considered them to be evidence that Baldi had seen the manuscripts of Leonardo. In any event, the engravings were made after Baldi's death, and the errors in them may not be his.

Before turning to Galileo, we may mention in passing a few writers who were connected with neither of the two chief schools of Italian sixteenth-century mechanics, as a means of showing the variety of topics that were considered of interest by men of widely different vocations.

Alessandro Piccolomini of Siena (1508–78) is remembered chiefly as a dramatist, poet, and popular educator. Among his writings are a treatise on astronomy which contains the first printed star maps, a commentary on the *Meteorology* of Aristotle in which Piccolomini attempted an explanation of the rainbow, and a treatise on the relative dimensions of the spheres of earth and water which drew critical comment from Benedetti for its strict Aristotelian orthodoxy. Piccolomini was the first to write an extended commentary on the *Questions of Mechanics,* which he did in the form of a paraphrase that included the entire text. He was dissatisfied with Leonico's translation, and examined Greek manuscripts at the principal libraries of Venice, Padua, Bologna, and Florence in quest of a correct text. Ultimately he concluded that he would perform a greater service by paraphrasing the text and adding explanatory comments. The most interesting section is a discussion of the conflicting doctrines of antiperistasis and impetus with regard to the problems of projectile motion and acceleration in free fall. Piccolomini's account shows the degree to which even an orthodox Aristotelian was aware by the mid-sixteenth century of the difficulties in this matter.

Oreste Biringuccio, translating Piccolomini's commentary into Italian in 1582, states that the author himself was so deeply devoted to the ideal of public education that he had expressed regret at having written the book originally in Latin. In vernacular form, Piccolomini felt, it would have been of value to engineers and builders. Piccolomini had also remarked on the curious fact that Leonico's 1525 edition of the *Questions* appeared to be the first Latin translation ever made of it. Another sixteenth-century translation had in fact been published at Paris in 1517, but no medieval translation is known.

A second writer on mechanics connected with neither of the main schools was the engineer Giuseppe Ceredi of Piacenza, who published at Parma in 1567 a book called *Tre*

Discorsi sopra il Modo d'Alzar Acque da' Luoghi Bassi (Three Discourses on Means of Raising Water from Low Places). Ceredi was interested in the construction and use of the Archimedean screw for the irrigation of fields and the draining of swamps. He had found that the devices in use were inefficient, and sought to discover the rules of design by which they might be improved. The results led him to specify a maximum length and optimum dimension for the water-channel, to suggest batteries of screws for lifts higher than the efficient maximum length, and to examine the design of cranks and other devices for turning the screws. Though not written in deductive form, Ceredi's investigations belong to theoretical mechanics; they are reminiscent of the experiential rules given by Philo of Byzantium for the construction of ballistae. Also worthy of note, though unrelated to our subject, is Ceredi's economic analysis of the probable gain in crop yield through irrigation as compared with the operating and capital costs of machinery and the expense of labor in harvesting and hauling to market the increased yield. Ceredi obtained a patent from Ottavio Farnese in 1566 for the development of his machines, a fact suggesting that at Parma he may have talked with Benedetti, who was then Farnese's adviser on engineering matters. Ceredi was familiar with the works of Archimedes and Pappus; among later writers he mentions Giorgio Valla, Girolamo Cardano, and Georg Agricola.

As a final example of the diversity of mechanical writings in sixteenth-century Italy, I shall mention the *Discorso* of the historian Domenico Mellini, published at Florence in 1583, "in which it is proved against the opinion of some persons that one cannot by art devise, nor give to any material body that is subject to decay, any movement that shall be continuous and perpetual." Mellini's treatment of perpetual motion is philosophical rather than mechanical; the chief argument is Aristotle's dictum that nothing violent can be permanent. It is of interest that Mellini discusses at length magnetic and electric phenomena as well as mechanical sources of motion.

Galileo

Galileo's two main sixteenth-century works on motion and on mechanics have been previously translated and commented upon by us in an earlier volume of the present medieval science series. For our views of the background and achievements of those works, the reader is referred to that volume, *Galileo Galilei On Motion and On Mechanics* (Madison, 1960). One remaining document of the same tenor, together with Galileo's notes relating to them all, is translated here.

After the publication of our previous volume, I arrived at a conclusion, with which Professor Drabkin concurred, concerning the origin of the inertial concept in Galileo's mind. That conclusion may appropriately be summarized here since Galileo's line of thought may in turn have been stimulated by a remark of Cardano's for which a definite medieval ancestry may be traced.

The introduction of the inertial concept has long been regarded as linked inseparably to the search for a better explanation than Aristotle's for the phenomena of projectile motion. Though the concept of inertia ultimately offered a rational explanation for projectiles, it appears to have first occurred to Galileo in a totally different connection. We have seen that, throughout the sixteenth century, impetus was widely accepted as a satisfactory substitute for the clumsy Aristotelian idea that projectiles were moved by some action of the medium after they had left the initial projecting instrument. Professor Alexandre Koyré believed that Galileo, finding that the supposed loss of impetus by projected bodies did not lend itself to mathematical treatment, was led to the idea of inertia, though in a confused form that included (or required) circular elements of motion. But the germ of Galileo's inertial concept is to be found in his *De motu,* a work in which he shows himself perfectly satisfied with impetus as an explanation of projectile motions. In conducting a general attack against Aristotelian physics, Galileo questioned the validity of Aristotle's basic dichotomy of natural and violent motion. Inquiring whether a motion might exist that was neither natural nor violent, he found an affirmative example in the rotation of a sphere situated at the center of the universe. As we have seen, Cardano assigned to similar exceptions the term "voluntary" motions. Galileo added to them the rotation of a homogeneous sphere anywhere in the universe. He suggested in *De motu* that all such motions, as well as the motion of any heavy body supported on a surface concentric with the center of the earth, should be called "neutral" motions, and applied that concept next to motions on the horizontal plane. From this concept of neutral motions, conceived quite apart from any consideration of projectiles, we can trace step by step Galileo's elaboration of an inertial concept as a state of rest or motion to which a body is indifferent and in which it will continue perpetually if undisturbed. The application of that concept to projectile motions in Galileo's published books was made decades after the original emergence of the concept in his unpublished writings.

The evolution of the inertial concept in Galileo's own writings throws much light on his habitual association of inertial motions with instances of the conservation of angular momentum. Under the usual hypothesis as to the origin of the inertial concept, that association of ideas has led to many misinterpretations of Galileo's thought, often in direct contradiction with his own considered statements in his published books.

Galileo, unlike the other writers whose works are translated here, was associated through a considerable part of his life with the universities both as a student and as a professor. It is therefore appropriate here to mention the Peripatetic tradition in mechanics and motion. Galileo was doubtless more familiar with that tradition than was Tartaglia, Benedetti, or Guido Ubaldo. Of those we have mentioned, only Cardano is likely to have had a considerable knowledge of it. Now, we have seen evidence that impetus theory, which belonged to the Peripatetic tradition, was quite widely known and accepted both inside and outside the universities during the sixteenth century. An important question still to be resolved is the fate of the mean-speed theorem during that same period. A thorough inquiry has yet to be made whether Italian university instruction continued to lay stress on the medieval and Renaissance commentators on Aristotle who concerned themselves with that theorem and with mathematical analyses of possible and actual motions.

When that question is studied, it should be remembered that a philosophical topic that was much debated in 1500 may have become dormant by 1600. No less than seventeen printed books before 1600 are reported to have set forth the mean-speed theorem and related discussions of motion; yet their influence on sixteenth-century writers on mechanics, including Galileo, is problematical, to say the least. It may be that the neglected Peripatetic tradition was the needed link for a viable science of mechanics, and that Galileo's familiarity with it made possible for him the synthesis that had eluded his predecessors. Evidences both for and against that idea can be found within his own early writings. To decide the issue, the probable strength of the Peripatetic tradition in mechanics at Pisa in the 1580's needs careful study.

Girolamo Borro, who was probably one of Galileo's teachers and whose *De motu gravium et levium* of 1576 was certainly read by Galileo, ignored in it the entire tradition of the "calculators," but he described an experiment of his own with falling bodies and a mistaken result adopted by Galileo in his own *De motu*. Perhaps experiment rather than calculation was the tradition at Pisa even among philosophers when Galileo studied there. The suggestion may sound preposterous, but Borro at least seems to have thought that a philosophical problem might be settled by experiment. From Galileo's own jottings, it would appear that his philosophical attention during the writing of *De motu* was given chiefly to Aristotle's own text and to the commentaries of Philoponus, Simplicius, Alexander of Aphrodisias, and Themistius rather than to Oresme, Buridan, Bradwardine, or the "calculators."

Benedetti and Galileo have in common chiefly the idea of an open and full-scale attack on the physics of Aristotle. The Archimedean revival seems to have influenced Benedetti mainly through the single key idea of buoyancy in fall; Galileo's thought is permeated with Archimedean mathematics, mechanics, and procedures, beginning with his earliest writings, *La Bilancetta* and the theorems on centers of gravity. It is Archimedes who links Galileo most closely with the work of Guido Ubaldo and the central Italian group, and likewise links him with Tartaglia and the rival northern group.

The Archimedean tradition was certainly essential to Galileo's mechanics, but so was the Aristotelian tradition. To the *Questions of Mechanics* Galileo owed his own acquaintance with the principle of virtual velocities, as he acknowledged in his book *On Bodies in Water* in 1612. That principle runs through all Galileo's writings on mechanics; he used it to explain the common level of fluids in communicating vessels, and it underlay his concept of moment. He lectured on the *Questions* at Padua and composed a commentary on it, now lost.

On the other hand, the Jordanus tradition appears to have had little if any influence on Galileo. The reason, as with Guido Ubaldo, was probably the sheer impact of the Archimedean revival. It is often said that Galileo's analysis of the inclined plane is similar to that of Jordanus, though in fact they have nothing in common beyond the correct result. The proof offered by Jordanus, even as improved by Tartaglia, lacked a definitively stated physical postulate other than that to be found in the statement to be proved. It was that circular character of the proof that rendered it wholly unacceptable to Guido Ubaldo in the sixteenth century and to William Whewell in the nineteenth. Galileo's proof linked the inclined plane with the balance, the lever, the pendulum, and the free descent of bodies supported on circular concave surfaces. In a sense that proof foreshadowed most of the mechanical investigations to which Galileo devoted a large part of his life.

The influence of the tradition of Hero on Galileo is debatable. In his early *Mechanics* he was critical of Pappus, from whose work he must have become familiar with Hero. One of Hero's devices is discussed by Galileo in a letter, but he showed little enthusiasm for such things. The large element of the technological tradition in Galileo is concerned rather with working techniques of his own time than with traditions of antiquity.

In the picture we have sketched here, the dynamic interests of the northern Italian school have been seen to be isolated in a sectarian way from the static investigations of the central group. Both were more or less oblivious to the Peripatetic tradition, except with regard to the concept of impetus. The universities paid little attention to either of the two main groups investigating mechanics, and still less to the technology of the time. The ultimate synthesis seems to have awaited a university man to whom both the Archimedean and the Aristotelian traditions were known and were seen to be of value, and who had also a knowledge of the philosophical and the technological traditions of mechanics. Such a man was Galileo, whose teachers and patrons urged on him from various sides these diverse sources of a new science. To what extent this picture is accurate with respect to the traditions named, and to what extent other elements played a role in Galileo's work, only future studies of the sixteenth century can determine.

For the Galileo translations in this book, Professor Drabkin left the following preface:

Galileo left in autograph manuscript form two Latin works on motion, probably written at the time he was teaching at Pisa (1589–92), but not published during his lifetime. Both works reflect the same basic doctrine and even include some identical passages, but they have different literary forms. One is a series of chapters that were obviously intended to form an essay or course of lectures; the other is in dialogue form. They will be referred to hereafter as "Essay" and "Dialogue," respectively.

Almost all of the "Dialogue" and parts of the "Essay" were first published in 1854 in Volume 11 of Eugenio Alberi's edition of Galileo's works. Antonio Favaro published the rest in 1883. The first volume (1890) of the National Edition of the works of Galileo, *Opere di Galileo Galilei,* edited by Favaro, includes, under the title *De motu,* the relevant contents of these manuscripts, as follows: (1) "Essay," first version (pp. 251–340); (2) "Essay," two separate reworkings of parts of the first version (pp. 341–43 and 344–66); (3) "Dialogue" (pp. 367–408); (4) a series of separate notes, reflections, outlines, etc., mainly on the subject of motion, referred to hereafter as "Memoranda" (pp. 409–19). All page and line references in the present translation will be to the Favaro *Opere,* Volume I.

Our earlier volume, *Galileo Galilei On Motion and On Mechanics* (Madison, 1960), included a complete translation of the first of the above items and brief discussions of the other three. The present work contains complete translations of the third and fourth items, the "Dialogue" and the "Memoranda," which it was not feasible to include in the earlier volume. For a general introduction to Galileo's early work on motion, including a summary of the doctrine it contains, the reader may refer to our 1960 volume. I add here a few remarks about the "Dialogue."

The locale of the "Dialogue" is the shore near the mouth of the Arno (*Opere,* I, 369.29; 370.23). There are also one or two indications of the time: the "Dialogue" mentions (*Opere,* I, 379.28) the construction of the *bilancetta,* and, if Favaro is right (*Opere,* I, 211) in placing Galileo's essay on this instrument in 1586, the composition of the "Dialogue" would belong after that date. Again, one of the problems listed for discussion in the "Dialogue" is attributed by Domenico to "my dear friend, Dionisio Font, most worthy knight" (368.19), a reference which seems to imply that Font was then living. According to Favaro (*Opere,* XX, 442), Font died on 5 December 1590. If that was so, the "Dialogue" is likely to have been begun, and possibly even abandoned, before that time.

At all events, the references to time and place are not inconsistent with the traditional view that Galileo's early writings on motion date from the years when he was teaching mathematics at Pisa, i.e., between 1589 and 1592 (see *Opere,* I, 245–49).

Of the two interlocutors of the "Dialogue," Alexander obviously speaks for Galileo. Any doubt on this point is eliminated by the reference to the *bilancetta.* As for Domenico, if Galileo had a particular person in mind, he does not help us to identify that person.

The "Dialogue" as we have it is quite incomplete. Of the six problems listed at the outset (*Opere,* I, 368) for discussion, three—the second, fifth, and sixth—are never dealt with. (The "Essay," incidentally, takes up all these questions, as well as others not treated in the "Dialogue," e.g., motion on the inclined plane.) Now there is, of course, the possibility that not all of the "Dialogue," as Galileo wrote it, has come down to us. But there is no reason to proceed on that assumption. And we may note that the Italian summary of the "Dialogue" on folio 36a of the manuscript, going back possibly to the eighteenth century, does not indicate that any more was present at the time it was made (see Favaro in *Bullettino di Bibliografia e di Storia delle Scienze Matematiche e Fisiche,* 16 [1883], 41).

Whether or not Galileo abandoned the "Dialogue" in the midst of its composition, he certainly did not publish it. Nor did he publish the "Essay" with which it is so closely related. In fact, the existing manuscripts of both works are in no sense "fair" copies. There are numerous deletions, alterations, marginalia, and indications for the insertion of separate sheets, and, in the case of the "Essay," the sequence of the unnumbered chapters is not altogether certain.

The omission of publication makes it impossible to set a terminal date for Galileo's work on these manuscripts. But the author's later discoveries on the subject of motion, the result of researches after he left Pisa, rendered much of the early work obsolete, and whatever was fruitful of that early work was developed over the years in the newer studies. There was less reason, as time passed, for Galileo himself to publish the earlier material as such.

Why did he not publish the material during the years at Pisa? And what is the chronological relation of the "Essay," the "Dialogue," and the brief notes on the subject of motion that I have called the "Memoranda"? I have discussed these questions in a paper, "A Note on Galileo's *De motu*" (*Isis,* 51 [1960], 271–77) and shall only summarize here.

I do not share the generally accepted view that the "Dialogue" was written after the "Essay." In fact, there is reason to suppose that the reverse may have been the case. To be sure, the works were written within a relatively short time of each other, if not actually together, so that the question of priority is perhaps not of large moment. But the unwillingness of Galileo to publish these writings or even, as it appears, to put them in final form and cause them to be circulated raises

some questions. Was it dissatisfaction with literary form or
with the artificiality of the Latin language (L. Olschki, *Galilei
und seine Zeit* [Halle, 1927], p. 204)? Was it the climate of
opinion at Pisa? Or the circumstance of his leaving Pisa and
turning to new tasks and new problems? Or the pressure of
personal problems at a difficult time in Galileo's life? Did
Galileo have misgivings about publishing material that was so
critical of powerful authorities?

I would suggest that a major deterrent to publication may
well have been Galileo's dissatisfaction with the content of
these writings. Certain observed phenomena were quite at
variance with the theory that Galileo was expounding. He
himself frankly admitted this and sought explanations (e.g.,
in the "Essay," *Opere*, I, 273, 302, 330, 333ff.; in the "Dialogue," *Opere*, I, 407), and I am inclined to believe that the
explanations were not wholly to his liking. At all events, the
writings remained in manuscript form when Galileo left Pisa
for Padua. And before he left Padua, he had already solved
the problem of falling bodies and related problems along quite
different lines from those laid down in the early works.

In presenting the translation, I have occasionally added
words in square brackets to clarify the meaning or to indicate
the Latin words that are being translated. The heading at the
top of each page gives the pagination of Favaro's edition to
facilitate reference to the Latin text.

I have also indicated, by italic page and line numbers in
brackets following the relevant passages of the "Dialogue,"
the more important parallels to be found in the Favaro edition
of the "Essay" and "Memoranda." A simple reference indicates a passage so similar as to imply copying between the
texts. The interested reader will thus be able, by turning to
the original passages, to form an opinion as to which was the
original and which the copied passage. Where the parallel
does not necessarily point to such copying or is merely a
reference to related material, I preface the reference with
"cf." —*I. E. Drabkin*.

I share Professor Drabkin's opinion that the Galilean dialogue on motion preceded the composition of the treatise on
motion that he translated in our previous volume in this series.
I think it likely that the "Dialogue" was drafted at Florence,
where Galileo was studying and teaching in the interim between his student days at Pisa and his return to the university
in 1589 as professor of mathematics. Quite possibly he con-

sidered the informal dialogue style unsuitable to the dignity of his new post, or even inadequate for the development of a serious treatise on the whole subject of motion.

In my opinion the most probable reason for Galileo's having withheld the treatise on motion from publication was his knowledge that the theorems concerning the speeds of bodies on inclined planes were simply incorrect. The fact that his correct rule for equilibrium on inclined planes contradicted the rule adopted by his patron, Guido Ubaldo, may have led him to test it to make sure it was right, particularly if he also knew of Cardano's still different rule. It is not a hard experiment to make. But neither is it hard to check whether the speeds of descent through a given height are roughly proportional to the lengths of the planes, and they are not. The failure of his rules of motion to meet even a rough experimental test would account for his holding back the product of so much labor of writing and rewriting.

Ultimately Galileo abandoned his belief that acceleration was but a temporary and accidental event in free fall, as well as his attempted explanation of acceleration, so vastly inferior to Benedetti's. It is hard to see why he forsook his earlier position unless he was forced to do so by some convincing observations. Reason alone would scarcely have changed his view, as he had previously arrived at it by reasons that were satisfactory to him. Perhaps the first relevant observations were those made on pendulums by Galileo about the year 1602, when he wrote of them to Guido Ubaldo. As has been noted, Galileo's proof of the rule for equilibrium on inclined planes linked tangential motions with pendulums—and with descent on concave circular surfaces, of which Guido Ubaldo had written to him. Thus shortly after 1600, when Galileo recognized the fundamental role of acceleration, his early treatise *De motu* became obsolete, and, though he kept it all his life, he never published it.

If my view is correct, the end of the sixteenth century was also the end of the incubation period for the modern science of mechanics.

Niccolò Tartaglia

Nova Scientia

Venice, 1537

Quesiti et inventioni diverse

Venice, 1546

Excerpts translated and annotated by
Stillman Drake

The Newly Discovered Invention of Nicolò Tartalea of Brescia, Most Useful for Every Theoretical Mathematician, Bombardier, and Others, Entitled

New Science

Divided into five books:[1] In the First is demonstrated theoretically the nature and effects of uniformly heavy bodies in the two contrary motions that may occur in them, and their contrary effects. In the Second is geometrically proved and demonstrated the similarity and proportionality of their trajectories in the various ways that they can be ejected or thrown forcibly through the air, and likewise the [proportionality] of their distances. . . .

Venice, 1537

Letter of Dedication to the Duke of Urbino

When I dwelt at Verona in 1531, Illustrious Duke, I had a very close and cordial friend, an expert bombardier at Castel Vecchio, an aged man blessed with many virtues. He asked me about the manner of aiming a given artillery piece for its farthest shot. Now I had had no actual practice in that art (for truly, Excellent Duke, I have never fired artillery, arquebus, mortar, or musket); nevertheless, desiring to serve my friend,

---

1. Only the first two books are here translated. The third book, originally published together with these two, deals with the determination of distances by sighting and calculation. The fourth and fifth books were not published; they are briefly described in the letter of dedication, where it is said that they are partly completed in manuscript. The fourth book was intended to give rules for the calculation of lengths of all shots made by a given cannon from the results of a single shot—a project far beyond Tartaglia's powers. The fifth book was to set forth detailed information concerning the manufacture of gunpowder for military purposes and of fireworks in general. Such information was published in 1540 by Vannocci Biringuccio in his *Pirotechnia,* which became rapidly known throughout Italy. Tartaglia had compiled similar material, which he published in the third book of his *Quesiti,* six years after the appearance of the *Pirotechnia.*

Pagination, shown in the running heads, is based on the 1537 edition, counting the engraved title page and including blank pages.

I promised to give him shortly a definite answer. And after I
had chewed over and ruminated on this matter, I concluded
and proved to him by physical and geometrical reasoning how
the mouth of the piece must be elevated in such a way as to
point straight at an angle of 45 degrees above the horizon. And
to do this most expeditiously, you must have a square made of
metal or hard wood that includes a quadrant with its vertical
pendant, as appears below in the figure; and placing a part of
its longer leg (that is, the part *BE*) in the barrel or mouth of the
piece lying straight along the bottom of the tube, elevate the
said piece so that the perpendicular *HD* cuts the curved side
of the quadrant *EGF* in two equal parts, that is, at the point *G*.

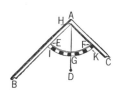

Now the piece will point straight at 45 degrees above the
horizon, because the curved side *EGF* of the quadrant is
divided by astronomers into 90 equal parts, each of which is
called a degree, and one-half of these, that is *GF*, will be 45
degrees. But in that which will be said, we have divided it into
12 equal parts, and in order that your Excellency may see in
the figure what we have described above in words, we have
drawn below a cannon with the square in its mouth arranged
according to the purpose as concluded by us for our friend.
This conclusion seemed plausible to him, yet he had some
doubt, since it appeared to him that the piece might be pointed
too high. This came about through his not understanding our
reasons, not being well grounded in mathematics; nevertheless with some particular experiments, it was in the end completely verified to be so.

Moreover in 1532, when the prefect at Verona was the
noble[2] Leonardo Giustiniano, a chief of bombardiers who was
very close to that friend of ours came there together with another man who is presently chief of bombardiers at Padua. And
one day it happened that the two of them took up the same
problem which our friend had proposed to us; that is, how a
cannon should be pointed in order to shoot as far as possible
over a plain. This friend of our friend, using a square, came to
the same conclusion we had reached, as we have said and
shown in the previous figure. But the other said that the gun
would shoot much farther at two points lower on the square[3]
(which was divided in 12 parts), as appears below in this
figure.

2. *Magnifico,* a word then often applied to any patrician of Venice. The
Giustiniani were, however, among the most distinguished of the ancient
Venetian families.
3. That is, at an elevation of 30 degrees.

Now on this point a certain sum of money was wagered, and finally they came to the trial, and a culverin of 20 [pound shot][4] was taken into the field at Santa Lucia. Each man shot according to his proposition, without any advantage in the powder or in the ball; and he that used our determination shot a distance (as it was reported) of 1972 perches [*pertiche*] of six feet per perch, while the other, who aimed two points lower, shot only 1872 perches. By this trial all the bombardiers and other people saw the truth of our determination, though before this experiment they were in disagreement, and most of them even had the contrary opinion, it having seemed to them that the piece was pointed too high.

Now I urgently wish your Excellency to know that one of these three things may have happened: namely, that the referees made an error in measuring, or that the truth was not brought back to me, or that the second gunner charged [the cannon] more carefully than the first. Because, your Excellency, reasoning shows that the second (that is, the one who aimed two points lower) shot somewhat farther in proportion to the first than he should have done, or else the first shot somewhat less than he should have done in proportion to the second, as may soon be known from the Fourth Book, where we shall treat of the proportions of shots.[5]

Now, your Excellency, having once gone into this matter, I thought seriously of a further trial, and I began (not without reason) to investigate the kinds of motion that can take place in a heavy body. I thus found that there are two such motions, the natural and the violent; and I found these to be totally contrary in events [*accidenti*] through their contrary actions [*effetti*]. I also found, by reasons evident to the intellect, that it is impossible for a heavy body to move with natural motion and violent motion mixed together. I then investigated, with demonstrative geometrical reasons, the character of the trajectories or violent motions of heavy bodies according to the various ways in which they may be ejected or thrown violently

4. In the first book of his *Quesiti* (Venice, 1546, f. 19v) Tartaglia describes a 20-pound culverin as being 10 feet in length, weighing 4300 pounds, and requiring seven oxen to move it. The diameter of the iron ball which it shot would have been about 4⅜ inches according to another of Tartaglia's tables (*Quesiti*, f. 36r). A cannon firing the same ball would have been seven to eight feet in length and about one-half the weight of the culverin.

5. This book was never published, and the topic is not discussed fully in the *Quesiti;* hence it is not known what method Tartaglia may have used in computing the theoretical ratios of shots with a 20-pound ball at 30 degrees and at 45 degrees. But see notes 9 and 28, below, and the last proposition of Book II.

through air. Beyond this, I made certain, by means of demonstrative geometrical reasons, that all shots with every kind of artillery, large and small, equally elevated above the plane of the horizon, or equally oblique, or along the plane of the horizon, are similar to one another and consequently proportional, as are their distances also. Next, I knew by Archimedean reasoning[6] that the distance of the aforementioned shot elevated at 45 degrees above the horizon was about ten times the straight carriage[7] of a shot made in the plane of the horizon, which is called "point-blank" by bombardiers. With such evidence, Excellent Duke, I found by geometric and algebraic reasons that a ball shot toward a point 45 degrees above the horizon goes about four times as far in a straight line[8] as it goes when shot in the plane of the horizon, or (as I said) "at point-blank." From which it is manifest how a ball shot by a given piece of artillery holds to a straight line farther [when shot] in one direction than in another, and consequently has greater effect. By calculation, I found the proportion of the increase or diminution in the shots made by every piece of artillery when it is raised or lowered over the horizontal plane. And similarly I found the method of reasoning out the variation of the said shots for every piece, large or small, by knowing only the result of a single shot, assuming that the charge is always equal.[9] From this, I investigated the proportion and the order of shots of a mortar, and likewise found the manner of calculating quickly the variation of shots merely by means of a single shot. Further, since by evident reasons I

6. The edition of 1558 and later editions say "by physical reasoning" *(con ragion naturale)*. I have not seen the edition of 1550. Since there is reason to believe that Tartaglia did not see the authentic works of Archimedes before 1539, the change suggests that tradition wrongly ascribed to him some writings on projectiles.

7. That is, the total distance traveled as measured horizontally from the point of discharge to the point at which the ball strikes the horizontal plane.

8. That is, the distance traveled "in straight violent motion" only, before the ball's trajectory begins to curve circularly, according to the theory set forth in the text and shown in the subsequent diagrams of Book I.

9. If the point-blank range is taken as unity, and the "straight" travel at 45 degrees is four times as great, then the horizontal distance traveled when "curved movement" begins, for the maximum trajectory, is $2\sqrt{2}$, or about 2.8; but if total horizontal travel in this trajectory is 10, the "curved movement" must be about 2.5 times as great horizontally as the initial "straight movement." The final proposition of Book II shows how Tartaglia attempted to reconcile his theory with the assumptions. But from the diagrams in Book I, it would appear that Tartaglia thought the observed horizontal motion to be about equally divided between the "straight" and "curved" portions. The difficulties implied here (empirical knowledge versus attempted mathematical representation) precluded his development of the promised general ballistic formulas.

knew that a cannon could strike in the same place with two different elevations or aimings, I found the way of bringing about this event, a thing unheard of and not thought by any other, ancient or modern.

But I then considered, Sire, that all these things would be of little use to a bombardier if the distance of the place he must hit were unknown to him. For example, suppose he must shoot at a visible place of which the distance is hidden to him. Of what avail would it be in this case, O Excellent Duke, for him to know that his piece would shoot 1356 paces at such and such an elevation, and at such another, 1468, and at still another, 1574? Reasoning thus from one degree to another would do him no good, for unless he knows the distance, he will not know the elevation at which to place his artillery piece to hit the desired place. It therefore follows that there are two principal things necessary to a loyal bombardier who wants to shoot by reason and not at a hazard, of which one is of no use without the other. First, he must know roughly how to find and calculate by sight the distance of the place at which he is to shoot; and second, he must know the extent of his artillery shots according to their various elevations. Knowing both these, he will not err much in his shots; but lacking either of them, he can never shoot by reasoning but only by his judgment. And if by chance on the first shot he hits the place or comes close to the place he wishes, it is rather by luck than by science, especially in long shots.

Therefore I found a new method of investigating quickly the heights, the hypotenusal (or diametral) distances, and also the horizontal distances of visible things. This is not completely a new thing, for indeed Euclid in his perspective shows it briefly, theoretically and in part. Likewise Johann Stoeffler, Orontio [Finé], Pietro Lombardo, and many others have given rules for such matters, here with the sun, there with a mirror, one with the quadrant, another with the astrolabe, some with two rods, others with a pole (called Jacob's staff), and in various other ways.[10] But, your Excellency, I have found a new and expedient way to investigate the said distances, which is quick and easily understood by anyone, and is less subject to error than any other; yet this has not been

---

10. Long before the invention of cannons, engineers had worked out practical methods for the determination of distances of inaccessible places; these were used in various surveying and construction practices as well as in war. To the writers mentioned by Tartaglia, many more might be added; but the matter is of no consequence to mechanics, and in any case Tartaglia added nothing of value to what had long been known about it.

put down by anyone else. Especially the hypotenusal or diametral distances, and also the horizontal, which are the most useful to the bombardier of all the dimensions, for to him it is not very necessary to know the height of a thing elevated perpendicularly above the horizon, nor the depth of anything, nor even the breadth of something wide. Only the hypotenusal and horizontal distances are very significant to him, as will be shown in the Fourth Book.

Moreover, out of curiosity, I undertook to run through the digests of Avicenna and many other excellent physicists[11] to learn the origin and the nature of different kinds of gums, salts, oils, and distilled waters and of various mineral simples and minerals not produced by nature but artificially manufactured. Also, I found some of their particular properties relating to the art of fireworks; similarly, I investigated which of the said materials were in agreement and which were not in accordance (so as to burn together), and as a result I found the method of compounding various and diverse species of explosives, not only very useful for the defense of walled places but also appropriate in many other pursuits.[12]

Through these discoveries, I was going to give rules for the art of the bombardier and bring this to the greatest possible detail through some particular experiments: for truly, as Aristotle says (in the seventh book of his *Physics,* text 20)[13] universal sciences are drawn from particular experiments. But then one day I fell to thinking it a blameworthy thing, to be condemned—cruel and deserving of no small punishment by God—to study and improve such a damnable exercise, destroyer of the human species, and especially of Christians in their continual wars. For which reasons, O Excellent Duke, not only did I wholly put off the study of such matters and turn to other studies, but also I destroyed and burned all my calculations and writings that bore on this subject. I much regretted and blushed over the time I had spent on this, and those

11. Later editions say, ". . . through the various modes observed by our ancient physicists, and also by moderns in the manufacture of fire[work]s, and among physicists [whose writings] I investigated."

12. It was this information that was to comprise the fifth book of the *Nova Scientia* and was published, at least in part, in the *Quesiti* of 1546.

13. Probably a reference to *Physics,* 247b.6–8: "It is when knowledge meets with the particular object that it knows in a manner the particular through its knowledge of the universal," on which Sir David Ross comments: "The point here is that knowledge of the universal includes a sort of knowledge of the particular, out of which it was originally built up." *(The Students' Oxford Aristotle* [Oxford, 1942], vol. II, *Physics.)*

details that remained in my memory (against my will) I wished never to reveal in writing to anyone, either in friendship or for profit (even though it has been requested by many), because such teaching seemed to me to mean disaster and great wrong.

But now, seeing that the wolf is anxious to ravage our flock, while all our shepherds hasten to the defense, it no longer appears permissible to me at present to keep these things hidden.[14] I have hence resolved to publish them partly in writing and partly by word of mouth, to every faithful Christian, so that each may be better fitted in offense as well as defense. And I am very sorry, Sire, that I ever abandoned this study, since I am certain that if I had kept on without pause I should have discovered things of more value, as I hope soon to do. But since the present is certain, time is short, and the future is always doubtful, I want to speed first that which I now have; and to carry this out in part, I have hastily composed the present little work. And like every river that flows to approach and unite with the sea, this will seek to approach and unite with your greatness, your Excellency being the greatest of mortals in warlike virtue. For just as the abundant sea, which has no need of water, does not disdain to receive a little stream, so I hope that your Excellency will not disdain to accept this, in order that the expert bombardiers of this our most illustrious ducal dominion, subjected to your Excellency, in addition to their fine and practical skill, may be better instructed by reason and able to carry out your mandates. And if in these three books I have not fully satisfied your Excellency together with the said expert bombardiers, I hope in a short time to do so with the practice of the fourth and fifth books, not indeed in print (for many reasons) but in writing or by word of mouth; to satisfy, in part, them and your Excellency, to whom I devotedly recommend myself.

Venice, at the new houses in San Salvatore
20th December, 1537
Your Excellency's humblest servant,
Nicolò Tartalea of Brescia.

---

14. The wolf was the Turkish emperor Suleiman, who was threatening invasion of Italy and was particularly menacing Venice. Francesco Maria della Rovere, Duke of Urbino, to whom this letter is addressed, was employed by the Venetian government to organize its defense.

The First Book of The New Science of Niccolò Tartaglia, Commencing with the Definitions, or Descriptions of Principles Known Directly from the Things Presumed

A body is called uniformly heavy *which, according to the weight of the material and its shape, is apt not to suffer noticeable resistance from the air in any motion.*

Every body (according to the physicists) is either simple or compound; the simple bodies are five—earth, water, air, fire, and sky. All other things are said to be composed of the foregoing, including men, animals, plants, stones, cloths, metals, and every other kind of body. Of the said five simple bodies, four are called elemental; that is, earth, water, air, and fire. The other, that is, the sky, is called the quintessence. Of the four elements (as Avicenna says in the second doctrine of the first *Fen* of his first book), two are light and two heavy. The light [elements] are fire and air. The heavy are earth and water, but Averroës in his fourth book on the heavens and world (text 19) will have it that all the said bodies in their own surroundings have some heaviness (except fire) or some lightness (except earth). Whence it would follow that air in its own surroundings would partake of weight. Thence it would follow that every body compounded of the four elements would have weight in air. Nevertheless, by *uniformly heavy body* we mean here only that which, according to the heaviness of the material and its form, is suited not to suffer sensibly the oppositions of the air in any motion. "According to the material"—that is, whether it is of iron, or lead, or stone, or any other material similar in heaviness. "According to its form"—that is, it shall have the quality of being not apt to suffer sensibly (by virtue of its form) the said opposition in any motion it possesses. Whence among the shapes or forms of bodies, the wedge or pyramidal shape would come first among those least apt to shrink before opposition from the air, supposing it designedly kept with the vertex or sharp point of the angle proceeding against the impetus of the air. But since if not kept this way it would not conform to the rule for being *uniformly heavy,* we shall assume the spherical shape to be unconditionally best suited of all forms to suffer least the opposition of the air in every kind of motion, being most ready to move in any direction, and *uniformly heavy* in all.

DEFINITION II

Uniformly heavy bodies *are called* similar and equal *when there is no essential or accidental difference in them.*

DEFINITION III

An instant *is that which has no parts.*

The instant in time and in motion is like the geometrical point in magnitude; that is, it has no parts, but is indivisible; and consequently it is neither a time nor a movement, though indeed it is the beginning and end of every time and of every finite movement. And it is inherently the end of time past and not a part of time future, as Aristotle shows us in the sixth book of his *Physics* (text 24).[15]

DEFINITION IV

Time *is a measure of movement and of rest, the endpoints of which are two* instants.

Time, for scientists, has been defined in different ways; that is, some say (as we have said above) that it is a measure of movement and of rest. Others determine it to be the inducer of motion of variable things. Some conclude it to be the changes of things, which are known in many ways by subtle investigation. And others say it is a moving epoch that is quickly gone. Of these definitions, we have taken the first by reason of its being best adapted to our purpose, saying that time is a measure of movement and rest. For by means of a material measure (in many lands called "perch," divided into six feet, with 12 inches to the foot) one comes to a knowledge of the length and breadth and height of material bodies. And similarly by means of a measure of time called a year, divided into 12 months, each month ordinarily of 30 days, each day of 24 hours, and each hour of 60 minutes, the difference of motion of bodies is known; that is, their speed and slowness. For how is the motion of one of the seven planets known to be faster than that of another? unless by the measure of these movements, namely a year with its parts (that is, months, hours, and minutes). This is clearly apparent in astronomical determinations. And the endpoints of this year, that is, its beginning and end, are two instants. The same must be understood of its other parts, and of every finite time.

15. *Physics*, 234a. 1–4.

DEFINITION V

The movement of a uniformly heavy body *is that trans-mutation which it makes occasionally from one place to another, the endpoints of which [movement] are two* instants.

All scientists, and especially Aristotle in the fifth book of his *Physics* (text 9),[16] have defined movement to be a change or transmutation. Some say the kinds of movement or trans-mutation are six; that is, Generation, Corruption, Augmentation, Diminution, Alteration, and Local Motion. But Aristotle in the place cited holds the mutations to be three, and no more; that is, change of quantity, of quality, and of place. Of these kinds, we have taken only the last (because the others are irrelevant), and say that the movement of a uniformly heavy body is that transmutation which it makes in time from one place to another—that is to say, from above to below and below to above, from here to there, from the right side to the left, or vice versa. And the endpoints of such movements (that is, the beginning and end) are two instants.

DEFINITION VI

Natural movement of uniformly heavy bodies *is that which they make by nature from a place above to another beneath, perpendicularly, without [the application of] any force whatever.*

DEFINITION VII

Violent movement of uniformly heavy bodies *is that which they make under [application of] force from below to above, from above to below, or from here to there, by reason of some motive power.*

DEFINITION VIII

The movements *of* uniformly heavy bodies *are said to be* equal *when the said bodies are similar and go with equal speed, that is, traverse equal intervals in equal times.*

DEFINITION IX

Any resting body is called a resistant *that by offering resistance to a* uniformly heavy body *is offended by it in any motion.*

16. *Physics,* 225a.34–225b.9.

DEFINITION X

Those bodies are called similar resistants *which are equally offended by similar* uniformly heavy bodies *in* equal movements, *and in unequal movements are unequally offended in such a way that that which resists the speedier* [*body*] *is the more offended.*

DEFINITION XI

The effect *that a* uniformly heavy body, *in any motion, causes in a* resistant *is called* offense, *or* percussion, *or* penetration.

DEFINITION XII

And when the percussions *or* penetrations *of similar* uniformly heavy bodies *are equal, they are called* equal effects, *and if unequal,* unequal effects.

DEFINITION XIII

Any artificial machine or material is called a motive power *that is suited to drive or shoot a* uniformly heavy body *forcibly through air.*

DEFINITION XIV

Motive powers *will be called* similar and equal *when there is no essential or accidental difference in them in driving similar and equal* uniformly heavy bodies. *But when there is any accidental difference in them, they are called* dissimilar and unequal.

FIRST SUPPOSITION

It is assumed that the uniformly heavy body in every movement goes more quickly when it makes, or would make (by the axioms [below]), a greater effect in a resistant.

SUPPOSITION II

It is assumed that two similar and equal uniformly heavy bodies traverse (or will traverse) equal spaces in equal times terminating in two instants, whenever the said bodies pass with equal speed.

SUPPOSITION III

It is assumed that, where similar and equal uniformly heavy bodies would (by the axioms) make equal effects in similar resistants, they would pass through such instants or places with equal speed.

SUPPOSITION IV

But where they would make unequal effects, it is assumed that they would pass with unequal speed, and that that which would make the greater effect would pass more swiftly.

SUPPOSITION V

The effects of equal and similar uniformly heavy bodies made in similar resistants at the last instant of their violent motions are assumed to be equal.

FIRST AXIOM

The greater the altitude from which a uniformly heavy body comes in natural motion, the greater the effect it will make in a resistant.

But it must be noted that the said altitude is to be understood with respect to the resistant.

SECOND [AXIOM]

If similar and equal uniformly heavy bodies come, in natural motion, from equal altitudes upon similar resistants, they will make therein equal effects.

THIRD [AXIOM]

But if they come from unequal altitudes, they will make therein unequal effects, and that which comes from the greater altitude will make the greater effect.

But it must be noted that the said altitudes must be understood with respect to the resistants.

FOURTH [AXIOM]

If a uniformly heavy body in violent motion encounters a resistant, the closer the said resistant is to the beginning of such motion, the greater the effect that will be made in it by the said body.

FIRST PROPOSITION

Every uniformly heavy body in natural motion will go more swiftly the more it shall depart from its beginning or the more it shall approach its end.[17]

---

17. The discussion of acceleration in free fall is of particular importance to its subsequent development by Benedetti and by Galileo. It should be noted

Example: Let there be the three different heights *A, B,* and *C,* in a straight line, as appears below. And from the height *A* there shall accidentally fall freely a uniformly heavy body; doubtless that body, not finding resistance, would go with natural motion to the earth, making its journey similarly to the line *DEFG.* Now I say that the movement of such a body would be of such condition that the more it departed from its beginning (that is, from the instant or point *D*) or approached its end (that is, the instant or point *G*), the more swiftly it would go. For the said body in such movement (by the First Axiom) would make greater effect in a resistant which lay beyond the height *C* than at the point *B.* It would therefore follow that the said body (by the First Supposition) would go more swiftly through the space *EF* than through the space *DE.* Similarly, the said body (by the First Axiom) would make a greater effect in a resistant if it were at the point *G* than if it were at the height *C.* It would therefore follow (by the same First Supposition) that the said body would go more swiftly through the space *FG* than through the space *EF;* and if it could pass the point *G,* the earth yielding place to it as does the air, it would go continually increasing in speed to the center of the earth, and then repose there (in the general opinion of philosophers), so that when the said body approached the center, it would be of a swifter motion than it had been in any space passed, which is the proposition.

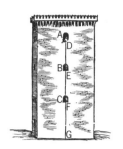

This same is also verified in anything that goes toward a desired place, for the more closely it approaches the said place, the more happily it goes, and the more it forces its pace, as appears in a pilgrim that comes from a distant place; for when he nears his country, he naturally hastens his pace as much as he can, and the more so, the more distant the land from which he comes. Therefore the heavy body does the same thing in going toward its proper home, which is the center of the earth; and when it comes from farther from that center, it will go so much the more swiftly approaching it.

[Added to later editions: The opinion of many is that if there were a tunnel that penetrated diametrically through the whole

---

that Tartaglia considers it a matter of indifference whether the acceleration be considered from the point of origin or from the point of termination; further, that he takes it for granted that the increase in speed is related to space traversed or to be traversed, saying nothing about the time elapsed. It would appear that most writers up to the time of Galileo, and Galileo himself until about 1610, simply assumed that there was no difference between the two concepts of space and time as measures of acceleration, and leaned toward the former as the easier to conceive.

earth, and through this there were let go a uniformly heavy body, as said above, then that body when it arrived at the center of the world would immediately stop there. But I say that this opinion (that it would stop immediately upon arriving there) is not true; instead, by the great speed which would be found in it there, it would be forced to pass by with very violent motion, running much beyond the said center toward the sky of our subterranean hemisphere; and thereafter it would return by natural motion toward the same center, and arriving there it would pass by once more, for the same reasons, with violent motion toward us; and yet again it would return by natural motion toward the same center, and pass it still again with violent motion, thereafter returning by natural motion, and so it would continue for a time, passing with violent motion and returning by natural motion, continually diminishing in speed, and then finally it would stop at the said center. By which it is a manifest thing that violent motion is caused by natural motion, and not the reverse; that is, natural motion is never caused by violent motion, but is rather its own cause.][18]

First Corollary. — From which it is manifest also in what manner every uniformly heavy body goes more slowly at the beginning of natural movement, and more swiftly at the end, than at any other place; and the longer the space it shall pass, the faster it shall go at the end.

Corollary II. — Also it is manifest that a uniformly heavy body in natural motion cannot pass through two different instants with equal speed.

PROPOSITION II

All similar and equal[19] uniformly heavy bodies leave from the beginning of their natural movements with equal speed,

18. From this later addition it appears that Tartaglia had adopted the idea of impressed force, or of self-exhausting impetus, associated with the name of Albert of Saxony, as opposed to the idea of a form of impetus diminishing only with external resistance (at least in certain kinds of motion), associated with the name of Jean Buridan. A similar discussion of the behavior of a body in a tunnel through the earth is given by Galileo in the *Dialogue,* where it is used to destroy the Aristotelian distinction of natural and violent motions as produced by internal and external forces (Galileo, *Dialogue* [Berkeley, 1953], trans. S. Drake, p. 236). Galileo, however, implies that the reciprocations would be perpetual (*Dialogue,* p. 227), though in another passage (*Dialogue,* pp. 135–36) he suggests that, in small oscillations about the center, a body would ultimately come to rest there.

19. Benedetti, who for a short time was Tartaglia's pupil, was first to declare the restriction of equality in weight to be irrelevant to the first part of

but, coming to the end of their movements, that which shall have passed through a longer space will go more swiftly.

Let there be the two pairs of different heights *AB* and *CD*, each placed in a straight line as appears below, with the height *A* as far from the height *B* as the height *C* is from the height *D*, and let a uniformly heavy body fall accidentally from the height *A* and another from the height *C*, the two bodies being similar and equal. I note that those bodies will go to the earth with natural motion, and their passages will be straight and perpendicular to the ground, resembling the two lines *GF* and *IE*. Now I say that these bodies would leave from the beginning (that is, one from the instant or point *G* and the other from the instant or point *I*) with equal speed, but, coming to the end of their movements, that is, to the two instants *E* and *F*, the one which came from the height *A* would go more swiftly than the other, because it would have passed through a longer space, which is the space *AF*. For the height *B* is as distant from the height *A* as the height *D* from the height *C* (by assumption); hence the body that falls from the height *A*, striking on a resistant projecting from the height *B*, would not make in it a greater effect (by the Second Axiom) than that which would be made by the other that should fall from the height *C* upon a similar resistant projecting from the height *D*. Whence (by the Third Supposition) one of the two said bodies would pass through the height *AB* at the point *H* and the other through the height *CD* at the point *K* with equal speed to that (by the Second Supposition) by which the said two bodies would pass, in equal times, the one the space *GH* and the other the space *IK*. Therefore the said two bodies would leave from the beginning of their movements (that is, one from the instant *G* and the other from the instant *I*) with equal speed, which is the first proposition. And since the body that came from the height *A* would make a greater effect in a resistant at the instant *F* (by the Third Axiom) than that which came from the height *C* would make on a similar resistant at the point *E*, then (by the First Supposition) the said body that came from the height *A*, arriving at the end of its movement (that is, at the instant or point *F*),

---

this proposition. G. B. Baliani noted the fact in 1611, when he was observing the behavior of different cannonballs at Savona. Tartaglia, who seems to have had similar opportunities to observe the fact, appears not to have noted it, or at least never to have discussed the effect (or lack of effect) of weight on speed in free fall.

would go more swiftly than that which came from the height *C*, arriving at its end, that is, at the instant or point *E;* and this is the second proposition. To show this second proposition in another way: Of the whole line or passage *GF*, take (by Euclid I.3) the part *GM* equal to the passage or line *IE;* and since all similar and equal uniformly heavy bodies from the beginnings of their natural movements leave with equal speeds (as was demonstrated above), then that body which left from the height *A* would go as swiftly through the space *GM* as that which parted from the height *C* would go through the space *IE;* that is, both would pass in equal times. And since the said body parting from the height *A* (by the preceding proposition) would go more swiftly in the space *MF* than in the space *GM*, by common consent it would go still more swiftly in the said space *MF* than the other in the space *IE*, which is the same second [part of the present] proposition.

PROPOSITION III

A uniformly heavy body in violent motion will go more weakly and slowly the more it departs from its beginning or approaches its end.

Example: Let there be a motive power at the point *A* that can drive a uniformly heavy body through air, and let the entire shot that can be made by the said power with this body be the whole line *AB*. I say that the farther that body departs from its beginning (that is, the instant *A*) or approaches its end (that is, the instant *B*), the more its speed is retarded. Which is demonstrated in this way. Let us divide the said line or passage *AB* into many spaces, and let these be *BC*, *CD*,

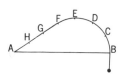

DE, EF, FG, GH, and *HA*. Now since the said body (by the Fourth Axiom) would make a greater effect on a resistant at point *C* than at point *B*, so that (by the First Supposition) the said body would go more swiftly through the point *C* than through the point *B*, and similarly through the space *DC* than through the space *CB*, then for the same reasons the said body would pass more swiftly through the space *ED* than through the space *DC*, and through the space *FE* than through the space *ED*, and through the space *GF* than through the space *FE*, and through the space *HG* than through the space *GF*, and through the space *AH* than through the space *HG*. And if the beginning of such violent motion were farther forward, so much the more swiftly would it go in the successive spaces, which is the proposition.

The same is verified in any person forcibly driven toward a detested place; for the more he approaches the said place, the sadder his mind becomes, and the more he seeks to retard his journey.

First Corollary.—Whence it is manifest how a uniformly heavy body goes more rapidly at the beginning of every violent motion, and at the end more slowly, than at any other place; and the longer the space it shall have to traverse, the more rapidly it must go at the beginning of the movement.

Corollary II.—Also it is manifest how a uniformly heavy body in violent motion cannot pass through two different instants with equal speed.

PROPOSITION IV

All similar and equal uniformly heavy bodies, coming to the end of their violent motions, will go with equal speed; but, from the beginning of such movements, that which shall have to pass through the longer space will leave more swiftly.

Example: Let there be two dissimilar and unequal motive powers, one at the point A and the other at the point C, that must drive two similar and equal uniformly heavy bodies violently through air, and let the entire shots that must be made by the said two powers with these bodies be, the one, the line AB and, the other, the line CD. I say that these two bodies reaching the end of these two violent movements, that is, the one at the instant or point B and the other at the instant or point D, will go with equal speed. But from the beginning of these movements of theirs (that is, the one from the instant A and the other from the instant C) they would leave with unequal speed, because that which must pass through the journey or space AB (being longer than the other) would leave more swiftly from the instant A than the other from the instant C, which will be demonstrated in this manner. If the said two bodies should find any resistant at the two instants D and B of similar and equal resistance, they would make thereon two effects (by the First Supposition), equal (by the Third Supposition) where they go with equal speed; this is the first proposition. To demonstrate the second, from the greater passage, or line AB, let us cut off in imagination the part BK equal to the lesser passage, or line CD; and since the said two bodies coming to the instants D and B would have equal speed (as has been shown above), they should have passed with equal speed through spaces equally distant from the afore-

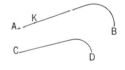

said two places or instants *B* and *D* (by the Second Supposition). Therefore the said two bodies would travel with equal speed, the one through the partial space *KB* and the other through the whole space *CD;* that is, they would traverse these in equal times. And since in violent motion the farther a heavy body shall depart from its beginning (by the Third Supposition) the more weakly and slowly it will go, therefore the body that came from the instant *A* would go more swiftly through the space *AK* than through any place in the partial space *KB*. Thence it follows (by common knowledge) that that body which came from the instant *A* would go through the space *AK* more swiftly than the other would go at any place in the space *CD*. Therefore the body that came from the point or instant *A* would leave from that instant more rapidly than that which came from the instant *C* would leave from that instant, which is our second proposition.

PROPOSITION V

No uniformly heavy body can go through any interval of time or of space with mixed natural and violent motion.

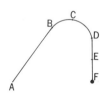

Example: Let there be a motive power at the point *A* which must shoot a uniformly heavy body violently through air, and let the entire passage that the said body should make by that shot be the whole line *ABCDEF*. I say that the said body will not pass through any part of that trajectory with violent motion and natural motion mixed together, but it will pass either entirely with pure violent motion through this [whole space] or with pure violent motion [through] part and with pure natural motion [through] part; and that the instant which terminates the violent motion will be the beginning of the natural motion. And if it were possible (for the adversary) for it to pass some part with violent and natural motion mixed together, let us assume that this part is *CD*. Then it would follow that the said body, passing from the point *C* to the point *D,* would go increasing its speed according to its share in the natural motion (by the First Proposition), and similarly that it would go diminishing in speed according to its participation in the violent motion (by the Third Proposition), which would be an absurd thing, for such a body would have to be increasing and diminishing in speed at the same time. Therefore, the contrary destroyed, the proposition remains.[20]

20. This line of reasoning, which remained the principal obstacle to an understanding of actual motions from the time of Aristotle to that of Galileo, is not so weak or implausible as it may seem to us. Under the assumption that

PROPOSITION VI

Any resistant will be least offended, by a uniformly heavy body ejected violently through air, at that instant which distinguishes the violent motion from the natural.

Example: Let there be a motive power at the point A, which shall drive a uniformly heavy body violently through air; and let the entire line $ABCDEF$ be passed by that body in that shot, and let the point D be the place of the instant wherein the violent is separated from the natural motion. I say that any resistant will be less offended at the point D than at any other place of the said passage. For the body would pass more slowly through the instant D than through any other place of the violent passage $ABCD$ (by the First Corollary of the Third Proposition) and consequently would make less effect. Similarly the body would pass more slowly through the instant D (by the First Corollary of the First Proposition) than through any other place of the natural passage DEF and consequently would make less effect. Therefore if the resistant were struck at the point C or at the point E by the body, it would be more offended than if struck at the point D, because the body would pass more swiftly through the point C (with violent motion) and through the point E (with natural motion) than through the point D; which is the proposition.

End of the First Book

The Second Book

FIRST DEFINITION

Straight movement of uniformly heavy bodies *is that which they make from one place to another directly, that is, by a straight line.*

As it would be to move from point A to point B so as to follow the line AB.

A⎯⎯⎯⎯⎯⎯⎯B

DEFINITION II

Curved movement of uniformly heavy bodies *is that which they make from one place to another curvedly, that is, by a curved line.*

------

deceleration is an essential (not an accidental) characteristic of violent motion, and acceleration an essential characteristic of natural motion, the argument is most compelling.

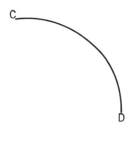

As it would be to move from the point *C* to the point *D* as along the line *CD*.

DEFINITION III

Movement partly straight and partly curved of uniformly heavy bodies *is that which they make from one place to another partly directly and partly curvedly; that is, along a line partly straight and partly curved.*

Such a way of motion would be from the point *E* to the point *G,* the line *EFG* being so situated, provided, however, the said two parts are directly joined — that is, the straight part *EF* with the curved part *FG* — so that no angle exists at the point *F.* For if there were an angle, one could not say that this was one continuous motion, but rather there would be two different motions; just as one could not say the whole quantity *EFG* was one straight line, but two lines, that is, one straight and the other curved. And this was what was to be explained.

DEFINITION IV

Any circular plane is called horizontal *which divides the lower from the upper hemisphere, or which divides into two equal parts any* uniformly heavy body, *even though it be violently ejected or thrown through the air; and the* horizontal plane *is concentric with the said body.*[21]

DEFINITION V

Radius of the horizon *is the name of that line which starts at the center and terminates in the circumference* [*of the horizon*], *going straight* [*but horizontally*] *in that direction in which the uniformly heavy body is thrown violently through the air.*

DEFINITION VI

Perpendicular of the horizon *is the name of that line which starts at the pole of the horizontal plane (called the* zenith*) and arrives perpendicularly at its center, and if continued would arrive finally at the center of the world.*

---

21. That is, "horizontal" means not only the plane tangent to the earth, but all planes parallel thereto. Tartaglia does not actually use this generalized definition, but it suggests that he had noticed the exceptional role of the horizontal plane in the theory of natural and violent motions. By "circular plane" Tartaglia here means "plane bounded by the circumference of the universe"; not (as with Galileo) a spherical surface concentric with the earth.

DEFINITION VII

But that part which goes from the center to the pole is called the perpendicular above the horizon, *and the other which is from the said center to the center of the universe is called the* perpendicular under the horizon.

DEFINITION VIII

The violent trajectory or motion of a uniformly heavy body *is said to be* along the plane of the horizon *if at its beginning it extends along the* radius of the horizon.

DEFINITION IX

The violent trajectory or motion of a uniformly heavy body *is said to be* elevated above the horizontal *when at the beginning it extends in such a way as to make an acute angle above the horizontal with the radius of the horizon; and it is said to be the* more elevated, *the greater an acute angle it makes; but when it makes a right angle, it is said to be* erect to the horizon.

DEFINITION X

The violent trajectory or motion of a uniformly heavy body *is said to be* elevated 45 degrees above the horizon *when at the beginning it goes in such a way as to divide into two equal parts the right angle made by the* perpendicular of the horizon *and the* radius of the horizon.

DEFINITION XI

The violent trajectory or motion of a uniformly heavy body *is said to be* oblique under the horizontal *when in the beginning it extends in such a way as to make an acute angle beneath the horizontal with the* radius of the horizon; *and it is said to be the* more oblique, *the greater the acute angle it makes. But when it makes a right angle, it is called* erect under the horizon.

DEFINITION XII

The violent trajectories or motions of uniformly heavy bodies *are said to be* equally elevated above the horizontal *when at the beginning they extend in such a way as to make equal acute angles above the horizontal with the* radius of the horizon, *and similarly* equally oblique *when in the beginning they make equal acute angles beneath the horizontal with the said radius.*

DEFINITION XIII

The violent trajectory or motion of a uniformly heavy body *is said to be* along the perpendicular of the horizon *when its beginning and end are in the said perpendicular; that is, when it is erect above or beneath the horizon.*

DEFINITION XIV

The distance *of the violent trajectory or motion of a* uniformly heavy body *is taken along that interval which goes in a straight line from the beginning to the end of such violent motion.*

FIRST SUPPOSITION

All the natural trajectories or movements of uniformly heavy bodies are parallel to one another and to the perpendicular of the horizon as well.

Truly two natural trajectories or motions of uniformly heavy bodies can never be perfectly parallel to one another, nor to the perpendicular of the horizon, for if the earth should give way to them as does the air, no doubt they would run together in the center of the earth; hence (by the last definition of Euclid I) they would not be, as I have said, parallel. Nevertheless, since this error is undetectable in a short space, we shall suppose them all parallel to one another and to the perpendicular of the horizon as well.

SUPPOSITION II

Every violent trajectory or motion of uniformly heavy bodies outside the perpendicular of the horizon will always be partly straight and partly curved, and the curved part will form part of the circumference of a circle.

Truly no violent trajectory or motion of a uniformly heavy body outside the perpendicular of the horizon can have any part that is perfectly straight, because of the weight residing in that body, which continually acts on it and draws it toward the center of the world.[22] Nevertheless, we shall suppose that part which is insensibly curved to be straight, and that which

22. It is often said that Tartaglia was unaware of this fact in the *Nova Scientia,* and corrected it only in the *Quesiti.* His differing treatments of it are better described as a progressive leaning away from strict Aristotelian theories of motion. See the portions of *Quesiti,* Bk. I, translated below.

is evidently curved we shall suppose to be part of the circumference of a circle, as they do not sensibly differ.

SUPPOSITION III

Any uniformly heavy body at the end of any violent motion
that is outside the perpendicular of the horizon will move with
natural motion which will be tangent to the curved part of the
violent motion.

For example, if a uniformly heavy body is ejected or thrown
violently through air, outside the perpendicular of the horizon,
I say that at the end of that violent motion (not encountering
resistance), it will move with natural motion which will be
tangent to the curved part of the violent motion, similarly to
the line *ABCD*, of which the part *ABC* will be the trajectory
of the violent motion, and the part *CD* will be the trajectory
made with natural motion, which will be continuous and tangent to the curved part *BC* at the point *C;* and this is what we
wish to conclude.

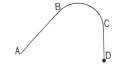

SUPPOSITION IV

The most distant effect from its beginning that can be made
by a uniformly heavy body in violent motion upon any plane,
or any straight line, is [in] that [violent motion] which terminates precisely in this plane (or in that line), [the body] being
ejected or thrown by a given motive force.

For example, let there be a motive force at the point *A* which
has ejected or thrown the uniformly heavy body *B* violently
through air, the trajectory of which let be the line *AEDB*, and
let us assume the point *D* to be the instant that distinguishes
the violent trajectory or motion *AED* from the natural trajectory or motion *DB;* and from the point *A* to the point *D*
extend the line *ADC*. Now I say that the point *D* is the most
distant effect from the point *A* that the said body *B* can make
upon the line *ADC* or upon that plane in which the line *ADC*
is situated, when aimed in that way. For if the force *A* should
throw the body *B* at greater elevation above the horizontal,
it would make its effect by natural motion upon the same line
ADC, as appears in the line or trajectory *AFG*, at the point *G;*
which effect *G*, I say, would be closer to the point *A*, that is
to say, to the beginning of that motion, than the effect *D* would
be. For the said body *B* will not come to terminate its violent
motion in the said line *ADC;* rather, that will terminate above
in the point *F*. And the greater the elevation of the throw, the

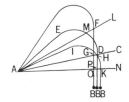

more its effect upon the line *ADC* will move toward the point *A*, for the end of the violent motion will depart so much the more from the said line *ADC*; that is, it will terminate so much the higher. Similarly if the same power should throw the same body *B* with less elevation than in its trajectory or line *AED*, as in the trajectory or line *AIHK*, it will make its effect by violent motion upon the said line *ADC* as at the point *H*; which effect *H*, I say, will be closer to the point *A* than that made at the point *D*. For that violent motion will go to end below the line *ADC* in the point *K*, and the more the power *A* shall be lowered in throwing the body *B*, the closer to *A* the body *B* will make its effect upon the line *ADC*. For the more it shall be lowered, the farther beneath the line *ADC* its violent motion will end. The same must be understood of any other shot; for example, drawing the line *AFL* from the point *A* to the point *F* (the end of the violent motion *AF*), I say that the body *B* thrown by the same power in another way will never be able to reach the point *F*, as is manifest in the trajectory *AEDB* which cuts the line *AFL* at the point *M*, which point *M* is much closer to the point *A* than is the point *F*. Similarly again, drawing a line from the point *A* to the point *K* (end of the violent motion *AIK*), which shall be *AKN*, I say that the body *B* shot in a different manner by the same force can never reach to the said point *K*, as for example appears in the other two higher shots that cut the said line *AKN* in natural motion in the two points *O* and *P*, each of which is closer to the point *A* than is the said point *K*; and this is what we wished to conclude.

FIRST PROPOSITION

The four angles of any rectilinear quadrilateral are equal to four right angles.[23]

PROPOSITION II

If from the center of a circle there shall be extended two lines to the circumference, the whole circumference of the circle will have the same ratio to the arc intercepted by these two lines which four right angles have to the angle at the center contained within the two said lines.

PROPOSITION III

If two straight lines forming an angle are tangent to a circle, and one of these is produced beyond the angle, the circum-

23. The proofs of the first three propositions are omitted, as being geometrical only.

ference of the circle will have that ratio to the intercepted arc which four right angles have to the exterior angle made by the extended line.

PROPOSITION IV

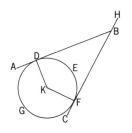

If the violent trajectory or motion of a uniformly heavy body shall be along the plane of the horizon, the curved part of this will be one-fourth the circumference of the circle from which it is derived.

Let the radius of the plane of the horizon be the line AB, and the perpendicular of the horizon the line CAD, and the violent trajectory of a uniformly heavy body the line AEF, the curved part of which shall be the arc EF; and the part FG shall be the trajectory made with natural motion. I say that the said curved part EF is one-fourth the circumference of the circle from which it is derived. Thus: produce the natural trajectory GF toward the radius of the horizon in such a way that it meets this at the point H; and since the trajectory FGH is parallel (by the First Supposition) to the perpendicular CAD, therefore the angle FHA (by Euclid I.29, 1) will be equal to the angle HAC, which is a right angle; therefore the exterior angle FHB (by Euclid I.13) will be a right angle; whence four right angles become the quadruple of the said exterior angle, for which reason the circumference of the circle from which the said curved portion EF is derived (by the foregoing Third Proposition) becomes quadruple the arc EF. Therefore the arc EF is one-fourth of the circumference of the circle from which it derives, which is the proposition.

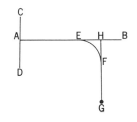

PROPOSITION V

If the violent trajectory or motion of a uniformly heavy body shall be elevated above the horizon, its curved part will be greater than one-fourth of the circumference of the circle from which it derives, and the more it is elevated, the more it will exceed one-fourth of the said circumference; but it can never reach one-half that circumference.[24]

Let the radius of the horizontal plane be the line AB and the perpendicular of the horizon the line CAD, the violent trajectory of a uniformly heavy body the line AEF, its curved part the arc EF, and the trajectory made with natural motion

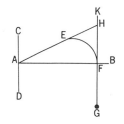

24. This proposition illustrates the hopelessness of Tartaglia's project of finding all trajectories from a single shot and reconciling them with empirical observations. See note 9, above, and compare the ensuing propositions.

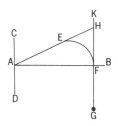

the part *FG*. I say the arc *EF* is greater than one-fourth of the circumference of the circle from which it derives. Thus: extend the natural trajectory *FG* and the straight part *AE* until they come together in the point *H,* and extend *FH* to *K,* making the exterior angle *EHK;* and since the angle *FHE* is equal (by Euclid I.29, 1) to the angle *EAC,* and the angle *EAC* (by the last axiom of the first book of Euclid) is less than a right angle, therefore the angle *EHF* (by common agreement) will be less than a right angle, wherefore the exterior angle *EHK* will be greater than a right angle (by Euclid I.13); and (by Euclid V.8, 2) four right angles will bear a less than fourfold ratio to the said exterior angle, and similarly the circumference of the circle from which the arc *EF* derives (by the Third Proposition above) will have a less than fourfold ratio to the said arc, and (by Euclid V.10, 2) the arc *EF* will be greater than one-fourth of the circumference of the circle from which it derives; which is the first [part of] our proposition. And since the more one elevates the straight part *AE* above the horizon, the smaller the angle that will be made by the line *AE* with the line *AC,* and consequently by the line *EH* with the line *FH,* the angle *EHK* will be continually enlarged, and the ratio of four right angles to it will be diminished from fourfold; and similarly the ratio of the circumference of the circle from which the arc *EF* derives will diminish from quadruple, for which reason the said arc *EF* (by Euclid V.10, 2) will continually grow beyond one-fourth of the circumference; which is the second [part of the] proposition. And since the exterior angle *EHK* can never equal two right angles (by Euclid I.32, 1 conjoined with Euclid I.17), therefore the proportion of four right angles to the said exterior angle can never be double. It therefore follows that the proportion of the circumference of the circle from which any arc or curved part of a violent motion derives can never be double the said arc or curved part, and consequently the said arc or curved part can never be one-half the circumference of the circle from which it derives; which is the third [part of the] proposition.

PROPOSITION VI

If the violent trajectory or motion of a uniformly heavy body shall be oblique under the horizontal, its curved part will be less than one-fourth the circumference of the circle from which it derives, and progressively less, the more oblique it shall be.

Let the radius of the horizon be the line *AB* and the perpendicular of the horizon the line *CAD* and the violent tra-

jectory of a uniformly heavy body the line *AEF*, the curved
part of which shall be the arc *EF*, and the part *FG* shall be
the trajectory made with natural motion. I say that the arc
EF is less than one-fourth of the circumference of the circle
from which it derives. Thus: extend the natural trajectory *FG*
and the straight part *AE* so that they shall meet in the point *H*,
and extend *FH* to *K* making the exterior angle *EHK;* and
because the angle *FHE* is equal (by Euclid I.29, 1) to the
angle *EAC*, and the angle *EAC* (by the last axiom of the first
book of Euclid) is greater than a right angle (that is, than the
angle *BAC* which is part of it), therefore the angle *EHF* will
be greater than a right angle; wherefore the exterior angle
EHK (by Euclid I.13) will be less than a right angle, and (by
Euclid V.8, 2) four right angles will have to it a more than
fourfold ratio; and similarly the circumference of the circle
from which the arc *EF* derives will have a more than fourfold
ratio (by the Third Proposition above), and (by Euclid V.10, 2)
the arc *EF* will be less than one-fourth of the circumference
of the circle from which it derives, which is the first [part of
the] proposition. And since the farther it shall go beneath
the horizon, the greater the angle made by the line *EA* with
the line *CA*, and consequently by the line *FH* with the line
EH, the exterior angle *EHK* will continually diminish, and the
ratio of four right angles to it will be augmented more than
fourfold; and similarly the ratio of the circumference of the
circle from which the arc *EF* derives to the said arc *EF* will
be augmented more than fourfold, from which fact the said
arc *EF* (by Euclid V.10, 2) will continually diminish to less
than one-fourth of the circumference of the circle from which
it derives; which is the second [part of the] proposition.

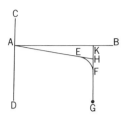

PROPOSITION VII

All the violent trajectories or motions of uniformly heavy
bodies, whether large or small, if equally elevated above the
horizon, or equally oblique [below it], or along the plane of
the horizon, are similar to one another, and consequently
proportional, as are their distances.

Let the radius of the plane of the horizon be the line *AB* and
the perpendicular of the horizon be the line *CAD*, and let the
trajectories of two different uniformly heavy bodies equally
elevated above the horizon be the two lines *AEFG* and *AHIK*,
of which the two parts *AEF* and *AHI* shall be the trajectories
made with violent motion and the two parts *FG* and *IK* the
trajectories made with natural motion, and let the two parts

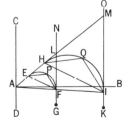

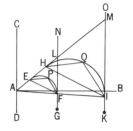

AE and *AH* be their straight parts, which straight parts, by being equally elevated, shall form together a single straightness, that is a single line, which will be the line *AEH*. And from the point *A* draw the line *AF*, which extended and continued straight will necessarily pass through the point *I*, because when the straight parts of violent trajectories or motions correspond, their distances also correspond, otherwise it would be very contradictory. Now I say that the trajectory *AEF*, made with violent motion, is similar to the trajectory *AEHI* also made with violent motion, and consequently they are proportional; and similarly the distance *AF* to the distance *AI*. Hence I shall extend their natural trajectories and their common straight part *AEH* until they run together in the two points *L* and *M*, and I shall extend the said natural trajectories to *N* and *O*, forming the two exterior angles *ELN* and *LMO*, and draw the two chords *EF* and *HI* to their curved parts. And since the two natural trajectories *GN* and *KO* (by the First Supposition) are parallel, the angle *ELN* (by Euclid I.29, 2) will be equal to the angle *LMO*, wherefore (by Euclid V.7, 2) four right angles will have the same ratio to each of them; and similarly the circumference of each of the two said circles from which the two arcs *EF* and *HI* derive will have the same proportion to the said two arcs, each to its related arc (by the Third Proposition). Hence the arc *EF* comes to be similar to the arc *HI* and similarly the segment *P* to the segment *Q;* wherefore, if we form in each of the said arcs an angle, which shall be *EPF* and *HQI*, these two angles (by the converse of the last two definitions of the third book of Euclid) will be equal to one another. Thus the angle *FEA* (by Euclid III.31) will be equal to the angle *IHE*, whence (by Euclid I.28) the chord *EF* will be parallel to the chord *IH*, so that the angle *EFA* will be equal (by Euclid I.29, 2) to the angle *FIH*. Therefore the triangle *AEF* will be equiangular with the triangle *AHI*, and consequently similar to it. Therefore the ratio of the straight part *AE* to the straight part *AH* will be that of the chord *EF* to the chord *HI*, and of the distance *AF* to the distance *AI*, and of the arc *EF* to the arc *HI;* which is the proposition. And in the same way, by the same method, such similarity will be demonstrated in violent trajectories or motions that are equally oblique below the horizon or are along the plane of the horizon, because the two exterior angles are always equal, and the arcs or curved parts of these will always be similar; for equal parts taken from the circumferences of circles are similar. And arguing as has been done above, we will show that the ratio of the straight part of one to the straight part of the other is that of the distance of one to the distance of the other, and of arc to arc; and by inverse proportion it is demonstrated that the ratio of the straight part of one to the distance of the same, or to the curved part of the same, is that of the straight part of the other to the distance, or to the curved part of that one; which is the proposition.

PROPOSITION VIII

If the same motive power shall eject or shoot similar and equal uniformly heavy bodies in different ways violently through air, that shot which shall have a trajectory elevated 45 degrees above the horizon will make its effect farther from its beginning on the plane of the horizon than one elevated in any other way.

To demonstrate this proposition, we shall use a physical argument, which is this: Anything that passes from the less to the greater, and through everything between, necessarily passes also through the equal. Or this other: Where there is found the greater, and also the lesser, than something, there is also found its equal. Now it is true that such arguments will not hold for, nor are they accepted or conceded by, the geometer; this is clearly demonstrated by the commentary on Euclid III.15 and III.30.[25] Nevertheless, such conclusions are verified in things that are truly unambiguous, even though in those that contain some equivocation they are sometimes erroneous. For example, one might say that a sector of a circle may be found in which the angle formed on the arc is less than a right angle, and this is any sector greater than a semicircle (by Euclid III.30); similarly there may be found another in which the angle is greater than a right angle, and this is any sector less than a semicircle (by the said Euclid III.30); therefore it would be possible, by the same argument, to find one in which will be formed an angle equal to a right angle. Now I say that in this case the argument will not be wrong; that is, it is possible to find a sector of a circle in which there really is formed on the arc an angle equal to a right angle, and this will happen because in the said angles there is no equivocation. But if anyone should say that, since there is a

25. The two propositions meant are numbered in standard modern editions of Euclid as III.16 and III.31. They state in effect that (1) the angle between a circle and its tangent is less than any angle contained between two straight lines, and (2) the angle formed at a point on the circumference by lines connecting it with the ends of a diameter is a right angle, whereas an acute angle is formed if the lines are drawn to the ends of a chord in a segment greater than a semicircle, and an obtuse angle is formed in a segment less than a semicircle. Tartaglia's numbering of Euclid's propositions here was that used in the translation by Campano, the earliest Latin version to be published.

sector of a circle, the angle of which sector [i.e., the "angle" between the arc and its chord] is less than a right angle (and this is the sector less than a semicircle, by Euclid III.30) and since similarly there is another [sector] of which the angle is greater than a right angle (and this is the sector greater than a semicircle, by the said Euclid III.30), therefore (by the previous reasoning) it would be possible to find a sector which would have its angle equal to a right angle, I say that in this case the said proposition and reasoning would be wrong, since the angle of the sector of a circle is not really unambiguous in the case of the right angle; for the right angle is contained between two straight lines, but the angle of the sector is contained between a straight line and a curved line; that is, between the chord and its arc.[26]

Nevertheless, I say that the proposition or reasoning that is true is always verified to the senses and to the intellect in the mean quality between those two diversities or contrary qualities, or between the sectors greater and less than the semicircle; which mean quality is precisely in that semicircle, as proved by Euclid III.30. But the incorrect reasoning is also verified to our senses in that same term, or mean quality (that is, in the semicircle), because its error is not perceptible and no sense by itself is fitted to recognize it in matter, so it is known only to the intellect. And that this is true is known, because the angle contained between chord and arc in the semicircle is so close to a right angle that it is not possible to form with two straight lines an acute angle closer than that to a right angle, or even as close (as proved by Euclid III.15). Therefore it follows that such propositions or arguments are always verified to the senses in that term, or mean quality, which is situated between two qualities contrary in properties or effects; that is, which shares equally in each of them.

And not to rely on this single example, let us take this other. The sun, turning continually through the zodiac, gives days sometimes longer and sometimes shorter than nights. Now by the above argument, it should follow that, at some time or place, the sun should give a day equal to the night; which is true, and verified to the senses and to the intellect, at that time

26. Tartaglia's argument is that, if we consider the mixed angle made between chord and arc for any chord other than a diameter, that angle will be acute or obtuse according as we consider the side of the arc lying in the lesser or in the greater segment of the circle, but that the limiting case of the diameter is not comparable, for in that case (by Euclid III.16) we have in fact a bona fide right angle between diameter and tangent; in this one case, no arc at all comes into consideration, and hence no mixed angle.

or in that mean place between the two times or places most contrary in such effects. These two most contrary places are the first degree of Cancer and the first degree of Capricorn, for when the sun enters the first degree of Cancer it gives the longest day with respect to night of all places or times, and when it enters the first degree of Capricorn it gives the shortest day with respect to night of all places. But the middle point between these two extremes in contrary effects will be in the first degree of Aries and the first degree of Libra. Now suppose the said reasoning in this case is incorrect; I say that, as above, it will still be verified to the senses in the aforesaid mean places, as we continually see that when the sun enters into one of the said places the day does equal the night, and even if not perfectly equal (as is rightly shown by the Reverend Cardinal Pietro Aliaco in his sixth question on Sacrobosco), the differences are imperceptible.

Returning therefore to our purpose, we clearly know that if a uniformly heavy body is ejected or shot violently along the plane of the horizon, it would complete its violent motion farther below the plane of the horizon than when elevated in any other direction; and if we elevate it gradually above the horizon, it would still terminate its violent motion below the horizon. But continuing such elevation, we evidently know that at some time this [violent motion] will end above the horizon; and then the more we go on elevating it, the higher, and thus the farther from the horizontal plane, this [violent motion] will end. Finally, coming to the perpendicular above the horizon (that is, where its motion or trajectory is straight above the horizontal), that [violent motion] will terminate higher, or farther, above the plane of the horizon than [it would] at any other elevation. Whence it would follow by the previous reasoning that there will be an elevation so conditioned as to make this [violent motion] terminate precisely in the plane of the horizon. Which reasoning being true, it is verified really to the senses and also to the intellect in that elevation which is midway between those which are most contrary in results (that is, between that which is along the plane of the horizon and that which is straight above the horizon), because the one will terminate the violent motion of the body lower, and the other higher, with respect to the horizon, than if [the shot were] elevated in any other way. And this mean elevation occurs when the violent trajectory or motion of a uniformly heavy body is elevated at 45 degrees above the horizon (that is, when the straight part of this [trajectory] divides into two equal parts the right angle formed by the perpendicular above the horizon and the radius of the horizon). And even if the said argument is false for our geometer adversary, it is still verified to the senses in the said mean elevation (that is, at 45 degrees above the horizon). Thus, the body ejected or shot in such a way as to make its trajectory elevated at 45 degrees above the horizon will terminate its violent motion precisely in the plane of the horizon. And the effect that it will make in this plane will be the most distant from its beginning (by the Fourth Supposition) that can be made at any elevation above the plane of the horizon and ejected or shot by the same power; which is the proposition.

Corollary. — From this proposition and the last [proposition] of our First [Book], it is manifest how a uniformly heavy body in violent motion elevated 45 degrees above the horizon makes less effect in the plane of the horizon than if elevated in any other way.[27]

PROPOSITION IX

If the same motive power shall eject or shoot two similar and equal uniformly heavy bodies, one elevated at 45 degrees above the horizon and the other along the plane of the horizon, the straight part of the trajectory of that which is elevated at 45 degrees above the horizon will be about four times the straight part of the other.

To demonstrate this proposition, let us take as an assumption that which we said in the beginning we had found, that is, that the distance of the violent trajectory or motion elevated at 45 degrees above the horizon is about ten times the straight trajectory made along the horizontal plane, which is commonly called point-blank. (This ratio will be seen to be true in the Fourth Book, wherein will be given numerically the order and proportion of growth and diminution of shots of all sorts of devices.) Then let the radius of the horizon be the line *AB* and the perpendicular of the said horizon be the line *CAD*, and the trajectory of a uniformly heavy body made along the plane of the horizon be the line *AEFG*, the straight part of which shall be the line *AE*, and the curved part the line *EF*, and the trajectory with natural motion the line *FG*. And let the trajectory of another body similar and equal to the first, and shot with the same power, elevated 45 degrees above the horizon, be the line *AHIK*, the straight part of which is the line *AH* and the curved part the arc *HI*, the trajectory with natural motion

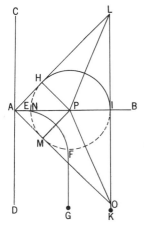

---

27. That is, its effect is weakest because it has reached the end of violent motion precisely on the horizontal plane, and natural motion has not begun at this moment.

the line *IK* and the distance the line *AI*, which distance is along
the radius of the horizon. I say that the straight part *AH* is
about four times the straight part *AE*. For I extend the natural
motion *IK* and the straight part [of violent motion] *AH* until
they intersect at the point *L*, and since the radius [of the ho-
rizon] *AB* is perpendicular to the natural trajectory *IK* in
the point *I* (by Euclid III.18), *AB* will pass through the center
of the circle from which the curved part derives.[28] Hence I
shall form (by Euclid III.24) the circle *HIMN*, which includes
the said curved part *HI*. From the point *A*, by Euclid III.16,
draw a line *AM* tangent to the said circle, and extend it straight
until it intersects the natural trajectory *IK* at the point *O*,
forming the triangle *ALO*. Now from the two points *H* and *M*
to the center of the circle, *P*, draw the two lines *HP* and *MP*,
which will be equal to one another (by definition of the circle).
Similarly the line *AH* (by Euclid III.35) will be equal to the
line *AM*, and the angle *PHA* will be equal to the angle *PMA*
because both are right angles (by Euclid III.17), and the base
AP is common to the two triangles *AHP* and *AMP*; whence
(by Euclid I.8) the said two triangles will be equiangular. And
since the angle *HAP* is half a right angle (being half the angle
CAP, by supposition), the angle *APH* (by Euclid I.32, 2)
will also be half a right angle. Therefore it follows that the
angle *MAP* of the other triangle is also half a right angle, so
that the whole angle *HAM* of the triangle *ALO* will be a right
angle, and since the angle *ALO* is half a right angle, being
equal to the alternate angle *LAC* (by Euclid I.29), it follows
(by Euclid I.32, 2) that the other angle *LOA* is also half a right
angle; whence (by Euclid I.6), the side *AL* will be equal to the
side *AO*, so that the said triangle *ALO* becomes half a square,
and the distance *AI* comes to be the altitude of the triangle
ALO and also comes to be equal to half the base *LO*; that is,
to *LI*. And since the said distance *AI* is assumed to be ten times
the straight line *AE* (that is, ten times as long as the straight
line *AE*), the area of the triangle *ALO* (by Euclid I.41) will
be 100; that is, 100 of the squares [measured by] the straight
line *AE*, which we here adopt as our unit of measure. And the
side *AL* would be the square root of 200 (by the penultimate
proposition of Euclid, Book 1), and similarly the other side *AO*.
Now, wishing to know numerically the quantity of the straight
line *AH*, we draw first from the center *P* two lines *PL* and *PO*,
and we proceed by algebra, assuming the radius of the circle to

28. That is, the circle of which *HI* is a part must have its center on the surface of the earth, making the trajectory predominantly circular—a new and unexpected development in the light of earlier textual discussions.

Tartaglia's trajectory diagram superimposed on a landscape, as shown by Walther H. Ryff in *Der geometrischen Büxenmeisterey,* included in *Der Architectür* . . . (Nürnberg, 1558). Ryff translated into German the first two books of Tartaglia's *Nova Scientia* and extensive sections of the *Quesiti et inventioni* without acknowledgment of their authorship. (From the Burndy Library copy, with kind permission of Mr. Bern Dibner.)

be the unknown. Since the said radius becomes the altitude of the triangle PLO on the base LO, and similarly of the triangle APL on the base AL, and similarly of the triangle APO on the base AO, which altitudes are PI, PH, and PM, we now find the area of each of the three triangles, by their rule, multiplying the altitude by half the base, or half the altitude by the whole base. Thus, multiplying PI, which is taken as the unknown, by the half of LO, which is ten, gives $10x$ for the area of the triangle PLO, which we put to one side.[29] Then from multiplying the altitude PH (which is also x) by half of AL, which will be the square root of 50, will come $\sqrt{50x^2}$ for the area of the triangle APL, which we put beside that other that we found. Then we shall find similarly the area of the other triangle APO, which will also be $\sqrt{50x^2}$ just as was the other, because their bases are equal; i.e., each base is $\sqrt{200}$. Now we add together the three areas, which will be in sum $\sqrt{200x^2}$ plus $10x$, and this sum will be equal to the area of the whole triangle ALO, which is 100. Whence, raising that $\sqrt{200x^2}$ and restoring the parts and reducing to x^2, we shall have x^2 plus $20x$ equal to 100; and following the rule we find x to be $\sqrt{200}$ minus 10, and this will be the radius of the circle, that is, the line PI or PM. And since the line AH is equal to the line HP (as shown above), it follows then that the said line AH is also $\sqrt{200}$ minus 10, or about $4\frac{1}{7}$; whence the straight line AH comes to be about four and one-seventh times the straight line AE; which is the proposition.

Corollary. — From this it is manifest how a uniformly heavy body ejected or thrown violently through air by a given power goes more by a straight line [if thrown] in one direction than in another, and consequently makes a greater effect.[30]

End of the Second Book

29. The modern notation x is used here for the unknown, then called *cosa* or "thing." Algebra was often called the "cossic art" after this Italian algebraic use of *cosa*.

30. The final corollary may appear to contradict Book I, Definition XI and Proposition VI, from which the "effect" of the horizontal shot at N is maximal and the elevated shot has minimal "effect" at I. See also the corollary to Book II, Proposition VIII. But what is meant above is that a horizontal shot aimed at E from the appropriate point on line AC would have less "effect" than the elevated shot shown, the same charge being used. This idea, developed in the *Quesiti*, was criticized by Benedetti in his letter to Cambiano, translated below, pp. 226–27.

Various Questions and Inventions
of Niccolò Tartaglia of Brescia

Venice, 1546

Extracts translated from Book One and Book Seven
Book Eight translated in full

To the Merciful and Invincible Henry VIII, by the Grace of
God King of England, of France, and of Ireland, etc.
Niccolò Tartaglia:

Inquiries, questions, or interrogations, your Majesty, that
are made by wise and prudent persons, often cause the man
interrogated to consider many things, and indeed to learn many
more, which, had he not been questioned, he would never have
known or considered. I say this of myself, who never made any
profession of or delighted in shooting of any kind — artillery,
arquebus, mortar, or pistol — and never intend to shoot. And
it was only a question asked of me by an expert bombardier
in the year 1531, at Verona, which induced me at that time to
consider and investigate theoretically the order and proportion
of shots far and near, according to the various elevations of
such gunnery machines; to which matters I should have given
small attention had not that gunner drawn me to them by his
question. But then, in the year 1537, having heard of the great
preparations being made by the Turkish emperor Suleiman to
attack our Christian religion, I composed with great haste a
little work on those matters and published it. Thus it came
about that those particular discoveries of mine had to be tried
out, in order to see and consider whether there might be drawn
out of them any good device for their improvement and defense; and even though nothing came of that (for various
reasons; nor did I care, since that war went up in smoke),
still that little work of mine provoked many people (and for
the most part not ordinary men, but men of high intellect)
to seek me out anew with various other questions or interrogations, and not merely about matters of artillery, ammuni-
tion, saltpeter, and powder. Indeed, on the contrary, they not

only made me consider anew those particulars about which
they asked but also led me to the knowledge and discovery
(as I said) of many others, which, without these questions and
interrogations, I might perhaps never have known or considered. Later, I reflected that no small blame attaches to that
man who, either through science or industry or through luck,
discovers some noteworthy things and wants to be their sole
possessor; for if all our ancients had done the same, we should
be little different from the irrational animals now. Therefore,
in order not to incur that censure, I have decided to publish
these questions or inventions of mine. And as a beginning in
the carrying out of this benevolent wish, I have made a collection, up to the present, from a part of the memoranda which
I have always kept by hand in order to preserve in memory
the more notable things; and this portion I have distributed
into nine books, distinguished according to the type of material
treated in the questions.[1] Then I remembered one day how

1. At this point, in the edition of 1546, Tartaglia inserted an abstract of the
contents of each of the nine books. In the 1554 edition, those abstracts were
transferred to the verso of the title page. A modified summary of the contents
follows.

Book I, comprising 30 questions, deals with the shooting of artillery pieces,
   recapitulating and expanding the conclusions reached in the first two books
   of the *Nova Scientia*. Included is a discussion of various sizes and types of
   gun. The translation here is limited to portions of the second and third questions in which the curvature of the entire path of a projectile is emphasized,
   in contrast with the treatment of that matter in the *Nova Scientia*.
Book II, comprising 12 questions, deals with cannonballs of various sizes
   and materials.
Book III, comprising 10 questions, deals with gunpowder.
Book IV, comprising 13 questions, deals with the ordering of infantry.
Book V, comprising 7 questions, deals with the mapping of terrain and emphasizes the importance of the use of the compass.
Book VI, comprising 8 questions in the original edition of 1546 and supple-
   mented by 7 additional questions in the edition of 1554, sets forth the principles and practice of fortification.
Book VII, comprising 7 questions, deals with the principles of the balance as a
   preparation for the discussion of statics. It questions the Aristotelian treatment and emphasizes the importance of the "science of weights." Only the
   first two and the final questions are here translated.
Book VIII, comprising 42 questions, sets forth the "science of weights,"
   based chiefly on the medieval work of Jordanus Nemorarius or his commentator; it is here translated in full.
Book IX, comprising 42 questions, deals with problems of arithmetic, geometry, and algebra, and in particular with the resolution of cubic equations.
   The pagination shown in the running heads is that of the 1554 edition, of
which a facsimile reprint was published at Brescia in 1959.

our honored friend Mr. Richard Wentworth,[2] gentleman of
your Majesty, had spoken to me of your magnificence, mag-
nanimity, liberality, generosity, humanity, and clemency, and
had told me also how your Majesty took great delight in all
matters pertaining to war. Which thought made me wish
(although I lack that eloquence and polish of speech which is
requisite to the hearing of your Majesty) that these questions
or inventions of mine, with their replies and solutions, might
be offered and dedicated—not as something necessary to
your Majesty (for indeed even things of profound learning,
set forth and explained in elegant and lucid style, could not
add to your Majesty's high perfection; let alone these of mine,
that are mechanical things, plebeian, and written as spoken,
in rough and low style) but only as new things—I offer them
and dedicate them to you, as is the custom to do with the first
fruits which are found at the beginning of the season; these,
though they be somewhat unripe and of little substance and
less flavor, are always presented to gentlemen and to persons
of high place, not for their quality, but for their novelty. For
new things naturally gratify the human intellect, and it is that
which makes me think that these inventions of mine should
not displease your Majesty, but rather please you; which
occurring (as I hope), I shall take heart to try further in the
future. And I present myself at your Majesty's feet, bowing
low with hands joined and head humbly inclined.

The First Book, on Artillery Shots and Various Things Relating to Them
(second and third questions)

Tartaglia: ... In a vertical shot, the entire path will be
along a straight line perpendicular to the horizontal, and this
will be more perfectly straight than any of those trajectories
previously mentioned; for truly the violent trajectory or motion
of a uniformly heavy body that is outside the perpendicular
of the horizon can never have any part that is perfectly straight,

2. Richard Wentworth was one of Tartaglia's pupils and friends, perhaps
the one most frequently mentioned. Tartaglia dedicated to him the Latin
translation of Archimedes (1543) and the first part of his great treatise on
mathematics (1566). Wentworth returned to England in 1541, according to a
passage in the *Quesiti*. The Bodleian Library has a manuscript in Italian,
given to it in 1602, which was once supposed to be Tartaglia's but which is
probably in the hand of Wentworth. Nothing else is known of him, but the
crest on the title page of Tartaglia's *Archimedes* bears his initials and might
make possible further identification.

as was said under the Second Supposition of the Second Book of our *Nova Scientia*.[3]

Duke:[4] Then why did you say "by a straight line," it being not perfectly straight?

Tartaglia: To be understood by the common people, because the part which is insensibly curved, they call straight; and that which is evidently curved, they call curved. . . .

Duke: But in your argument you have put me into great difficulty or doubt; for if I recall correctly, you said that the ball shot from a cannon never makes any part of its motion by a straight line, except when shot vertically toward the sky.

Tartaglia: Or vertically down toward the center of the world.

Duke: This I grant; that is, that when the ball is shot either straight up toward the sky or straight down toward the center of the world, its trajectory or motion is completely straight. And I also concede that in those two directions the ball goes much farther in a straight line than it does at any other elevation or in any other direction. But that in every direction other than the said two it makes no part of its motion straight—that is, along a straight line—seems to me something not to be believed, nor do I believe it. For if I remember correctly, you said before that in those two shots made at Verona, you found that the 20-pound culverin, when shot point-blank (that is, sensibly straight), carried about 200 paces level. Now if, through reasoning, you found that passage of 200 paces to be not entirely straight, I can believe it and concede it; but if that piece could not shoot 200 paces along a straight line, would you not concede that such a device might shoot that way at least half the distance; that is, 100 paces, or if not 100, then at least 50?

Tartaglia: Not only would the ball fail [to travel] the said 50 paces along a perfectly straight line, but it would not do so even a single pace.

Duke: You are joking.

Tartaglia: It is reason that satisfies the mind, for thereby the true is discerned from the false.

Duke: Quite so.

Tartaglia: Then as to your Excellency's opinion that the ball shot by that cannon point-blank should go a part of its violent trajectory or motion along a straight line and the

3. See note 22 to the *Nova Scientia*.
4. The interlocutor is the Duke of Urbino to whom the *Nova Scientia* was dedicated.

remainder along a curved line, supposing this to be true, I should like to have you tell me the cause for which the ball goes along a straight line in that part where you assume that it does, and similarly the reason for its going by a curved line where your Excellency supposes it to do that.

Duke: The extreme speed that is found in the motion of the ball when issuing from the mouth of the gun is the true cause that this ball, for some little time or space, goes straight through the air. But later, losing somewhat its vigor there, it then commences to slow down and to drop successively toward the earth, and so it goes continuing to the end, when it strikes the ground.

Tartaglia: Certainly your Excellency could not have responded better than you have done, saying that the great speed is the true cause that holds the ball's motion to straightness, if that is possible. And similarly, the loss of speed in the air is the true cause that makes it tend and decline in its motion curvedly toward the earth, and the more it loses its speed there, the greater becomes its declining curvature. And all this happens because, for any heavy body driven violently through air, the faster it goes, the less heavy it is in that motion, and therefore the straighter it goes through air, because the air more easily sustains a body the lighter it is. Also, in making its effect in such motion, it assumes[5] much greater heaviness than its own, and therefore the faster a heavy body goes in violent motion, the greater effect it makes in any resistant. Likewise, the more speed it loses, the more its heaviness grows in such motion, which heaviness continually acts on it and draws it toward the earth. But for making its effects in such motion, it assumes greater lightness (or less heaviness), and hence makes less effect.

Duke: This reasoning of yours does not disturb me; proceed.

Tartaglia: I say, then, that from the things said and approved by physical reasoning is born this conclusion: that

5. The evident contradiction of the two ideas given here was not apparent to writers in the sixteenth century. The ball is supposed to travel farther in a straight line with greater speed because great speed diminishes weight; at the same time, it is supposed to strike harder because it assumes greater weight with speed. The Italian *assumere* here is not a usual word, but it must be understood to mean "take on" rather than "pretend to." The established distinctions in statics between absolute weight, positional weight, and specific weight may have helped only to confuse the writers on dynamics who had to deal with problems of weight, momentum, work, and energy. The linking of weight with speed, in place of the transfer of properties of one to the other, was slow in coming. Tartaglia's small step in that direction is pointed out in note 39, below.

where there is greater speed in the ball thrown violently through air, there is less heaviness in it; also, conversely, where its speed is diminished, there is more heaviness in it.

Duke: Very true.

Tartaglia: Also I say that where there is greater heaviness in it, there is greater stimulation to draw the ball toward the center of the world; that is, toward the earth.

Duke: This is easy to believe.

Tartaglia: Now to conclude our proposition, let us suppose that the whole trajectory or journey that must be made, or has been made, by the ball shot from the said cannon is this whole line *ABCD* and, if possible, that some part of it is perfectly straight. Let us assume that part to be *AB,* which can be divided into equal parts at the point *E.* Then the ball will  go more swiftly through the space *AE* (by the Third Proposition of the First Book of our *Nova Scientia*)[6] than through the space *EB.* Therefore the said ball will go straighter (for the reason given above) through the space *AE* than it will through the space *EB;* whence the line *AE* would be straighter than *EB.* But this is impossible, for if the whole of *AB* is assumed to be perfectly straight, one half of it cannot be less straight than the other half. And if the one half shall be more straight than the other, it necessarily follows that that other half is not straight, whence it follows that the part *EB* is not perfectly straight. And if anyone should still hold the opinion that the part *AE* was perfectly straight, that opinion would be reproved as false by the same method and manner, that is, by dividing the part *AE* again into two equal parts at the point *F;* and for the same reason as above it will be manifest that the part *AF* is straighter than the part *FE;* hence the part *FE* necessarily will be not perfectly straight. Likewise one may divide *AF* into two equal parts, and with the same reasoning it is manifest that the half of it toward *A* is more straight than that toward *F,* and thus one may divide that half into still two other equal parts, and the same will follow: that is, that the part ending at *A* is more straight than the other. And since this procedure is infinite, it follows of necessity not only that the whole line *AB* is not perfectly straight, but that not the least part of it can be perfectly straight; which is the proposition.

It is therefore seen how the ball shot from the cannon in such a direction does not go along a perfectly straight line for even the least distance of its motion or trajectory, though

6. I.e., that bodies in violent motion continually diminish in speed.

shot with the most enormous velocity you please; for the speed, however great, is never sufficient in such a direction [i.e., any way but vertically] to make it go in a straight line. It is true that the faster it goes in such a direction, the more its motion approximates straight motion, and it may even hit its target; hence it is best to say in this case that the faster the ball goes, the less curved its motion is. . . .[7]

The Seventh Book, on the Principles of Aristotle's "Questions of Mechanics"

First Question
Raised by Don Diego Hurtado de Mendoza,
Imperial Ambassador at Venice[8]

Mendoza: Tartaglia, since we took a vacation from the reading of Euclid, I have found some new things relating to mathematics.

Tartaglia: And what has your Excellency found?

Mendoza: Aristotle's *Questions of Mechanics* in Greek and in Latin.[9]

Tartaglia: It is quite a while since I saw these, particularly the Latin.

Mendoza: What did you think of them?

Tartaglia: They are very good, and certainly most subtle and profound in learning.

Mendoza: I, too, have run through them and I understood most of them; yet many questions remained with me, which I should like to have more fully explained.

7. The discussion continues with remarks on observed phenomena in certain types of shots and an additional theoretical proof that no part of an actual trajectory can be geometrically straight.

8. Mendoza was ambassador of Charles V of Spain at Venice from 1539 to 1546. It is probable that he acquired in Italy the manuscript, now preserved at Madrid, from which it is evident that Tartaglia copied the Latin translation of Archimedes that he published in 1543, as well as the text of Jordanus published posthumously in 1565 as "corrected" by Tartaglia; the same manuscript contains two other medieval treatises (traditionally but incorrectly attributed to Archimedes and Euclid) used by Tartaglia in his account of the "science of weights" in the *Quesiti*. It would be natural for Mendoza to discuss the latter topic with Tartaglia if he was then the owner of the manuscript, and his ownership would account for its presence in Madrid as well.

9. The work is now generally attributed to Strato, but in Tartaglia's time it was considered a genuine work of Aristotle. Thus Tartaglia's attack on its correctness is perhaps the first open onslaught against the reliability of Aristotle on a matter of fact rather than of philosophical opinion. References to the text in these notes are to the translation by W. S. Hett in *Aristotle: Minor Works* (Harvard, 1936), pp. 330–411, identified hereafter as *Aristotle.*

Tartaglia: Sir, should you wish me to explain them to you properly, many of the problems would require that I first explain to your Excellency the principles of the Science of Weights.[10]

Mendoza: It appears to me that Aristotle proves everything without using, or so much as knowing about, the Science of Weights.

Tartaglia: It is true that he proves each of his problems partly by physical reasons and arguments and partly by mathematical. But some of his physical arguments may be opposed by other physical reasoning, and others can even be shown to be false through mathematical arguments by means of the said Science of Weights. And besides that, he omits or remains silent about a problem of no little importance concerning the balance, because (so far as I can judge) one cannot assign the cause for that problem by physical reasoning, but only through the Science of Weights.

Mendoza: I do not believe this is true, that is, that any of his arguments can be contradicted; for Aristotle was no fool. Nor do I believe that he omitted anything or was silent on any problem of importance concerning the balance.

Tartaglia: Yet it is only too true; for if, as a physicist, one wishes to consider, judge, and prove the cause of his first problem, using physical arguments that he adduces for the material balance or scale, then one can equally prove with physical arguments (as I said before) that things are quite the opposite of what he concludes or assumes in that problem. And if one wishes then to consider and judge this problem as a mathematician, Aristotle's arguments can similarly be proved false by means of the Science of Weights.

Mendoza: How are things judged and considered as physical, and how as mathematical?

Tartaglia: The physicist considers, judges, and determines things according to the senses and material appearances, while the mathematician considers and determines them not according to the senses, but according to reason, all matter being

10. Here, as in his translation of Euclid (1543), Tartaglia speaks of the *scientia di pesi* as a recognized discipline familiar to his readers. The phrase originated in medieval treatises and appeared in the opening sentence *("Cum scientia de ponderibus sit subalternata . . .")* of the *Liber Iordani . . . de ponderibus* etc., which had been published at Nürnberg in 1533 in an edition by Peter Apianus. In this translation the phrase is capitalized and designates not statics as such, but the mixture of statics and dynamics to which mathematical reasoning had been applied with varying degrees of success by medieval writers, a portion of which science Tartaglia tried to codify in the *Quesiti.*

abstracted—as your Excellency knows that Euclid was ac-
customed to do.

Mendoza: On this I can say nothing, because offhand I do
not recall the subject of [Aristotle's] first problem. Pray tell
me what it says.

Tartaglia: It is worded precisely thus: "Why large balances
or scales are more sensitive than small ones." [11]

Mendoza: Good; what would you say about this problem?

Tartaglia: Considering it as a mathematician, in abstraction from all matter, I should say this: Without doubt the
statement is universally true, whether for the many reasons
prefaced by Aristotle[12] or for many others that may be brought
in from the Science of Weights. For that line whose moving
extremity is farther from the center of a circle, being moved
by a given force or power at that extremity, is more easily
moved, driven, or carried, and with greater speed, than another
at its extremity less distant from the center. And for that
reason, larger scales or balances are found to be more sensitive
than smaller ones.

But next, wishing to consider and test that statement materially and with physical arguments (as he does at the end)
by the sense of sight and with a material balance, I say that
by this sort of argument the problem is not generally verified,
and even that the opposite occurs; that is, smaller balances
are found to be more sensitive than larger ones. That this is
true in material balances, experience makes manifest; for if
we have a worn ducat and want to see by how many grains it
is too light, using a large balance such as one of those used to
weigh spices, sugar, ginger, cinnamon, and such materials,
we shall get a poor result; but if we use one of those small
balances employed by bankers, goldsmiths, and jewelers, no
doubt we can be quite certain of the result. This is just the
contrary of that which was concluded in this problem; for
here, small balances are more sensitive than large ones because they more thoroughly and more subtly show the difference of weights. And the cause of this contradiction stems
simply from matter; for things constructed or fabricated thereof
can never be made as perfectly as they can be imagined apart
from matter, which sometimes may cause in them effects quite
contrary to reason. And for this and other reasons, the mathematician does not accept or consent to proofs and demonstra-

11. *Aristotle,* p. 337.
12. *Aristotle,* pp. 333–37, where all such problems are explained in terms of
properties inherent in circles.

tions made on the strength and authority of the senses in matter, but only those made by demonstrations and arguments abstracted from all matter. Consequently, the mathematical disciplines are considered by the wise not only to be more certain than the physical, but even to have the highest degree of certainty. And therefore those questions which can be demonstrated with mathematical arguments cannot be suitably proved by physical arguments. Likewise those which have already been demonstrated by mathematical arguments (which are the most certain) should not be subjected to attempts to certify them still better by physical arguments, which are less certain.

Mendoza: It seems to me that you wish this first problem [of Aristotle] to be given the greatest clarity of truth by reasons and arguments adduced and demonstrated in advance, which reasons or arguments are all mathematical, and not physical, for part of them are verified by Euclid VI.23 and part by the fourth book of Euclid.

Tartaglia: Your Excellency well says, with Aristotle, that that problem is made manifest by the reasons he prefaced [to the problem], and I myself affirmed this before, because such antecedent arguments are proved mathematically by him. But at the end of those good arguments, he subjoins two other conclusions, the first of which is precisely this: "And certainly there are some weights which, placed in the small balance, are not manifest to the senses, and in the larger balance are manifest." [13] Which conclusion when considered, judged, and tested as physical—that is, by the strength and authority of the sense of sight in material scales—will doubtless suffer much opposition; for in such material scales or balances the exact opposite will be found to occur most of the time. That is, there are some weights which, placed in large scales or balances, make no tilting manifest to the sense of sight, but which will do so in little balances (that is, will make a visible tilting); and all this is shown by experience. For if, on one of those great spice scales mentioned above, there shall be placed a grain of wheat, it is obvious that on most of them it will make no visible tilting, while on most small bankers' balances it will make a quite evident tilting. But since we wish to consider, judge, and demonstrate this problem or conclusion of Aristotle's as mathematicians, that is, without any material, doubtless the conclusion will be false, since every little weight

13. *Aristotle,* p. 347.

placed in any scale will make it continually incline to the last or lowest place it can go. And all this I shall make manifest to your Excellency in the principles of the Science of Weights.

Aristotle also adds this other conclusion, and in this form: "And certainly there are some weights which manifest themselves in both sorts of scales (that is, the large and small), but much more in the larger, a far greater tilting being made there by the same weight."[14] Now if we consider, judge, and test this conclusion as physicists—that is, by the strength and authority of the sense of sight—then, as was said of the other, it will certainly suffer no less opposition in the said material scales than will the other [conclusion], and for the same reasons. And similarly if we consider, judge, and test it as mathematicians (that is, apart from any matter), this conclusion will still be false, because every sort of weight placed in any sort of scale will make it tilt continually until it comes to the last and lowest place it can. And all this is demonstrated in the said principles of the Science of Weights.

Mendoza: Although all these objections and physical arguments of yours are probable, I cannot believe that there are not other arguments and reasons, both physical and mathematical, by which [Aristotle's solution of] this problem can be saved and defended together with his two additional conclusions. Indeed, I am of the firm opinion that anyone who would study this matter diligently would discover all the special properties of matter which give rise to [the effects mentioned in] that problem as well as those conclusions that are not verified materially, as the author concludes and says. And once these were discovered and known, I think it would be easy to remedy them and to make everything verifiable in material precisely as the author proposes.

Tartaglia: Your Excellency is not mistaken, for in fact all those things that are known by the mind to be true, and particularly by abstraction from all material, should reasonably be verifiable in matter also by the sense of sight; otherwise mathematics would be wholly vain and useless and devoid of profit to man. And if it happens that they are not verified in the aforesaid scales or in large and small balances, as questioned, then it is to be believed and even held for certain that all this proceeds from the disproportionality and inequality of the material parts and members that make one scale differ more than another from balances considered apart from all matter. So if we want to defend and save this problem of

14. *Aristotle*, p. 347.

Aristotle—that is, make it verified in matter and in every kind of balance or scale, large or small—it is necessary to make all the parts or members of each balance uniform, in such a way that all are equally applicable to those considered apart from all material. This done, we shall not only verify sensibly in matter this problem of his for material scales and balances, but will also verify those other two conclusions he adds at the end.

Second Question

Mendoza: I am glad to hear my opinion confirmed. But since I did not entirely understand your reasons, I should like them repeated more clearly.

Tartaglia: I say, Sir, that the cause that the larger and smaller balances do not behave as the author concludes and proves has its roots in the difference between the material parts or members of which they are composed, which parts or members are the two arms and the pivot (that is, the axis or center on which the arms turn in both cases). For the said arms and pivot in the larger scale or balance are much more gross and bulky than in the smaller. And since the arms of those scales or balances are to be considered mathematically, that is, apart from all material, they are considered and assumed to be as simple lines, without breadth or thickness; and the pivot or axis [of support] is assumed to be a simple indivisible point. Such a scale or balance, as much as possible, would be given as in fact despoiled and naked of any sensible material, as is considered by the mind, and would doubtless be agile and responsive far beyond all material scales or balances of the same size, for it would be completely free of any material hindrance. And thus I say in conclusion that the more the parts or members of a material scale or balance resemble or approach the parts or members of an immaterial one (which is the original or ideal of all material ones), so much the more agile and responsive will it be than those which less resemble or approach this, the sizes being the same. And the parts or members of those small scales used by bankers and jewelers, as mentioned above, much more resemble and approach the parts or members of their said ideal than do the parts or members of those larger scales or balances used by merchants; for the little arms of the smaller balances are very thin, and those of the larger ones are gross. Wherefore the pivot of the smaller balance much more resembles and approaches to its ideal pivot, which is an indivisible point, than does the pivot of the large balance by reason of its gross size. And this is the principal reason why the aforementioned small balances are sensibly more accurate than the large ones, which is completely contrary to the Aristotelian view in the problem under discussion.[15]

.   .   .   .

Seventh Question

Mendoza: Now let us come to the third part, which is still lacking here; that is, how it comes about that, when the support of a scale is precisely in its center, neither above nor below, but in the center, as is the case with most of our ordinary scales, and one of the arms is pushed down either by some weight or by our hand, and the weight or hand is then removed, this arm immediately ascends again and returns to its original place, as does the arm of a scale whose support comes from above. For in fact the cause of this seems to me farther removed from common sense than for either of the two usual cases.

Tartaglia: I have told your Excellency that in order to demonstrate the cause of this effect, it will be necessary for me first to define and explain to your Excellency some of the terms and principles of the Science of Weights.

Mendoza: Is this something lengthy, these principles you must explain?

Tartaglia: So far as it concerns simply the demonstration of this particular, it will be quite short; however, if your Excellency wants to learn in an orderly manner all the principles of the Science of Weights, that will be quite lengthy.

Mendoza: You know very well that I should like to learn the whole thing, and in proper order.

Tartaglia: It is getting rather late to accomplish this.

Mendoza: Well, you may go, then, and return tomorrow morning.

Tartaglia: I shall return, your Excellency.

End of the Seventh Book

15. Questions 3 to 6 have not been translated here; in them, Tartaglia follows the medieval argument regarding material and ideal mathematical balances. That distinction was supported by Benedetti but was rejected by Guido Ubaldo in his statics (see pp. 16v–26v); it was, however, taken up by him in his attempt to deal with dynamic problems (pp. 308 *et seq.*). He was particularly (and justly) critical of the theorems of Jordanus regarding the location of the point of suspension of a material balance, repeated by Tartaglia in the section omitted here. Tartaglia's primary purpose in the Seventh Book was to excite interest in the Science of Weights as a mathematical discipline that afforded a basis for correction of the Aristotelian *Questions of Mechanics*.

The Eighth Book, on the Science of Weights

First Question

Mendoza: Now, Tartaglia, I want you to start explaining in due order that Science of Weights of which you spoke to me yesterday. And since I know that it is not a simple science in itself (there being no more than seven liberal arts), but rather that it is a subordinate science or discipline, I want you first to tell me from which others it is derived.

Tartaglia: Sir, part of this science is derived from geometry and part from natural philosophy; for part of its conclusions are demonstrated geometrically and part are tested physically, that is, through nature.

Second Question

Mendoza: Now I understand this point. But tell me what constructs can be drawn from that science.

Tartaglia: The constructs which can be drawn from that science would be almost impossible to impress upon your Excellency, or to enumerate; nevertheless, I shall repeat those which are manifest to me at the moment. Hence I say that first, by the power of this science, it is possible to know and to measure by reason the force and strength of all those mechanical instruments that were discovered by the ancients to augment the strength of a man for raising, carrying, or driving forward all heavy weights, in whatever size they are constituted or fabricated. Second, by virtue of that science it is possible not only to be able to know and measure by reason the force of a man, but also to find how to augment this infinitely, and in various ways, and thus it is possible to know the order and proportion of such augmentation in any manner, as finally, by means of various mechanical instruments, I shall make your Excellency see and know.

Third Question

Mendoza: This I should like very much; now proceed as you wish with this science.

Tartaglia: To proceed in an orderly fashion, we shall today define only some terms and ways of speaking that occur in this science, in order that the fruit of the understanding of this may be the more easily apprehended by your Excellency. Then, tomorrow, we shall proceed to explain the principles of that science, that is, those things which cannot be demonstrated in the science; for as your Excellency knows, every science has its indemonstrable first principles, which, being conceded or assumed, afford the means to discuss and sustain the whole science. Then we shall go on by setting forth various propositions or conclusions concerning the science; and part of these we shall demonstrate to your Excellency by geometrical arguments, and part we shall test by physical reasons, as I said before. And after this, your Excellency, you shall put forward those doubts or questions that occur to you concerning things mechanical, and especially the miraculous effects of the said material instruments that augment the force of a man; and by the things said and tested in the Science of Weights, all will be resolved.[16]

Fourth Question

Mendoza: This orderly procedure of yours suits me very well. Therefore go on with the said definitions, in order.

FIRST DEFINITION

Tartaglia: Bodies are said to be of *equal size* when they occupy or fill equal spaces.[17]

Mendoza: Give me some material example.

Tartaglia: For instance, two spherical bodies cast or shaped in the same mold, or in equal molds, will be said to be of the same size even though of different materials, as when one is of lead and the other of iron or of stone. And the same is to be understood of any other variety of form.

Mendoza: I understand; go on.

Fifth Question. DEFINITION II

Tartaglia: Similarly the bodies are said to be of different or *unequal size* when they occupy or fill different or unequal

16. To give the reader an approximate idea of the sources of Tartaglia's Science of Weights, his propositions will be correlated with those of the three medieval treatises on which he drew, as edited and translated in E. Moody and M. Clagett, *The Medieval Science of Weights* (Madison, 1953), hereafter cited as *Medieval*. Following each page reference to that book, the imputed author of the appropriate medieval treatise will be identified, as well as the proposition (in the numbering assigned by the English translators). If the medieval proposition has been materially modified by Tartaglia, the abbreviation "cf." will be prefixed. Alternative references could be given, but a complete collation would serve no purpose here. It should be remembered that the imputed authors are not necessarily the real ones, but merely those traditionally assigned in medieval writings to the respective treatises.

17. *Medieval*, p. 27: Euclid (1).

spaces, and *greater* means that which occupies more space.[18]

Mendoza: I understand; proceed.

Sixth Question. DEFINITION III

Tartaglia: By *force* of a heavy body is understood and assumed that power which it has to tend or go downward, as also to resist the contrary motion which would draw it upward.[19]

Mendoza: When I say nothing to you, continue, for by my silence I denote that I have understood and wish you to continue.

Seventh Question. DEFINITION IV

Tartaglia: Bodies are said to be of *equal force* or power when in equal times they run through equal spaces.[20]

Eighth Question. DEFINITION V

Bodies are said to be of *different force* or power when in different times they move through equal spaces, or when in equal times they traverse unequal intervals.[21]

Ninth Question. DEFINITION VI

The *force or power* of different bodies is said to be *greater* in that which traverses the same space in less time, and *less* in that which employs more time; or [greater in that] which in equal time traverses greater space.[22]

Tenth Question. DEFINITION VII

Those bodies are said to be *generically the same* when they are of equal size and also of equal force or power.[23]

Eleventh Question. DEFINITION VIII

Bodies are said to be *generically different* when they are of equal size and are not of equal force or power.

---

18. *Medieval,* p. 27: Euclid (2) (3).
19. *Medieval,* p. 175: Jordanus R1.001.
20. *Medieval,* p. 27: Euclid (4).
21. *Medieval,* p. 27: Euclid (5).
22. *Medieval,* p. 27: Euclid (6); cf. p. 175: Jordanus R1.002.
23. *Medieval,* p. 27: Euclid (7).

Twelfth Question. DEFINITION IX

Those bodies are said to be *simply equal in heaviness* which are actually of equal weight, even though of different material.[24]

Thirteenth Question. DEFINITION X

A body is said to be *simply heavier* than another when it is actually more ponderous, though of different material.[25]

Fourteenth Question. DEFINITION XI

A body is said to be *specifically heavier* than another when its material substance is more ponderous than the material substance of the other, as is lead than iron, and similarly other materials.[26]

Fifteenth Question. DEFINITION XII

A body is said to be *more* or *less heavy in descent* than another when the straightness, obliquity, or pendency of the place or space where it descends makes it descend more or less heav[il]y than the other, and similarly more or less rapidly than the other, though both are simply equal in heaviness.[27]

Sixteenth Question. DEFINITION XIII

A body is said to be *positionally more* or *less heavy* than another when the quality of the place where it rests and is located makes it heavier [or less heavy] than the other, even though both are *simply equal in heaviness*.[28]

Seventeenth Question. DEFINITION XIV

The *heaviness of a body* is said to be known when one knows the number of pounds, or other named weight, that it weighs.[29]

---

24. Cf. *Medieval,* p. 43: Archimedes, Definition 9.
25. Cf. *Medieval,* p. 43: Archimedes, Definition 10.
26. Cf. *Medieval,* p. 43: Archimedes, Postulate 3.
27. Cf. *Medieval,* p. 175: Jordanus R1.003. The curious and ungrammatical use of the word *grave* in the text suggests that the speed and power at the end of the descent are intended.
28. Cf. *Medieval* p. 175: Jordanus R1.006.
29. Cf. *Medieval,* p. 41: Archimedes, Definition 5.

Eighteenth Question. DEFINITION XV

The arms of a scale or balance are said to be in the *level position* (or *place of equality*) when they stand parallel to the plane of the horizon.[30]

Nineteenth Question. DEFINITION XVI

The *line of direction* is a straight line imagined to come perpendicularly from above to below and to pass through the support or axis of any kind of scale or balance.

Twentieth Question. DEFINITION XVII

The descent of a heavy body is said to be *more oblique* when for a given quantity it contains less of the line of direction, or of straight descent toward the center of the world.[31]

Mendoza: I do not understand this very well; therefore give me an example.

Tartaglia: To exemplify this definition, let there be the body A, and its straight descent toward the center of the world shall be the line AB; and let there be also the descents AC and AD; and of these two, let there be two designated equal quantities or parts AE and AF. From the points E and F, draw the two lines EG and FH parallel to the plane of the horizon. Since the part AH is less than the part AG, the descent AFD will be said to be more oblique than the descent AEC, because it contains less of the straight descent, that is, of the line AB, in a given quantity. And the same is to be understood for all descents that could be made by the body A, or any similar body, hung from the arm of any balance. That is, that descent will be said to be more oblique which, in the above way, contains less of the line of direction in a given quantity of descent.

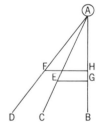

Mendoza: I have sufficiently understood; therefore proceed, if you have anything else to define.

Tartaglia: Sir, this is the last thing that we have to define concerning this subject. Tomorrow we shall explain the principles of this science, according to our promise.

Twenty-first Question

Mendoza: Now, Tartaglia, continue with your principles.

Tartaglia: Some say that the principles of any science

30. *Medieval*, p. 175: Jordanus R1.007.
31. *Medieval*, p. 175: Jordanus R1.005.

should be called axioms [*dignità*], because they prove other propositions but cannot be proved from others; some call them suppositions, because they are supposed to be true in the given science; others prefer to call them petitions, because, if we wish to debate such a science and sustain it with demonstrations, we must first request the adversary to concede them. For if he does not concede them but denies them, the entire science would be denied, nor could one debate further. And since this last opinion pleases me somewhat more than the other two, let us call them petitions and set them forth in the form of requests.

Twenty-second Question. FIRST PETITION

We request that it be conceded that the natural movement of any heavy and ponderable body is straight toward the center of the world.[32]

Mendoza: This is not to be denied.

Twenty-third Question. SECOND PETITION

Tartaglia: Likewise we request that it be conceded that that body which is of greater power should also descend more swiftly; and in the contrary motion, that is, of ascent, it should ascend more slowly — I mean in the balance.[33]

Mendoza: Give me a material example for this petition if you wish me to understand it.

Tartaglia: For example, let there be the two equal scales *ABC* and *DEF*, with the two arms *AB* and *BC* equal to the two arms *DE* and *EF*, and their supports or centers *B* and *E;* and at the extremity of the arm *BA* let there be hung the body *A*, say, of two pounds weight; and at the extremity of the other arm, that is, at the point *C*, let there be no other weight. And at the extremity of the arm *ED* let hang the body *D*, say, of a single pound weight; and at the extremity of the other arm, that is, at the point *F*, let there be no other weight. And let the two said bodies, so conjoined, be elevated by hand to equal heights,

---

32. Cf. *Medieval*, p. 175: Jordanus R1.001, first part. It is noteworthy that Tartaglia felt the need for physical assumptions in addition to definitions and separately stated from them. Such a procedure characterized the pseudo-Archimedean work, but not that of Jordanus.

33. Cf. *Medieval*, p. 175: Jordanus R1.002. The afterthought is Tartaglia's, and seems to imply that he construed the Jordanus proposition as general for the Aristotelian law of fall, whereas Tartaglia wanted it restricted to connected weights. Guido Ubaldo criticized his predecessors for the same confusion; see *Mechaniche* (16v), below.

as appears below in the figure. Now I request that it be conceded to me that, when both the said two elevated bodies are released, the body *A* (being heavier) will descend more swiftly than the body *D;* that is, the body *A* will take less time to run through the curved space *AG* than will the body *D* to run through the curved space *DH*, which spaces will be equal because the arms of the scales are assumed equal, whence the said two curved spaces or descents are circumferences of equal circles. And the converse happens when the said bodies shall have descended to their lowest places, that is, one to the point *G* and the other to the point *H*. I ask that it be conceded that the force or power which shall be hung at the other arm of the scale at the point *C,* in order to elevate the said body *A* to the place where it is presently shown in the figure, will be able to raise the body *D* more swiftly when hung from the other arm of its scale at the point *F*.

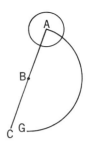

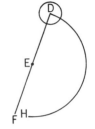

Mendoza: This I concede, because experience gives me good evidence of it.

Tartaglia: But your Excellency knows that what we have said and supposed of the two said bodies, of which one is simply more powerful than the other, we suppose of two bodies simply equal in power but unequal by force of their position or placement on the arms of the same balance. For example, on the arm *AB* of the balance *ABC* let there be hung the two weights *A* and *D,* simply equal in power, that is, one at the point *A* and the other at the point *D,* as appears below in the figure. Although they are equally powerful, nevertheless the body *A* in that position (by the Thirteenth Definition) will be said to be heavier than the body *D,* as will later be made manifest. For at this time the reason cannot be given for the things said, but later it will be proved that the body *A* in such a position is heavier than the body *D*. Nevertheless, these being raised, one to the point *E* and the other to the point *G,* and both then released, I say that the body *A* will descend more swiftly than the body *D;* and conversely, if both have descended to their lowest points, that is, one to the point *F* and the other to the point *H,* the power that, at the point *C,* will be able to elevate the body *A* from the point *F* to the point *E* will be able, in the same place, to elevate much more swiftly the body *D* from the point *H* to the point *G*.

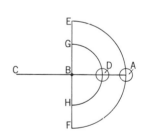

Mendoza: This is also clear, but I should like to hear from you two things. First, I wish to know why you do not draw the above figure of the scale with its two pans hung one from one end and one from the other (as is usual in actual scales), where we place weights and samples of things to be weighed. Second, I should like to know if this example of the scale should be understood of those that have their support above, or of those that have it below, or of those having it neither above nor below, but in the scale itself?

Tartaglia: As to the first, I shall reply that by the ideal scale is intended the mere length that forms the two arms on both sides of the support, whether such arms are equal or unequal, and that those two pans of which your Excellency speaks are not part of the scale, but are added for the convenience of the weigher in placing the weights and samples that are to be weighed — just as the saddle is not part of the horse, but something added for the convenience of him who must ride. And just as a horse is better seen and recognized bare of saddle than with saddle, so is a balance denuded of those pans seen better than with them; thus without pans we illustrate it. As to the second matter, I say that the present scale, as well as all those we shall later propose (unless we specify otherwise), should be understood to have the support within, as is usual with actual balances.

Mendoza: I understand; proceed.

Twenty-fourth Question. THIRD PETITION

Tartaglia: Also we request that it be conceded that a heavy body in descending is so much the heavier as the motion it makes is straighter toward the center of the world.[34]

Mendoza: Give me some material example of this new petition, if you want me to understand it.

Tartaglia: For example, let there be the heavy body *A*, and assume that the four lines *AB, AC, AD, AE* are four places or spaces by which the said body *A* can descend, and let us also assume that the line *AB* is the straight and perpendicular descent toward the center of the world. Thus *AD* becomes more direct toward the center of the world than the line *AE*, 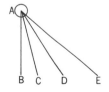 and hence in this case we request that it be conceded that the said body *A* is heavier in descending by the line *AD* than by the line *AE* (the former going more directly than the latter to the center of the world), and similarly heavier in descending by the line *AC* than by the line *AD*, because the line *AC* is more direct to the center of the world than the line *AD*. And thus the more the said body shall approach the line *AB* in its descent, it is assumed so much the heavier in descent, because

34. *Medieval,* p. 175: **Jordanus R1.003.**

that trajectory or descent which forms the more acute angle
with the line *AB* at the point *A* is understood to be more direct
toward the center of the world than one which forms a less
acute angle. Whence along the line *AB* it comes to descend
most heavily of any direction. And what we have said and requested of the said body *A* separated from any balance, we
request of those [bodies] which descend when hung from the
arm of any scale. For example, let there be also the body *A*
hung onto the arm of scale *ABC* that turns on the support or
center *B* or onto the arm of scale *ADE* that turns on the support or center *D;* and let the perpendicular descent toward
the center of the world be the straight line *AF;* and the descent
which the said body *A* would make with the arm *AB* of the
scale *ABC* on the center *B* will be the curved line *AG*, while
the descent which the same body *A* will make with the arm *AD*
of the scale *ADE* on the center *D* will be the curve *AH*. Now
I request it to be conceded that the said body *A* is heavier in
descending by the descent *AH* than by the descent *AG*, because the said descent *AH* is more direct toward the center
of the world than the descent *AG*. For the descent *AH* forms
a more acute angle with the line *AF* (which is the angle *HAF*
of tangency) than that made by the descent *AG*.

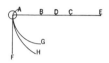

Mendoza: I understand very well, and that petition is not
to be denied. Now go on to the next.

Twenty-fifth Question. FOURTH PETITION

Tartaglia: Also we request that it be conceded that those
bodies are equally heavy positionally when their descents in
such positions are equally oblique, and that will be the heavier
which, in the position or place where it rests or is situated, has
the less oblique descent.[35]

Mendoza: This also is manifest by what was said of the
foregoing, and also of the second, petition; therefore proceed.

Twenty-sixth Question. FIFTH PETITION

Tartaglia: Similarly we request that it be conceded that
that body is less heavy than another positionally when, by the
descent of that other on the arm of the balance, a contrary
motion would follow in the first; that is, the first would thereby
be elevated toward the sky; and conversely.[36]

Mendoza: This is quite clearly to be conceded.

35. Cf. *Medieval*, p. 175: Jordanus R1.004.
36. *Medieval*, p. 175: Jordanus R1.006 (cf. Definition 13, above).

Twenty-seventh Question. SIXTH PETITION

Tartaglia: Also we request that it be conceded that no body is heavy in itself.[37]

Mendoza: This petition of yours I do not understand.

Tartaglia: That is, that water in water, wine in wine, oil in oil, and the air in air have no heaviness.

Mendoza: I understand, and this is something that may be conceded because experience makes it manifest; hence go on.

Tartaglia: There are no more petitions to be made to your Excellency.

Mendoza: There will be propositions enough.

Tartaglia: Not too many, Sir.

Mendoza: Do you think we can get through them to-morrow?

Tartaglia: I doubt, Sir, that we can finish them tomorrow and the next day.

Mendoza: Well, you may go, and return early tomorrow.

Twenty-eighth Question

Mendoza: Now continue, Tartaglia, with your propositions or conclusions in order, one after another, and briefly.

FIRST PROPOSITION

Tartaglia: The ratio of size of bodies of the same kind is the same as the ratio of their power.[38]

Mendoza: Give me an example.

Tartaglia: Let there be the two bodies *AB* and *C* of the same kind; let *AB* be the greater, and let the power of the body *AB* be [represented by the line] *DE,* and that of the body *C* [by the line] *F.* Now I say that that ratio which the body *AB* bears to the body *C* is that of the power *DE* to the power *F.* And if possible (for the adversary), let it be otherwise, so that the ratio of the body *AB* to the body *C* is less than the ratio of the power *DE* to the power *F.* Now let the greater body *AB* include a part equal to the lesser body *C,* and let this be the part *A;* and since the force or power of the whole is composed of the forces of the parts, the force or power of

---

37. *Medieval,* p. 43: Archimedes, Postulate 1. Here Tartaglia reverses the position taken at the beginning of his *Nova Scientia.*
38. Cf. *Medieval,* pp. 175–77: Jordanus R1.01.

the part A will be D, and the force or power of the remainder B will necessarily be the remaining power E; and since the part A is taken equal to C, the power D (by the converse of Definition 7) will be equal to the power F, and the ratio of the whole body AB to its part A (by Euclid V.7, 2) will be as that of the same body AB to the body C (A being equal to C), and similarly the ratio of the power DE to the power F will be as that of the said power DE to its part D (D being equal to F). Therefore [by the adversary's assumption] the ratio of the whole body AB to its part A will be less than that of the whole power DE to its part D. Therefore, when inverted (by Euclid V.30), the ratio of the body AB to the residual body B will be greater than that of the whole power DE to the remaining power E, which will be contradictory and against the opinion of the adversary, who wants the ratio of the greater body to the less to be smaller than that of its power to the power of the lesser body. Thus, the contrary destroyed, the proposition stands.

Mendoza: Very good; continue.

Twenty-ninth Question. SECOND PROPOSITION

Tartaglia: The ratio of the power of heavy bodies of the same kind and that of their speeds (in descent) is concluded to be the same; also that of their contrary motions (that is, of their ascents) is concluded to be the same, but inversely.[39]

Mendoza: Illustrate this proposition for me.

Tartaglia: Let there be, again, the two bodies AB and C of the same kind but different size, and let AB be the larger, and let the power of AB be DE, and that of C be F; and since the body of greater power or heaviness descends more swiftly (by the Second Petition), let the speed in descent of the body AB be GH and that of C be K. Now I say that the ratio of the power DE to the power F is the same as that of the speed GH to the speed K, while that of their contrary motions is the same but inversely; that is, the ratio of the speed of the body AB to the speed of the body C in contrary motion (that is, in ascending) is as that of the power F to the power DE, or as that of the body C to the body A. This [first part of our propo-

39. Though related closely to the foregoing proposition, the comparison of powers to speeds of descent here appears new and less in line with the medieval tradition than with that of the pseudo-Aristotelian *Questions of Mechanics*. It is a definite step toward the unification of dynamics with statics. Jordanus left the relationship undemonstrated, as a postulate; see note 33, above.

sition] is demonstrated in the same way as the foregoing. [Demonstration omitted, consisting of the previous argument applied to speeds and powers.]

Now for the second part of our conclusion, I say that the ratio of the speeds of the descents and of the contrary motions (that is, of the ascents) of the said bodies is the same, but inversely; that is, the ratio of the speed of the body *AB* in being raised by some other power imposed on the other arm of the balance (say, to the line of direction) to the speed of the body *C* raised also by the same power to the same line of direction will be as that of the speed *K* to the speed *GH*, or of the power *F* to the power *DE*, or of the body *C* to the body *AB*. For that force or power that a heavy body has by descending, it has also for resisting the contrary motion against anyone who wants to draw it or lift it up. Therefore the power of the body *AB* to resist whatever would raise it will be as much as the said *DE*, and that of the body *C* will be as much as the said *F*. Hence that force which, on the other arm of the scale, will be barely able thus to elevate the said body *AB* to the line of direction will be able to raise the said body *C* so much the more swiftly to the line of direction as its resistance shall be proportionately less than that of the body *AB*. And since the said resistance of the body *C* is as much less than the resistance of the body *AB* as the power *F* than the power *DE*, the speed of the body *C* in contrary motion will be to the speed of the body *AB* as the power *DE* to the power *F*, or as the body *AB* to the body *C;* which is the proposition.

Corollary. — From this it is manifest how the ratio of the sizes of bodies of the same kind, and that of their powers, and that of their speeds in descent, is one and the same ratio. And similarly that of their speeds in contrary motion is the inverse ratio.

Mendoza: I understood this; continue.

Thirtieth Question. THIRD PROPOSITION

Tartaglia: If there are two bodies simply equal in heaviness, but unequal positionally, the ratio of their powers and that of their speeds will necessarily be the same. But in their contrary motions (that is, in ascent) the ratio of their powers and that of their speeds is affirmed to be inversely the same.

Mendoza: Demonstrate this for me.

Tartaglia: Let there be the two bodies *A* and *B*, simply equal in heaviness, and the balance *CD*, whose center of support is the point *E;* and at the end of the arm *EC*, that is, at the point *C*, let there be hung and sustained the body *A*, while at another place closer to the support on the same arm, say, at *F*, there is sustained the body *B*. And though these two bodies are simply equal in heaviness, nevertheless (by the Fourth Petition) the body *A* will by positional force be heavier than the body *B*, because its descent will be *CH*, less oblique than the descent of the body *B*, which shall be *FG* (by the Third Petition). Hence the body *A*, being positionally heavier than the body *B*, will also be more powerful; and being more powerful, it will (by the Second Petition) fall more swiftly than the body *B* in descents, and in the contrary motion, of ascents, it will rise more slowly. I say therefore that the ratio of their speeds in descents is similar to that of their powers, and that of their ascents is also the same, but inversely. And to demonstrate the first part, let the power of the body *A* be *L* and that of the body *B* be *M*, and let the speed of the body *A* in descents be *N*, and that of the body *B* be *O*. I say that the ratio of the speed *N* to the speed *O* is as that of the power *L* to the power *M*, which is demonstrated as in the preceding [proposition]. [Demonstration omitted, as formally the same as that omitted above.]

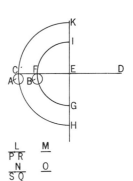

The second part is resolved or argued just as before; that is, that that power which in the other arm of the scale (let us put it at the point *D*) will be able to lift the body *A* to the line of direction, that is, to the point *K*, will be able to raise the body *B* to the point *I* as much more swiftly as the power of the body *B* (which is *M*) is less than the power of the body *A* (which is *L*), because by whatever amount the power of a body is less, by that much less it resists contrary motion, and conversely. Therefore the speed of the body *B* in ascents will be to the speed of the body *A* as the power *L* is to the power *M;* which is the second [part of this] proposition.

Mendoza: This is a very pretty proposition, but proceed.

Thirty-first Question. FOURTH PROPOSITION

Tartaglia: The ratio of the power of bodies simply equal in heaviness, but unequal in positional force, proves to be equal to that of their distances from the support or center of the scale.

Mendoza: Give me an example.

Tartaglia: Let there be the two bodies *A* and *B* of the preceding figure, simply equal in heaviness, and let the scale be *CED*, whose center or support is at the point *E;* and let there be hung the body *A* at the point *C* and the body *B* at the

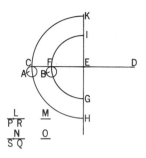

point *F*, as in the figure. I say that the ratio of the power of the body *A* (which is *L*) to the power of the body *B* (which is *M*) is like that of the distance or arm *EC* to the distance or arm *EF;* and this is proved according to the order of the preceding. [Demonstration omitted, as formally identical with those preceding.]

Corollary. — From the things said and demonstrated not only is manifest the sameness of the ratio of the distances from the support along the arms of the scale, and that of the powers of bodies simply equal in heaviness in such sites or places, and likewise [that] of their speeds in descent; but also both their descents and their ascents observe the same [rule]; for the ratio of the arm *EC* to the arm *EF* is that of the curved descent *CH* to the curved descent *FG*, and likewise of the curved ascent *CK* to the curved ascent *FI*. For the said descents and ascents are in each case one-fourth the circumference of the two [respective] circles, of which the radius of the larger is that of the arm or distance *EC*, and of the smaller, that of the arm or distance *EF*.

Mendoza: This also has been a pretty proposition. Continue.

Thirty-second Question. FIFTH PROPOSITION

Tartaglia: When a scale of equal arms is in the position of equality, and at the end of each arm there are hung weights simply equal in heaviness, the scale does not leave the said position of equality; and if it happens that by some other weight [or the hand] imposed on one of the arms it departs from the said position of equality, then, that weight or hand removed, the scale necessarily returns to the position of equality.[40]

Mendoza: This is that problem which you told me Aristotle omitted in his *Questions of Mechanics.*

Tartaglia: So it is, Sir.

Mendoza: I look forward to hearing the cause of that effect; therefore go on.

Tartaglia: Let there be, for example, the scale *ACB*, the center of which is at the point *C*, and let the arm *AC* equal the arm *BC*, and let it be in the position of equality as assumed. And at each extremity let there be hung a body (the bodies *A* and *B*) which are simply equal in heaviness. I say that the said scale, by the imposition of the said bodies, will not leave

40. Cf. *Medieval*, pp. 177–79: Jordanus, R1.02.

the position of equality; and if it is separated from that position
of equality either by the imposition of some other weight or
by hand, that imposed weight or hand being removed, the scale
will of necessity return to the position of equality. The first
part is manifest because the said two bodies are simply equal
in heaviness (by assumption), and similarly they are of
equal positional force by the Fourth Petition, their descents
being equally oblique. Hence, being equal in weight and power
both simply and positionally, neither of them will be able to
raise the other, that is, to make it ascend with contrary motion; and so they will rest in the same position of equality.

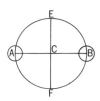

Mendoza: This I believe and would have conceded it
freely without any demonstration, it being a natural thing.
But go on to the second part, which appears to me much more
abstract, or remote from our natural intellect, than the other.

Tartaglia: For the second part, let there be also the scale
ACB of equal arms, and at its extremities let there also be
hung the two bodies *A* and *B*, simply equal in heaviness, which
scale by the reasons adduced above will stand in the position
of equality as appears in the figure. Now the arm *AC* having
been driven down by hand or by the imposition of some weight
on the body *A*, if we take away the hand or weight, the arm
will rise again and return to its first position of equality.

And to assign the immediate cause of that effect, let there
be described about the center *C* the circle *AEBF* for the journey that the two bodies will make in rising or falling with the
arms of the scale; and draw the line of direction *EF*, and divide
the arc *AF* into as many equal parts as you like (say, into four
parts at the three points *Q, S, V*); and into as many parts divide
the arc *EB* at the three points *I, L, N;* and from the said three
points *I, L, N* draw the three lines *NO, LM,* and *IK* parallel
to the position of equality, that is, [parallel] to the diameter
or line *AB*, which [three lines] shall cut the line of direction
EF at the three points *X, Y, Z.* Similarly, from the three
points *Q, S, V* are drawn the three lines *QP, SR,* and *VT,* also
parallel to the same line *AB*, which shall cut the same line of
direction *EF* at the three points *W, J, D.* And now let the body
A be depressed by hand (or by the imposition of some other
weight) to the point *V*, and the other body *B* (opposite to that)
will be found to be raised with contrary motion to the point *I.*
Now with things arranged this way, we have come to divide
the whole descent *AV* made by the body *A* in descending to
the point *V* into three equal descents or parts, which are *AQ,
QS,* and *SV;* and similarly the whole descent *IB* which the

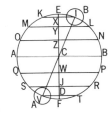

body *B* would make in descending or returning to its original place (that is, the point *B*) will come to be divided into three equal descents or parts which are *IL*, *LN*, and *NB;* and each of these three-plus-three partial descents includes one part of the line of direction; namely, the descent from *A* to *Q* partakes of or contains the part *CW* of the line of direction, and the descent *QS* contains the part *WJ*, and the descent *SV* contains the part *JD*, and the other descent that remains to the said body *A*, that is, the descent *VF* contains the line or part *DF*. Likewise the descent of the body *B* from the point *I* to the point *L* contains the part *XY* of the same line of direction, and in the descent from the point *L* to the point *N* it contains the part *YZ*, and from the point *N* to the point *B* it contains the part *ZC*, and all these parts are unequal; that is, the part *CZ* is greater than *ZY*, and *ZY* is greater than *YX*, and *YX* than *XE;* and similarly the part *CW* is greater than the part *WJ*, and *WJ* than *JD*, and *JD* than *DF*, and all this can be easily proved geometrically; and also the part *DF* can be proved equal to the part *EX*, and *JD* to *XY*, and *WJ* to *YZ*, and *CW* to *ZC*.

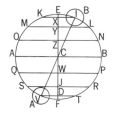

Now to resume our proposition, I say that the body *B* standing in this point *I* comes to be positionally heavier than the body *A* standing at the point *V* (as appears in the figure), because the descent of the body *B* from the point *I* to the point *L* is more direct than the descent of the body *A* from the point *V* to the point *F* (by the second part of the Fourth Petition), because it contains more of the line of direction. That is, the body *B* in descending from the point *I* to the point *L* contains the part *XY* of the line of direction, and the body *A* descending from the point *V* to the point *F* contains the part *DF* of the line of direction, and since the part *XY* is greater than the line or part *DF*, the descent (by Definition 17) from the point *V* to the point *F* will be more oblique than that from the point *I* to the point *L*. Whence (by the second part of the Fourth Petition) the body *B* in that position will be positionally heavier than the body *A;* and [body *B*] being thus heavier, when the imposed weight or hand is taken away from the body *A*, it will (by the converse of the Fifth Petition) make the said body *A* re-ascend with contrary motion from the point *V* to the point *S*, and it [*B*] will descend from the point *I* to the point *L;* and it [*B*] will come to be found still positionally heavier than the body *A*, because the said body *A* standing at the point *S* will have the descent *SV* more oblique than the descent *LN* of the body *B* because it contains less of the line of direction; that is, the part *JD* is smaller than the part *YZ*. Whence for the reasons adduced above, the body *B* will raise the body *A* when it is at the point *Q*, and *B* will descend to the point *N*, at which point *N* the same body *B* will yet be found positionally heavier than the body *A* because the descent from *Q* to *S* is more oblique than the descent from the point *N* to the point *B*, the part *ZC* being greater than the part *WJ*. And hence (by the reasons adduced above) the body *B* will make the body *A* re-ascend to the point *A* (its first and proper place) and will itself descend to the point *B* (also its first and proper place), that is, to the position of equality, in which position the said two bodies will be found (for the reasons adduced in the first part of this proposition) equally heavy positionally. And since they are also simply equally heavy, they will remain in the said place, as was said and proved above; which is our proposition.

Mendoza: This was a pretty demonstration, but, if I recall correctly, you said also, of the first mechanical problem of Aristotle, that those two conclusions of his that he adduces at the end are false.

Tartaglia: So they are.

Mendoza: Why?

Tartaglia: The reason for this objection will be verified in the next proposition, through some corollaries that are manifest from the things said and demonstrated in the above, of which the first is this:

[*First*] *Corollary.* — From the things said and demonstrated above, it is manifest how a heavy body, whenever parted or removed from the position of equality, is made positionally lighter, and the more so, the more it is removed from that position.[41]

For example, the body *A* will be found lighter at the point *V* than at the point *S*, and more at *S* than at the point *Q*, and at *Q* than at the point *A*, the position of equality, by reason of the various descents being one more oblique than another. That is, the descent *VF* becomes more oblique than the descent *SV* because the part *FD* of the [line of] direction is less than *DJ* [and so on; balance of demonstration omitted as repetitive.]

Second Corollary. — Also by the things said and demonstrated, it is manifest that the said two bodies being removed from the position of equality, that is, one downward and the other upward, both are made positionally lighter, and yet the

---

41. *Medieval*, pp. 179–81: Jordanus R1.03.

one that is lifted up is found to be less light than that which is pressed down; and this is manifest by the argumentation adduced above. That is, the body *B* at the point *I* is heavier than the body *A* at the point *V*, and so at the other higher points it will be heavier than at the corresponding lower points.

Mendoza: I understand; continue.

Thirty-third Question. SIXTH PROPOSITION

Tartaglia: Whenever a scale of equal arms is in the position of equality, and at the end of each arm are hung weights simply unequal in heaviness, it will be forced downward to the line of direction on the side where the heavier weight shall be [see note 40, above].

Mendoza: To me it does not appear that this proposition of yours can be universally true, and I think you have confessed this to me yourself, since you know that in the preceding corollary you have concluded that if the two bodies *A* and *B* (in the figure for the foregoing proposition) are removed from the position of equality, that is, one downward and the other upward, then, although both are made positionally lighter, yet in every position that one which is lifted up will be less light than that which is pressed down.

Tartaglia: True.

Mendoza: If this is true, it is to be believed (or rather held certain) that, if one should impose on the body *A*, pressed down, another little body which equaled in heaviness that difference by which the upper body exceeded positionally the heaviness of the lower, then each would remain in its own place where it was. That you may better understand me, you know that, the body *B* of the figure in the preceding proposition being lifted to the point *I* (as shown there) and the body *A* being depressed to the point *V*, it was proved by you that the body *B* was heavier than the body *A* in that position.

Tartaglia: Sir, this is true.

Mendoza: Therefore I conclude that, if one should impose on the body *A* in that position another small body of precisely as much heaviness as the difference between the said two bodies *A* and *B* in that position, the two bodies would remain fixed and stable in that position; for in that position they would be equally powerful. That is, the body *B* would not be sufficient to cause the body *A* to re-ascend to the position of equality, the said body *A* being (by the force of

that added little body) as heavy and powerful as it [B]. For
by the amount that the body B is positionally more powerful
or heavier than the body A, the body A is heavier than the
body B by force of the simple heaviness of that little body
added to it; whence the body B will not be able to make the
body A re-ascend to the position of equality; and still less
will the body A be able to raise the body B from the position I.
So neither can leave its place; that is, the body A with that
other body added cannot re-ascend to the position of equality,
nor can it descend to the line of direction, that is, to the point
F, as concluded in your proposition. Yet the said body A
together with that other little body added to it would be simply
heavier than the body B, so you cannot deny that your proposition is in general false; though it is true that, if the heaviness
of that little body that was added to the body A were greater
than the heaviness by which the body B was positionally
heavier than the body A, that which is concluded in your proposition would follow. And if it happened that the heaviness
of that little body were less than the said difference, the body
B would make the body A ascend to another place higher than
the point V, according to the greater or less deficiency in
heaviness of that little body with regard to their said difference
in positional force.

Tartaglia: This objection of yours, Sir, is certainly a very
pretty speculation. Nevertheless, I note that although the body
B in that place I is heavier than the body A in the place V, yet
the difference of those two unequal heavinesses is so small or
minute that it is impossible to find so small or minute a difference between two [corporeal] unequal quantities.

Mendoza: What you have just said seems to me a most
absurd thing to say and not to be believed; for, a continuous
quantity being infinitely divisible, it is a quibble to say that it
is impossible to have a body of so little quantity and heaviness
as is the difference between the heaviness of the body B at the
place I and that of the body A at the place V.

Tartaglia: Reason, Sir, is the means of clarifying doubts
and distinguishing the true from the false.

Mendoza: Very true.

Tartaglia: If this is true, then before your Excellency
forms an absolute opinion of my proposition, hear first my
reasons.

Mendoza: Go on and say how it seems to you.

Tartaglia: Let there be, for example, the same scale *ABC*
of the preceding proposition, at the ends of which are hung

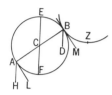

the bodies *A* and *B*, equal in simple heaviness; and let the hand
depress the body *A* and lift the body *B* as shown in the next
figure. I say that in this position the body *B* is positionally
more ponderous or heavy than the body *A*, and that the difference between the heavinesses of these two bodies is impossible to give or find between two unequal quantities. And
to demonstrate this proposition I draw two straight lines, *AH*
and *BD*, perpendicularly to the center of the world, and I
also draw two lines *AL* and *BM* tangent to the circle described
by the arms of the scale at the points *A* and *B*. I describe also
a part of the circumference of a circle touching the same circle
AEB at the point *B*, this being a similar and equal circle, *BZ*,
such that the arc *BZ* is similar and equal to the arc *AF* and
similarly placed (that is, in like position), and the line *BM*
which touches or is tangent to this. Since the obliquity of the
arc *AF* (by what was said about the Third Petition) is measured by means of the angle contained by the perpendicular
AH and the circumference *AF* at the point *A*, and the obliquity
of the arc *BF* is measured by the angle contained by the perpendicular *BD* and the circumference *BF* at the point *B*, the
body *B* in that position will be as much heavier than the body
A as the said angle (contained by the perpendicular *BD* and
the circumference *BF* at the point *B*) will be less than the
angle contained by the perpendicular *AH* and the circumference *AF* at the point *A*. And since the angle *HAF* is precisely equal to the angle *DBZ*, and the said angle *DBZ* is as
much greater than the angle contained by the said perpendicular *BD* and the circumference *BF* at the point *B* as the
angle of contact of the two circles *BZ* and *BF* at the point *B*,
and since this angle of contingency is more acute than any
of the acute angles made by straight lines (as proved by
Euclid III.16), then the difference or ratio between the angle
HAF and the angle contained by the perpendicular *BD* and
the circumference *BF* at the point *B* is less than any difference
or ratio you please which can occur between any large and
small quantities. And thus (by the Third Petition) the difference of the obliquity of the descent *AF* and the descent *BF*,
and consequently the difference of positional heaviness of
the two bodies *A* and *B*, is less than any you wish between
two unequal quantities. Therefore any small corporeal quantity that is added to the body *A* will necessarily be heavier
in any position than the body *B*, and hence it will not cease to
descend continuously as far as the line of direction, that is, to
the point *F;* and thus it will continue to raise the body *B* as far
as the line of direction, that is, to the point *E*.

And if this would take place in the position that is shown in the figure, it appears that it would happen so much the more at the position of equality, in which position there neither is nor will be any difference of positional heaviness of the descents—that is, in that position they would be equally heavy—and so any small quantity of weight, however minimal, that should be imposed on either side of any scale (that is, with equal arms, whether large or small) will immediately tilt the scale down on that side, and the arm will continue its declination, for the reasons adduced above, as far as the line of direction, that is, to the point *F*. Now this would be contrary to those two conclusions which Aristotle adduces concerning the first of his mechanical problems, of which I spoke with your Excellency once before. In one conclusion he says that there are some weights which, imposed on little scales, do not make themselves manifest to our senses by any tilting, while on large scales they do make themselves manifest. This conclusion, looked at mathematically, that is, abstracted from all matter, would be quite false for the reasons adduced above, because a small balance as well as a large one will be forced to tilt down on that side where such a weight is placed, however small it be, and to tilt as far as the line of direction. Thus in the tilting of small and large [balances] there will be no proportionate difference, and in one as in the other the tilting will continue to the line of direction.

The same would follow as to his other conclusion, that is, when he says that there are some weights which are manifest in both sorts of scales, large and small, but much more [manifest] in the larger. That conclusion would also be false for the reasons adduced above. For, as remarked, in both they will make that arm of the scale decline as far as the line of direction.

Mendoza: These reasons and arguments of yours are fine and good; nevertheless in actual or material scales it is seen that for the most part things happen as Aristotle says and concludes. For if on any scale you please (large or small) there is placed a grain of poppyseed or some other small quantity, few are the scales that will make a sensible tilting from so little heaviness. Some indeed are found which will make some sensible sign of tilting, but this does not go so far as the line of direction. And not only will the said grain of poppyseed fail to make any scale tilt as far as the line of direction, but so will a grain of wheat that is much more ponderous. And all this is demonstrated by experience. So that I do not know what to say, since on the one side, by your reasons and arguments,

I see and understand that you speak the truth, and on the other I find by experience that the opposite happens.

Tartaglia: Sir, all this comes about from matter, because in the scales considered by the mind, apart from all material, the support or axis is assumed to be an indivisible point. But in material scales that support or axis has always some corporeal thickness of its own, and the greater that thickness is, the more it reduces the sensitivity of the scale. Likewise the arms of the imagined (that is, ideal) scales are assumed to be lines, without breadth or thickness, but in material scales the arms are of some metal or of wood, and the bigger they are, the more they reduce the sensitivity of the scale.

Mendoza: I understand. Continue, then, if you have further propositions concerning this subject.

Thirty-fourth Question. SEVENTH PROPOSITION

Tartaglia: If the arms of the scale are unequal, and at the ends thereof are hung bodies simply equal in heaviness, the scale will tilt on the side of the longer arm.[42]

Mendoza: This is a physical matter.

Tartaglia: Although it is physical, if we wish to proceed correctly, we must assign the cause of this effect.

Mendoza: Go ahead.

Tartaglia: Let there be the rod or scale *ACB*, with the arm *AC* longer than *CB*. I say that, there being hung bodies simply equal in heaviness at the two points *A* and *B*, the scale will tilt on the side of *A*. For, when the perpendicular *CFG* (that is, the line of direction) is drawn, and the two quarter-circles, which shall be *AG* and *BF*, are traced on the center *C*, and when two tangent lines *AE* and *BD* are drawn from the points *A* and *B*, it is manifest that the angle of tangency *EAG* is less than the angle *DBF*. Hence the descent made along *AG* is less oblique than the descent made along *BF*. Therefore (by the Third Petition) the body *A* will be heavier than the body *B* in this position; which is the proposition.

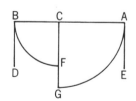

Mendoza: This I understand; continue.

Thirty-fifth Question. EIGHTH PROPOSITION

Tartaglia: If the arms of the balance are proportional to the weights imposed on them, in such a way that the heavier

---

42. *Medieval*, pp. 181–83: Jordanus R1.05.

weight is on the shorter arm, then those bodies or weights will be equally heavy positionally.[43]

Mendoza: Give me an example.

Tartaglia: First let there be the bar or balance *ACB* and the weights *A* and *B* hung thereon, and let the ratio of *B* to *A* be as that of the arm *AC* to the arm *BC*. I say that this balance will not tilt to either side. And if (for the adversary) it is possible for it to tilt, let us assume it to tilt on the side of *B* and to descend obliquely as the line *DCE* in place of *ACB*, and [let us] take *D* as *A* and *E* as *B;* and the line *DF* falls perpendicularly, and the line *EH* rises similarly. Now it is manifest (by Euclid I.16 and I.29) that the two triangles *DFC* and *EHC* have equal angles. Whence (by Euclid VI.4) they will be similar, and consequently will have proportional sides. Therefore the ratio of *DC* to *CE* is as that of *DF* to *EH;* and since the weight *B* is to the weight *A* as *DC* is to *CE* (by our assumption), the ratio of *DF* to *EH* will be as the weight *B* to the weight *A*. Hence, if we take from *CD* the part *CL*, equal to *CB* or *CE*, and put *L* equal in heaviness to *B* and descending along the perpendicular *LM*, then, since it is manifest that *LM* and *EH* are equal, the proportion of *DF* to *LM* will be as the simple heaviness of the body *B* to that of the body *A*, or as the simple heaviness of the body *L* to that of *D* (because the two bodies [*A* and *D*] are supposed to be one and the same, as are likewise the bodies *B* and *L*, the heaviness of the body *L* having been assumed equal to that of the body *B*). Hence I say that the ratio of all *DC* to *LC* will be as the heaviness of the body *L* to that of the body *D*. Whence if the said two heavy bodies, that is, *D* and *L*, were simply equal in heaviness, standing then in the same positions or places at which they are presently assumed to be, the body *D* would be positionally heavier than the body *L* (by the Fourth Proposition) in that ratio which holds between the whole arm *DC* and the arm *LC*. And since the body *L* is simply heavier than the body *D* (by our assumption) in the same ratio as that of the arm *DC* to the arm *LC*, then the said two bodies *D* and *L* in the level position would come to be equally heavy, because by as much as the body *D* is positionally heavier than the body *L*, by so much is the body *L* simply heavier than the body *D;* and therefore in the level position they come to be equally heavy. Hence that power or heaviness that will be sufficient

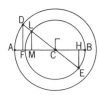

---

43. Cf. *Medieval*, p. 183: Jordanus R1.06; the converse is not proved.

to lift the body *A* from the level position to the point at which it is at present (that is, to the point *D*) will be sufficient to lift the body *L* from the same level position to the place where it is at present. Therefore if the body *B* (for the adversary) is able to lift the body *A* from the level position to the point *D*, the same body *B* would also be able and sufficient to lift the body *L* from the same level position to the point where it is at present, which consequence is false and contrary to the Fifth Proposition; that is, the body *B* (which is supposed equal in heaviness to the body *L*) would lift the said body *L* out of the level position [though they are] in equal places, that is, equally distant from the center *C*, which is impossible by the said Fifth Proposition. Thus, the adversary's position destroyed, the thesis stands.

Mendoza: This is a very pretty proposition, but it seems to me (if I recall correctly) that Archimedes of Syracuse has a similar one, and I believe he does not prove it in this way of yours.[44]

Tartaglia: Your Excellency is right. Indeed, of this proposition he makes two, and these are the fourth and fifth in that book of his wherein he deals with the centers of gravity of heavy bodies; and in fact he proves those two propositions succinctly by principles of his set forth and demonstrated previously. And since those principles and arguments of his would not be suitable in this treatise, it being of somewhat different subject, it appeared best in this place to prove those propositions with other principles or arguments more appropriate here.

Mendoza: I see. Proceed.

Thirty-sixth Question. NINTH PROPOSITION

Tartaglia: If there are two solid rods or beams of the same length, breadth, and width, hung on a balance in such a way that one is horizontal and the other vertical, with the distances equal from the center of the balance to the point of suspension of the latter and the center of the former, then they will be positionally equally heavy.[45]

Mendoza: I do not understand this, so give me an example.

Tartaglia: For example, let there be the ends of the balance arms *B* and *D* and the pivot or center at the point *C*,

44. Archimedes, *On Plane Equilibrium*, Propositions 6 and 7, in the translation of Sir Thomas Heath (*Works of Archimedes* [Dover Press, n.d.], pp. 192–94.) The same propositions are differently numbered by Tartaglia in the ensuing sentence.

45. *Medieval*, p. 185: Jordanus R1.07.

and let there be attached the two similar equal solids, of
which one shall be attached along the balance arm horizontally, called *FE*, whose midpoint is *D*, while the other shall be
attached hanging perpendicularly as *BG*, the point of attachment being *B*. And let the distance from the point *B* to the
point *C* (center of the balance) be as much as that from the
midpoint of the other solid (that is, the point *D*) to the same
point. I say that the two solids in that place or position are
equally heavy, and this can be demonstrated in several ways.
The first of these is this: it is manifest by the things demonstrated by Archimedes in his *Centers of Gravity* that the solid
FE weighs as much in that position on the balance as if it
were hung perpendicularly at the point *D*, because at that
point *D* is situated the center of gravity of the solid; and the
two solids being equal in weight by hypothesis and hung
equally distant from the central point *C*, then by the Fifth
Proposition they will not depart from the level position; which
is the proposition.

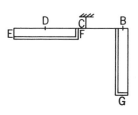

Also this proposition can be demonstrated in another way,
which is more suitable because it depends on its own principles
rather than imported ones. It is manifest that, when two simply
equal bodies, *H* and *K*, are suspended, the one at the point *E*
and the other at the point *F*, and two others which shall be *L*
and *M*, equal to them, are hung at the point *B*, these weights,
I say, will weigh equally at those points, because the ratio of
the weight *L* to the weight *K* is as that of the arm *BC* to the
arm *FC* (by the Fourth Proposition); for the body *L* will be
positionally as heavy at the point *D* as where it is at present,
that is, at the point *B* (since *CD* is equal to *CB* by hypothesis).
Therefore, by the said proposition, this ratio will be that of the
positional heaviness of the body *L* to the body *K*, which will
be that of the arm *DC* or *BC* to *CF;* and for the same reasons
this ratio will be that of the heaviness of the body *M* to the
heaviness of the body *H* positionally. And this will be the ratio
of the same arm *CD* or *BC* to the arm *CE*. Therefore the positional heaviness of both the bodies *L* and *M*, together, to the
positional heaviness of the other two bodies *H* and *K*, together,
will be as double the arm *CD* or *BC* to the two arms *CE* and
CF together. And since the said two arms *CE* and *CF*, together,
are precisely as much as the double of the said arm *CD* or
BC, it follows also that the heaviness of the said two bodies
L and *M* is equal to the positional heaviness of the two bodies
H and *K;* which is the proposition. For if the said solid *FE*
were made into two equal parts, one of those hanging at the
point *F* and the other at the point *E*, they would separately

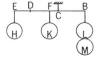

weigh as much thus at those points as [they would] if elon-
gated and joined in the manner supposed before. Similarly, if
the solid *BG* also were in two parts, both hung at the same point
B, they would thus weigh as much separated as conjoined (as
supposed above); hence from the things said and alleged the
proposition follows.

Mendoza: I should like to have you demonstrate to me that
the arm *CF* together with *CE* is as much as double the arm
DC or *BC*.

Tartaglia: Sir, it is manifest that the whole arm *CE* is
greater than the arm *CD* by the part *ED*, which part *ED* is

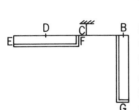

equal to *DF*. Therefore let us say that the whole of *CE* is equal
to *CD* added to its part *FD*, and if to part *FD* we add the arm
FC, these two parts together also equal the same *CD*. Therefore the whole *CE* together with *CF* is precisely the double of
CD; and since the said *CD* is equal by hypothesis to *BC*, it
follows that the whole *CE* together with *CF* is equal to the
double of *CB;* which is the proposition.

Mendoza: I understand very well, so continue.

Thirty-seventh Question. TENTH PROPOSITION

Tartaglia: If there is a solid rod or beam of the same
breadth, thickness, substance, and heaviness in every part, and
[if] its length is divided into two unequal parts, and at the end
of the shorter part there is hung another solid or heavy body
which makes the said rod, beam, or balance stay parallel to
the horizon, then the proportion of the heaviness of that body
to the difference between the heaviness of the longer part
of the rod (or beam or balance) and the heaviness of the shorter
part will be as the ratio of the length of the whole rod, beam,
or balance to the double of the length of its shorter part.[46]

Mendoza: Give me an example, if you want me to com-
prehend.

Tartaglia: Let *AB* be a solid rod (beam or balance) of
uniform breadth, thickness, substance, and heaviness throughout (that is, at every point), and divide it mentally into two
unequal parts at the point *C*, and mark *CD* equal to *CA;* then
DB becomes the difference between the longer part *CB* and
the shorter *CA*, of which difference find the center, which
shall be the point *E*. Now the said solid beam *AB* being sus-

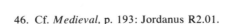

46. Cf. *Medieval*, p. 193: Jordanus R2.01.

pended at the point *C*, and there being attached or suspended
at the end of the shorter part another solid, which we shall
call *F*, which makes the first solid beam *AB* stand parallel to
the horizon, I say that the proportion of the heaviness of the
solid *F* to the heaviness of the difference *DB* is that of the
whole length *AB* to *AD*, the double of the length of the shorter
part *AC*. For the said difference *DB* weighs as much in that
position where it stands at present as it would if it were suspended perpendicularly at the point *E*, and therefore (by the
converse of the Eighth Proposition) the ratio of the heaviness
of the solid *F* to the heaviness of the partial solid beam *DB*
will be as the ratio of the distance *CE* to the distance *CA*. And
that ratio of *CE* to *CA* (by Euclid V.15) will be the same as
[that of] the double of the distance *CE* to the double of the
distance *CA*. And because the double of the said distance *CE*
is the whole length of the solid *AB*, and the double of the
distance *CA* is the whole of *ACD*, it follows (by Euclid V.11)
that the ratio of the heaviness of the solid *F* to the heaviness
of the difference *DB* is as the ratio of the whole length of the
solid rod *AB* to the double of the length of the shorter part
AC (which is the said *ACD*); which is the proposition.

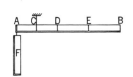

Mendoza: Why is double the distance *CE* equal to the
whole length of the beam *AB?*

Tartaglia: Because the said distance *CE* becomes equal to
precisely half of that length *AB*, for the part *DE* is the half of
the part *DB*, and *DC* is the half of the other part *DA;* therefore the two parts *DE* and *DC* joined together become the half
of the two parts *DB* and *DA* joined together.

Mendoza: I understand; therefore go on to the next.

Thirty-eighth Question. ELEVENTH PROPOSITION

Tartaglia: If the proportion of the heaviness of a solid
suspended at the end of the shorter part of a similar rod (beam
or balance) divided into two unequal parts, to the difference
between the heaviness of the longer part and that of the shorter,
shall be as the ratio of the whole length of the solid rod or staff
to the double of the length of its shorter part, such solid rod
(beam or staff) will necessarily be horizontal.[47]

Mendoza: I well believe that the preceding proposition
may have its converse; yet do not fail to give me the proof.

47. *Medieval*, p. 67: *De canonio* (anon.), II.

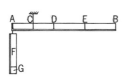

Tartaglia: This being the converse of the preceding, for its exemplification let us assume the same arrangement or figure. That is, let us suppose the ratio of the heaviness of the solid *F* to the difference of heaviness between the longer part and the shorter, that is, of *DB*, to be as the ratio of the whole length of the solid rod *AB* to the double of the length of the shorter part *AC*, which will be *AD*. I say that this solid rod *AB* will of necessity remain horizontal. If it is possible (for the adversary) that it must or might tilt from either side, let us assume that it tilts toward *B*. To the solid *F*, we add mentally such a part (which we shall call *G*) as will cause the said solid rod or staff to stand parallel to the horizon. Therefore (by the preceding), the proportion of the whole heaviness of the combination of the two bodies *F* and *G* to the difference between the weight of the longer part *BC* and that of the shorter part *AC* (which will be that of *DB*) shall be as the ratio of the whole length *AB* to the double of the length of its shorter part *AC*, which double would be *AD;* and since the simple solid *F* has that same ratio to the same difference (by what has gone before), it would follow (by Euclid V.9) that the heaviness of the simple solid *F* would equal the heaviness of the whole combination of the two solids *F* and *G* — which is impossible, for the part would be equal to the whole. The same contradiction would follow if the adversary should assume that it tilted toward *A,* because cutting away from the solid *F* such a part that the remainder would make the solid *AB* rest parallel to the horizon and arguing as above would make it follow that the heaviness of the same remainder was equal to the heaviness of the whole solid *F*. Therefore, being unable to tilt from either side toward *A* or *B,* it necessarily stands parallel to the horizon; which is the proposition.

Mendoza: Very good; now go on.

Thirty-ninth Question. TWELFTH PROPOSITION

Tartaglia: If there is a solid rod, beam, or staff, as in the two preceding [propositions], which is similar and equal in thickness, breadth, substance, and heaviness in every part and of which the heaviness as well as the length is known, and if it be divided into two unequal parts which are also known, it is possible to find a weight which, when suspended at the end of its shorter part, will make the said solid rod, beam, or staff stay horizontal.[48]

---

48. Cf. *Medieval*, pp. 197–99: Jordanus R2.06.

Mendoza: This operation I should like better explained to me by means of a material example, for I want to understand it thoroughly.

Tartaglia: For example, let there be the solid rod (beam or staff) *AB* as proposed, that is, equal and similar in breadth, thickness, substance, and heaviness on every side or in every part; and let us assume the heaviness of the said solid rod to be known, that is, let the length be two paces or ten feet; and let us also assume that the rod is divided into two unequal parts at the point *C* and that the [lengths of] said parts are known, it being assumed that the shorter part *AC* is two feet and the longer *CB* is eight feet. Now I say that it is possible to find how many pounds that body must be which, suspended at the point *A* (end of the shorter part), will make the said rod or beam stand parallel to the horizon. For (by the things demonstrated in the two previous propositions) it is manifest that the ratio of the heaviness of that body to the heaviness of that difference which exists between the longer part *CB* and the shorter *AC* (which difference becomes *DB*) will be as the length of the whole rod or beam *AB* (which is ten feet) to the double of the shorter part *AC* (which is two feet), and this double comes to be four feet. Let us call this *AD*. Then the heaviness of that body [at *A*] will be to the heaviness of the partial rod *DB* as the whole length of *AB* (which is ten feet) is to the length of *AD*, which is four feet. Whereby, arguing conversely, let us say that the ratio of *AD* (which is four feet) to the whole *AB* (which is ten feet) will be as the heaviness of the partial rod *DB*, which (at the rate of 40 pounds to all *AB*) is 24 pounds. Now the weight of the body we seek is that which, hung at the point *A*, should maintain the rod or beam parallel to the horizon. Whence in order to find this, we shall proceed by the rule ordinarily called the rule of three, founded on Euclid VII.20; multiplying ten by 24 gives 240, and this we shall divide by four, obtaining 60. I say that that weight which I called the body *F* will be 60 pounds; and this is the proposition.

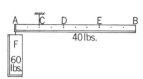

Mendoza: This problem pleased me very much and I understood it well; therefore go on to the next.

Fortieth Question. THIRTEENTH PROPOSITION

Tartaglia: If you shall have a rod, beam, or staff (as above) of which the length, as well as the heaviness, is known, and also a heavy body of which the weight is known, it is possible to determine the place at which the said rod, beam, or staff must be divided in order that the said heavy body, when hung at the end of the shorter part, will cause the rod, beam, or staff to remain parallel to the horizon.

Mendoza: Give me an example of this problem.

Tartaglia: To illustrate this problem, let us assume that there is a rod, beam, or staff as the above, ten feet long and weighing 40 pounds (as assumed before). And let us assume also that there is a body weighing 80 pounds. I say that it is possible to determine the place at which the rod may be divided so that the said heavy body, when hung at the end of its shorter part, will make the rod stay parallel to the horizon. And any such problem may be solved by ratios; nevertheless, it may be more easily solved by algebra, the shorter part of the rod being put as x, whence the longer part is 10 minus x. Double the shorter part (that is, x times 2) — this gives $2x$ — and subtract it from the whole length of 10 feet. There remains 10 minus $2x$, and this will be the difference between the longer part and the shorter. To find the weight of this difference, I multiply it by four, because, the whole rod weighing 40 pounds, each foot comes to weigh four pounds. Multiplying by four, as I said, the result is 40 minus $8x$ pounds. And since the ratio of the whole rod (which is ten feet) to the double of its shorter part (which double is $2x$) is as the weight of our heavy body (which is 80 pounds) to the weight of the above difference (which is 40 minus $8x$), by Euclid VII.20 the product of the first [term], or 10 feet, into the fourth, which is 40 minus $8x$, would be 400 minus $80x$, [and this] will equal the product of the third, which is 80 pounds, into the second, which is $2x$, [and this] will be $160x$. Thus we have $160x$ equal to 400 minus $80x$; and restoring the parts by rule we shall find x to be $1\frac{2}{3}$. Hence $1\frac{2}{3}$ feet will be the shorter part of the said rod or beam, whence the longer will be $8\frac{1}{3}$ feet; which was our problem.

Mendoza: This was a pretty solution. Now continue, for today and tomorrow I want to finish all that you have to say on this science, after which I should like to have you clear up for me some questions I have for you. . . .

Forty-first Question. FOURTEENTH PROPOSITION

Tartaglia: The equality of slant is an equality of [positional] weight.[49]

Mendoza: Give me an example.

Tartaglia: Equality of slant is preserved only in a straight path. Therefore let us assume that the said straight path is the

49. *Medieval,* p. 189: Jordanus R1.09.

line *AB*, and from the point *A* let the perpendicular *AC* be drawn, and let us also suppose some different places along the said slanted line *AB*. Let one of these be the point *D* and another the point *E*. Now I say that any heavy body in descending, whether at the point *D* or at the point *E*, will have the same positional weight as at any of the other said places. For let us take under *D* and *E* two equal parts in the path or line *AB;* let one be the part *DE* and the other *EG*. I say that the said equal parts partake equally in the [line of] direct[ion], that is, the line *AC*, so designated. From the two points *E* and *G* let there be drawn the two lines *EH* and *GL*, perpendicular to the line *AC*, and from the two points or places *D* and *E* the two lines *DK* and *EM*, perpendicular to the same *EH* and *GL*, which two perpendiculars *DK* and *EM* will be equal, wherefore the said heavy body, at point *D* as at point *E*, in equal quantities or descents [along *AB*] will share equally in the direct, and hence will be of the same positional heaviness in either of these places; which is the proposition.

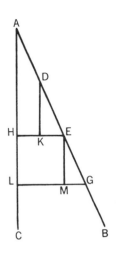

Mendoza: I have understood this; therefore continue.

Forty-second Question. FIFTEENTH PROPOSITION

Tartaglia: If two heavy bodies descend by paths of different obliquities, and if the proportions of inclinations of the two paths and of the weights of the two bodies be the same, taken in the same order, the power of both the said bodies in descending will also be the same.[50]

Mendoza: This proposition seems to me beautiful, and therefore give me a clear example, that I may be better satisfied.

Tartaglia: Let there be the line *ABC* parallel to the horizon, and upon this there is perpendicularly erected the line *BD*, and from the point *D* there shall descend on either side the two paths or lines *DA* and *DC*. Let *DC* be the more oblique. Then by the ratio of their tilts, I do not mean that of their angles, but of the lines to the parallel cut in which we take an equal part of the [line of] direct[ion]. Then let the letter *E* represent a heavy body placed on the line *DC*, and the letter *H* another on the line *DA*, and let the ratio of the simple heaviness of the body *E* to that of the body *H* be the ratio of *DC* to *DA*. I say that the two heavy bodies in those places are of the same power or force. And to demonstrate this, I draw *DK* of the same tilt as *DC*, and I imagine on that

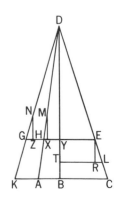

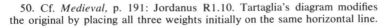

50. Cf. *Medieval*, p. 191: Jordanus R1.10. Tartaglia's diagram modifies the original by placing all three weights initially on the same horizontal line.

a heavy body, equal to the body *E*, which I letter *G*, in a straight line with *EH*, that is, parallel to *CK*.

Now if possible (for the adversary) that the said two bodies *E* and *H* are not the same in power and equal in force, assume that *E* is of greater force, and hence able to descend and thus to draw up the body *H*. Now let us suppose (if possible) that the said body *E* descends as far as the point *L*, and that it makes the body *H* ascend to the point *M*. Make or draw *GN* equal to *HM*, which becomes also equal to *EL*. And from the point *G*, draw *GH*, which will be perpendicular to *DB*, the said three points or bodies *G*, *H*, and *E* being assumed in line and parallel to *KC*. And similarly from the point *L*, let *LT* be drawn parallel to *CB*, which will also be perpendicular to the same *DB;* and from the three points *N*, *M*, and *E* draw the perpendiculars *NZ*, *MX*, and *ER*. Since the ratio of *NZ* to *NG* is as that of *DY* to *DG*, it is as that of *DB* to *DK* (the said three triangles being similar). Likewise the ratio of *MX* to *MH* is as that of the said *DB* to *DA* (the two triangles being similar). Also the ratio of *MX* to *NZ* will be as that of *DK* to *DA;* and (by hypothesis) that is the same as that of the weight of the body *G* to the weight of the body *H*, because *G* is supposed to be simply equal in heaviness with the body *E*. Therefore, by however much the body *G* is simply heavier than the body *H*, by so much does the body *H* become heavier by positional force than the said body *G*, and thus they come to be equal in force or power.[51] And since that same force or power that will

---

51. There are several interesting differences between Tartaglia's proof and that of Jordanus. The most serious objection to the medieval proof is that it failed to give a specific postulate or other physical reference relating it to the theorem (other than a repetition within the proof of the physical fact to be proved). Tartaglia attempted to rectify this defect, with some success, by his choice of definitions and postulates and by the addition of the sentence noted here. Other logical modifications and improvements of the medieval proof are to be noted in his manner of stating the assumption to be used in the *reductio ad absurdum,* and in the two final sentences of Tartaglia's proof.

Benedetti ("On Mechanics," Ch. 8, below) curtly dismissed both this and the preceding theorem on grounds which, taken literally, would render a science of mechanics impossible. He offered no alternative theorem or proof.

Guido Ubaldo ignored this theorem and adopted that of Pappus; see *Mechaniche* (110r–121v), below.

Galileo's proof proceeds on different lines; it relates each inclined plane to the vertical (not two inclined planes directly to one another); it considers the virtual movement at a point (not an assumed movement through an interval); and it relates a circular to a tangential motion at a point, both being related back to the lever or balance. None of these essential ideas is used by Jordanus or Tartaglia, nor does Galileo start from the same assumptions and definitions that they invoked.

be able to make one of the two bodies ascend (that is, to draw it up) will be able or sufficient to make the other ascend also, [then], if (for the adversary) the body *E* is able and sufficient to make the body *H* ascend to *M*, the same body *E* would be sufficient to make ascend also the body *G* equal to it, and equal in inclination. Which is impossible by the preceding proposition. Therefore the body *E* will not be of greater force than the body *H* in such places or positions; which is the proposition.

Mendoza: This was a beautiful speculation and satisfied me well. And since I see it is now late, I do not want you to proceed further today.

End of the Eighth Book

Giovanni Battista Benedetti

Resolutio
omnium Euclidis problematum aliorumque
ad hoc necessario inventorum
una tantummodo circini data apertura

Venice, 1553
(Letter of dedication only)

Demonstratio
proportionum motuum localium
contra Aristotelem et omnes philosophos

Venice, 1554

Diversarum speculationum

mathematicarum et physicarum liber

Turin, 1585

Excerpts translated and annotated by
I. E. Drabkin

The Resolution of All Geometrical Problems of Euclid and Others with a Single Setting of the Compass

Formulated by
Giovanni Battista Benedetti

Venice, 1553[1]

Dedicatory Epistle to the Abbot Gabriel Guzman

. . . Once, when we were still together, you eagerly begged and besought me to write something on the subject of natural motions based on sound theory and also, as far as possible, supported with mathematical demonstrations. Since your repeated request was a just one, I gladly took this heavy task upon myself. And so, after the month of September, when I was in the country to avoid the distractions caused by all the excitement in the city and by my large number of friends (for, when the occasion requires, it is ungenerous and ungracious not to give one's attention to one's friends), I fixed my thoughts on the loftiest ideas of natural philosophy, and on mathematics and other such branches of knowledge. To avoid waste and the frustration of my purpose, and to comply with your request, I have, with the Lord's help, composed three books.

One work contains demonstrations of many unknown things pertaining to nature and mathematics, which it is not my intention to publish in the near future, since additions of material that was lacking are being made to it daily. I do not speak of the second work, for, with good luck, it is possible that it will shortly go to press. The third work is the one you now see, dedicated to your name, as I have said.

Now it is in the first book that I carried out the task which you required of me. But, since it is not my intention to publish that work now for the reasons I have given, I shall here briefly give my solution of the problem.

---

1. [Pages are unnumbered in the original edition. The numbers shown in the running heads here have been assigned by counting from the title page and including blanks. —S. D.]

Understand, then, that the ratio [of the volume] of one body to another (it being assumed that they are homogeneous and uniform) is the same as the ratio of their respective forces [weights].

Suppose, for example, that there are two lead bodies of unequal size, designated with the letters *A* and *E*, of which body *A* is three times as large as body *E*. Now I infer that body *A* will have weight three times that of body *E*. Indicate the weight of body *A* with letter *B*, and that of body *E* with letter *F*. Now

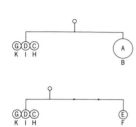

imagine body *A* divided into three equal parts, namely, *C*, *D*, and *G*, with weights *H*, *I*, and *K*, respectively. It is clear, in view of our assumption, that the separate parts *C*, *D*, and *G* will each be equal to body *E*, and obviously each will have a weight equal to *F*. If this were not the case, it could not be said that each of the parts of *A* was homogeneous with body *E;* thus our assumption would be contradicted. And since the sum of *H*, *I*, and *K* is equal to *B* alone, then also the ratio of *B* to *F* will obviously be equal to the ratio of *H*, *I*, and *K*, together, to *F* (Euclid V.7). But *H*, *I*, and *K* together are three times *F*. And by this reasoning our proposition is clear.

Next, I assume that the ratio of the [speed of the natural] motion of bodies alike [in shape] but of different material, in the same medium and over an equal distance, is equal to the ratio of the excesses (I mean, of weight or lightness) of the bodies over the medium.[2] (Note that those bodies must have the same shape.) And, conversely, I assume that the ratio between the respective excesses over the medium, as indicated, is equal to the ratio between the [speeds of the] motions of those bodies.

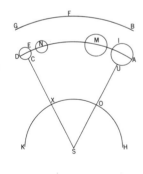

This proposition will be made obvious in the following way. Let there be a uniform medium *BFG*, say, water, and suppose that two bodies, each homogeneous but different, i.e., each made of a different material, are in that water. For example, suppose that body *DEC* is of lead and body *AUI* of wood, but that each of them is heavier than a body of water equal in size [to the lead body and the wood body, respectively]. And let the given bodies be spherical, and let the corresponding aqueous spheres be designated *N* and *M*, respectively. Let us imagine the center of the universe at *S*, the *terminus ad quem* of the motion a point on the line *HOXK*, and the *terminus a quo* a point on the line *AMD*, the latter being everywhere equidis-

2. As the sequel shows, Benedetti is concerned with specific weight rather than with total weight; "excess" here refers to the arithmetical difference between the specific weights of the moving body and of the medium.

tant from line *HOXK*, and both being circular arcs about the center of the universe *S*.

Then if lines *SO* and *SX* are extended to the line of the *terminus a quo*, the segments [on the extensions of *SO* and *SX*] intercepted by those terminal lines will be equal to each other, by Euclid's third axiom. (For by his definition all straight lines drawn from the center of a given circle to the circumference are equal to each other.) Let us also imagine the center of body *AUI* situated at the point of intersection of line *SO*, produced, with line *AMD;* and the center of body *DEC* at the point of intersection of *SX*, produced, with line *AMD*. Futhermore, let the aqueous body equal [in size] to body *AUI* be *M*, and the other aqueous body equal [in size] to body *DEC* be *N*. And suppose that body *DEC* is eight times as heavy as body *N* and body *AUI* twice as heavy as body *M*.

Now, then, I say that the ratio of the [speed of the natural] motion of body *DEC* to that of body *AUI* (on my hypothesis) is equal to the ratio of the excesses of bodies *DEC* and *AUI* over bodies *N* and *M*, that is, that the time in which *AUI* will move [the given distance in the given medium] will be seven times as long as the time required by body *DEC*.[3] For it is clear (see Archimedes, *On Floating Bodies*, [Bk.I,] Prop. 3) that if bodies *AUI* and *DEC* were of equal weight with bodies *M* and *N*, respectively [*AUI* with *M* and *DEC* with *N*], they would not move at all, either up or down; and also (*ibid.,* Prop. 7) that bodies heavier than the medium move downward. Therefore bodies *AUI* and *CED* will move downward. And the resistance[4] of the fluid medium (that is, of the water) to body *AUI* is in the ratio of 1 to 2 (as is obvious to ordinary understanding), but to body *DEC* is in the ratio of 1 to 8. Hence the time in which the center of body *AUI* will traverse a given distance will be seven times as long as the time in which the center of the body *DEC* will traverse the same distance.[5] (And I mean in natural motion, for nature acts in all things along shorter lines, i.e., along straight lines, unless something impedes.) For, as one may gather from the work of Archimedes cited above, the ratio of motion to motion [i.e., of the speeds of the motion of *AUI* and *DEC*] is not equal to the ratio of the

---

3. Thus, time of *AUI*: time of *DEC* = $(8 - 1) : (2 - 1) = 7:1$.

4. What is here called resistance is merely the effect of the buoyancy of the medium. Contrast the usage in the Padua version of the 1554 *Demonstratio,* translated subsequently.

[Benedetti's concept is easier to grasp if we read "1 part in 2" for "in the ratio of 1 to 2," etc. —*S. D.*]

5. In this sentence Benedetti had inadvertently transposed *AUI* and *DEC*.

weights of *AUI* and *DEC*, but to the ratio between the weight of *AUI* (relative to that of *M*) and the weight of *DEC* (relative to that of *N*).[6] And the converse of this assumption is quite evident, since what has been said is clear.

I now say that, if there are two bodies of the same shape and the same material, whether equal [in size] to each other or unequal, they will, in the same medium, move over an equal distance in an equal time. This proposition is quite obvious, for, if they did not move in equal times, the bodies would necessarily have to be made of different materials, by the converse of the previous assumption; or else the medium would not be uniform, or the distances traversed would be unequal—all of which would be contrary to the hypothesis.

But, by way of demonstration, suppose there are two bodies, *G* and *O*, similar in shape (spherical) and of the same material, a uniform medium *BDF*, and circular arcs representing the terminal lines, everywhere equidistant from each other [i.e., concentric] about center *S*, arc *PIQ* passing through the *terminus a quo*, and arc *RMUT* through the *terminus ad quem*. Now I infer that bodies *G* and *O* will, in natural motion, traverse the given distance in the given medium in equal times.

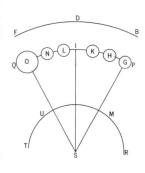

For example, suppose that body *O* is four times the size of body *G*. From what has been said it is clear that body *O* will also weigh four times as much as body *G*. (Indeed, if they were equal in both respects [size and weight], they would then, no doubt, move in equal times.) Now I shall imagine body *O* divided into four equal parts, each part similar to the whole (i.e., of spherical shape). Let these parts be *H*, *K*, *L*, and *N*. I shall place their centers on arc *PQ*, so that the distance between *H* and *K* is the same as that between *L* and *N*, and I shall also bisect arc *KL* at point *I*. This point will obviously be the center of gravity of bodies *H*, *K*, *L*, and *N* (see also Archimedes, *On Centers of Gravity*, Prop. 3).[7] Besides, it is clear that each of bodies *H*, *K*, *L*, and *N* will move from *PIQ* to *RMUT* in the same time as will body *G*. (For each of them is of the same size, shape, and weight as body *G*, by an axiom of Euclid.) By the first axiom, therefore, all the bodies, *H*, *K*, *L*, and *N*, if let fall together at the same instant, will move equally, i.e., in equal time, and the arc passing through their

6. Specific weight, not total weight, is controlling. Implicit in the formulation is the equality of the speed of natural motion (in a given medium in which such motion can take place) of bodies of the same shape and density, regardless of size and total weight. The next proposition makes this explicit.
7. *On Plane Equilibrium*, Prop. 4, in our modern editions.

centers will at all times be everywhere equidistant from arc
RMUT.

Now, if the arc connecting the center *I* with the center of
body *O* is thought of as bisected, the point of bisection will,
by the foregoing, be the center of the weight of *H, K, L, N*,
and *O*. But now if we imagine that that arc moves under the
force of those bodies, being allowed to descend from arc *PQ*
and remaining everywhere equidistant from it (and since the
moving arc would then obviously be everywhere equidistant
from *MURT*, it will always be equidistant from *MURT*), body
O will move in natural motion over the given distance in the
same time as [that in which] bodies *H, K, L*, and *N* will
move. (For the resistance of the medium to bodies *H, K, L*,
and *N* is the same as to body *O*,[8] according to what was said
above and also Euclid V.16, since the weight is the same and
the material is the same.) But that is obviously the same time
as that in which *G* moves. And this is what we set out to prove.

I can also, by this demonstration, prove a part of the assumption made above, namely, that if there are two bodies
of the same shape but of different material and unequal size,
each of them heavier than the medium through which they
move—the smaller of them being of heavier material than the
larger, but the total weight of the larger exceeding that of
the smaller—then I say that the above assumption is true.[9]

For example, suppose the two bodies are *M* and *N*, of the
same shape, but of different material and also of different size.
(For there will be no doubt about bodies of equal size.) Now
suppose that the larger body is *M*, but that the material of
body *N* is heavier than the material of body *M*. And let the
[total] weight of body *M* be greater than that of body *N;* and
let each of them be heavier than the medium through which
they move. I say now that our assumption is true.

First consider a body *AUI* of the same size and shape as
body *M* but of the material of body *N*. Then with respect to
bodies *AUI* and *M* our assumption is clearly true.[10] But by
our previous proof, body *N* moves [in natural motion over a
given distance] in the same time as body *AUI*. Therefore our

8. The assumption is made—wholly without proof—that the resistance of
the medium is merely a matter of buoyancy and therefore proportional to
the weight of the body. Note that the surface of *O* is not equal to the sum
of the surfaces of *H, K, L*, and *N*.
9. Presumably that $v_1 : v_2 = (d_1 - d_m) : (d_2 - d_m)$, where *v* is speed of natural
motion, *d* is density, and the subscripts refer to the two bodies and the medium.
But note the divergent discussion of the same matter in the Padua version
of the 1554 *Demonstratio*.
10. See the previous note.

proposition [with respect to bodies *M* and *N*] is also clearly true.

From this it is clear that swifter [natural] motion is not caused by an excess of the [total] weight or lightness of the swifter body in comparison with the slower (given bodies of the same shape), but by the difference in heaviness or lightness of the material[11] of the one body in comparison with that of the other. Now this was not Aristotle's thought or the thought of any of his commentators whose works I have chanced to see and to read or to discuss with professors.

Understand also that the same principle underlies forced motions, provided that the force of the movers is of corresponding strength, etc.[12]

The same is true when [the motive forces] make equal angles above or below the horizontal,[13] but the order is opposite that of natural motion, for forced motion is swifter at the beginning than at the end. And forced motion takes place in a way different from that of natural motion: for in every case of forced motion natural motion is mingled to some extent. If the forced motion is horizontal, or if it is oblique, whether above or below the horizontal, natural motion operates insofar as to impose some end upon the forced motion. But if the forced motion takes place perpendicularly to the horizon and toward the place which the body seeks in accordance with its nature [i.e., vertically downward], then nature [i.e., natural motion] cannot prevent force from always accompanying it, beginning at the point of departure.[14]

I should have been happier to write all this to you privately, in answer to your requests, than to divulge it by publication.

11. I.e., by difference in specific weight (or density).

12. Violent motion produced by application of external force to the body is analogous to natural motion caused by an excess of the body's specific weight (or lightness) over that of the medium. And the speed of that violent motion would be proportional to the force. For example, bodies of the same material and shape, but of different size, move with the same speed in violent motion, if the force applied to them is proportional to their size.

[Violent motion is to be taken as synonymous with forced motion, in opposition to natural motion. Professor Drabkin indicated a preference for the term "violent motion." — S. D.]

13. The meaning seems to be that if two bodies of the same shape and material, but of different size, or of the same shape and size, but of different material, are projected obliquely (upward or downward) at the same angle, with forces proportional to their weights, the speeds will be equal, i.e., over the initial part of the trajectory, which is considered to be a straight line. On the trajectory cf. Tartaglia's view.

14. Virtually the same language is used in the *Demonstratio proportionum motuum localium* of 1554. The passage contains a refutation of Tartaglia's view that natural and violent motion can never be commingled (*Nova*

But I feared that someone would get hold of my letter (as we see often happens) and, whoever it was, would steal my ideas and publish my demonstration in his own name or in the name of one of his friends, while I would have wasted my time and effort, an experience that has befallen many, and in our own time. For there is a type of person—men who may be called "pretenders," who love praise and attention, but avoid hard work, arrogating to themselves by fair means or foul a glory that was attained by the great effort and exertion of others.[15] For this reason I was compelled to publish this demonstration, just as you see it, though I do not intend at present to publish the volume of which I spoke: for many other more important things will be contained in that volume. . . .

Scientia I. Prop. 5). See also Galileo, *De motu* (*Opere di Galileo Galilei,* ed. A. Favaro [Ed. Naz., 1890], vol. 1), p. 300. Note that in his dialogue *De motu* (*Opere,* I, 408) Galileo holds that the impressed force (even in a vertical downward projection) is ultimately dissipated and that the body thereafter undergoes purely natural motion. Benedetti's view is different.

15. [The remark was general, but probably alluded to the recent behavior of Cardano toward Tartaglia. —*S. D.*]

A Demonstration of the Ratios of Local Motions in Opposition to Aristotle and All Philosophers

Formulated by
Giovanni Battista Benedetti

Venice, Ides of February, 1554[1]

To his ever revered Lordship, the Most Reverend and
Noble Gabriel de Guzman, Most Worthy Abbot
of Pontelongo, Giovanni Battista Benedetti
sends greetings.

Last summer the learned doctor Peter Arches, who (as
your most Reverend Lordship well knows) is highly trained
not only in theological matters but in those that pertain to
philosophy, made a trip to Venice and graciously came to pay
me a visit. I remember that, after conversation on various
subjects between us, he reported to me that many at Rome,
having examined that proposition of mine which I sent to
your Reverend Lordship along with the others, generally
expressed great surprise that I added that my proposition was
not in agreement with Aristotle's thought. They expressed
this feeling before giving any further thought to my demonstration (though after they understood it, they did, as I have
said, approve it). But they wanted to avoid admitting that
Aristotle was in any way mistaken, for they consider Aristotle not subject to human limitations, but rather as having
some divine power, and they hold it a sin if one differs even a
nail's breadth from him. And so they held and still hold the
opinion that I failed to attain the true and essential meaning
of Aristotle's views based on that author's thought.

But to avoid having this error imputed to me any longer
and to avoid suppressing or withholding any thoughts I enter-

1. Since the Vatican version of the *Demonstratio* more closely repeats
the doctrine of the *Resolutio,* translated above, it seemed better to present
here the Padua version, which introduces a different concept of the resistance
of the medium. On this difference in doctrine, see *Isis,* 54 (1963), 260.
[Pages are unnumbered in the Padua version. The numbers shown in the
running heads are here assigned, counting from the title page. — *S.D.*]

tain on this subject, I have decided to publish this new treatise and in it to express my thought more clearly. In this way all may realize that from the beginning I understood Aristotle aright and that I do not disagree with him capriciously on this matter. For, however reluctant I am to differ with him — and I certainly am not eager to differ, since I know how pre-eminent he was in every field of learning — still, I must have a higher regard for the truth. And this is what he himself asserted when in the *Ethics* he said: "Plato is my friend, and Socrates is my friend, but the truth is even more my friend." . . .

It is my intention in this brief work to show the errors of Aristotle in his treatment of the ratios or comparisons of local motions, and consequently the errors of all his commentators, all other philosophers, etc.

We must first understand that, for bodies of one and the same material, the ratio between their volumes [*quantitates*] is the same as the ratio between their heavinesses (or lightnesses).

Suppose, for example, that there are two lead bodies of unequal size, *A* and *E*, body *A* having a volume three times that of body *E*. (On the multiplication of volume, let there be no misunderstanding. Some have thought that a sphere is doubled [in volume] when its diameter is doubled. But this is a grave error, for it is proved in Euclid XII.15 that the ratio of [the volumes of] two spheres is equal to the cube of the ratio of their diameters. A similar proposition is proved in Euclid XI.37 for similar solids having surfaces respectively parallel. In addition, Albert Dürer treats the matter adequately in Book IV of his work on geometry,[2] when he takes up the doubling of the cube, etc.)

Now I assert that body *A* is three times as heavy as body *E*. Suppose that the weight of *A* is indicated by letter *B* and that of *E* by letter *F*. Imagine body *A* divided into three equal parts, *C*, *D*, and *G*, the weights of these parts being *H*, *I*, and *K*. Now I say, in accordance with our assumption, that the several parts, *C*, *D*, and *G*, will each be equal to body *E*, i.e., will obviously each equal *F* in weight. If this were not so, each of the parts of *A* would not be homogeneous with body *E*. This would conflict with our assumption.

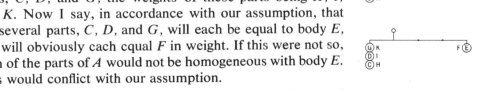

2. Latin translations of Durer's *Underweysung der Messung* had appeared in 1532 and 1535.

Thus, as is obvious, *H*, *I*, and *K* together weigh the same as *B* alone. Also (by Euclid V.7), the ratio of *B* to *F* is the same as the ratio of *H*, *I*, and *K* [taken together] to *F*. But the latter ratio is 3 to 1. Therefore weight *B* will be three times weight *F*. And by this reasoning our proposition becomes clear.

I shall now show that bodies in natural motion do not move according to the same ratio as Aristotle holds but according to one quite different.[3]

To begin with, in *Physics* IV, Ch. 8,[4] dealing with the void, Aristotle says that the ratio of the [speeds of the] motions of a given body through different elements is the same as the ratio of the [rarenesses of the] elements: so that if air is twice as rare as water, the body will traverse medium *B* in twice the time it traverses medium *D*, and time *C* will be double time *E*, etc. But this is false.

Suppose, first, that a body moves through water in natural motion, let us say, traversing a given distance. Let us assume that that body has twice the heaviness of water, and that water has twice the heaviness of air. Thus the given body will have four times the heaviness of air. And the body will therefore be three-halves[5] as heavy in air as in water (Archimedes, *On Floating Bodies*, [Bk. I, Prop.] 7). It therefore follows that the [natural] motion of the body will be three-halves as swift in air as in water. For example, if the distances traversed are equal, the length of time in which the given body moves through water will be one and one-half times as great as that in which it moves through air, and not twice as great, as Aristotle holds.

Moreover, in the same chapter he writes as follows:[6] "For we observe that bodies which have more tendency to movement [*momentum*] due to heaviness or lightness, provided they have the same shape, move more swiftly through an equal interval, and in the same ratio that the [heavinesses or lightnesses of the] bodies have to each other. Therefore they will so move through a void."

Later he reaffirms this, saying (*Physics* VI, Ch. 2)[7] that the magnitude[8] is divided in the same ratio as the time is divided;

3. Note that Benedetti's purpose here is chiefly to refute Aristotle rather than to build his own theory.

4. *Physics*, 215a.31–215b.12.

5. The text reads "three" for "three-halves" (or "one and one-half") here and in the following two sentences of this paragraph. The error is corrected in the *Diversarum speculationum* (see "Disputations," Ch. 5, below).

6. *Physics*, 216a.13–16.

7. *Physics*, 233a.4–5.

8. Aristotle in the passage cited speaks of dividing the magnitude of the distance traversed; but Benedetti seems to refer to the magnitude of the moving body.

and this he repeats at four places. In Chapter 4 [of Book VI][9] he writes thus: "Since everything that moves moves in some place and some time, and the motion is of the whole body that moves, there must be the same divisions of time and of [the quantity of] motion, and of the very act of motion, and of that which moves, and also of that in which the motion takes place." He says substantially the same thing in Chapter 7 [of Book VI].[10]

And in *De Caelo* I, Ch. 6,[11] he says that the times [of natural motion] will have the inverse ratio of the weights: for example, if half the weight moves in a given time, the whole weight will move in half the time. He also says the same thing at four places in *De Caelo* III, Ch. 2.[12] And in *De Caelo* III, Ch. 5,[13] he says that the larger a body is, the swifter must its [natural] motion be: as in the case of fire, too, whose [natural] motion is swifter, the greater the amount of fire. And he makes exactly the same assertion in *De Caelo* IV, Ch. 2. But I shall now show that this is not so.

Suppose that bodies *M* and *N*, as seen in the figure, are of the same material, *M* being four times the size of *N;* and that body *AUI* is equal to *M* in size, but to *N* in weight. In that case, body *M* will weigh four times as much as body *AUI*, by the first proposition of this book with the help of the first axiom in Campanus on Euclid, etc. And suppose that the medium through which those bodies move is so rare that practically no resistance can be detected, e.g., if the medium is air and *M* is of gold. In that case, there will be no doubt[14] that body *M* moves about four times as swiftly as body *AUI*. But body *N* obviously moves more swiftly than body *AUI*.[15] Therefore the motion of body *M* will not be four times as swift as that of body *N*.

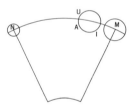

In the same chapter on the void [*Physics* IV, Ch. 8][16] Aristotle also says that, if there were motion in a void, the same ratio would obtain as in a plenum, which would be impossible.[17]

9. *Physics,* 235a.13–17.
10. Cf. *Physics,* 238a.22–27.
11. *De Caelo,* 273b.30–274a.2.
12. E.g., *De Caelo,* 301a.28.
13. *De Caelo,* 304b.15–18.
14. [Because weight is the cause of speed. Cf. p. 370*n*62. −*S. D.*]
15. Because the resistance of the medium, even if small, retards the larger of the two bodies (of the same weight and shape) more than it does the smaller one.
16. *Physics,* 216a.16–17.
17. It is not clear whether Benedetti intended this last clause to be part of the quotation from Aristotle. Cf. "Disputations," Ch. 9, below.

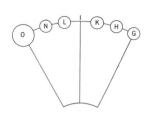

Let us imagine two bodies *O* and *G*[18] moving in a void and a body equal to *O* divided into four equal parts. In that case each of the parts will be equal [in weight and volume] to body *G*, by the first proposition of the present work. Let these parts be *N*, *L*, *K*, and *H*, all of which together are equal in weight to body *O*, as is obvious, since they are of the same material. I then join them by a line passing through their centers. Obviously bodies *N*, *L*, *K*, and *H* will move with the same speed as body *O*, for they encounter no resistance and their [total] weight is the same [as that of *O*].[19] Furthermore, body *G* will move with the same speed as [that with which] each of the bodies *N*, *L*, *K*, and *H* moves, but also with the same speed [as that] with which all the bodies *N*, *L*, *K*, and *H*, joined together by a line passing through their centers, move. Therefore *G* will move with the same speed as *O* (by the first axiom of Euclid).

It is therefore clear that bodies of the same material, though of unequal size, move [in natural motion in a void] with equal speed.

Furthermore, I say that, if those bodies were of different material and different[20] weight, they would still move with the same speed of [natural] motion [in the void]. And they would move not instantaneously,[21] but in a determinate interval of

18. The two bodies are of the same material and shape, *O* having four times the volume of *G*. Benedetti seems inadvertently to have omitted these necessary conditions or to have been thinking of bodies *M* and *N* of the previous discussion.

19. [Benedetti has not yet raised the question of any resistance other than buoyancy. −*S. D.*]

20. The 1553 *Resolutio* does not specifically deal with the speed of natural motion in a void. But if the void be considered as a medium of density zero, that speed would be proportional to the specific weight of the body. Such a relation was also implied earlier in the present treatise. But this sentence and the next two paragraphs now contradict that proposition. (Note that Aristotle holds [*Physics*, 216a.20] that bodies of different material and weight would move with the same [infinite] speed of natural motion in a void; he considers the absurdity of such a result an argument for the rejection of the possibility of a void.)

One might wish to read here *aequalis* for *diversi*, since in what follows Benedetti seeks to prove that the speed of natural motion in a void is the same for bodies having different size and material but the *same total weight*. But this proposition, which indeed happens to be true, whether the weight be the same or different, is not consistent with the discussion of motion in a plenum. [Or better: does not apply in the plenum. −*S. D.*]

Note that in his early work *De motu* (*Opere di Galileo Galilei*, ed. A. Favaro [Ed. Naz., 1890], I, 279–80) Galileo adopts a unified (though erroneous) treatment of natural motion in plenum and void, making the speed proportional to the specific weight of the body (diminished by the specific weight of the medium, a void being considered a medium of density zero).

21. In contrast with Aristotle's view (put forth to refute the possibility of a void) that bodies moving in a void would traverse any distance instan-

time. (For no motion could take place in an instant, since in that case an instant would have an earlier and a later part, and would consequently be divisible. Hence it would be [an interval of] time, and thus not a terminus or end-point [of time], but a portion of definite time, etc.)

But for the sake of clearer understanding, suppose there are three bodies in a void, *M*, *N*, and *AUI*. Suppose that body *M* has the same [total] weight as body *N*, but is of different material and (consequently) of different volume. Now, those bodies [*M* and *N*], since they encounter no resistance, will, no doubt, move in equal time over an equal distance.[22]

Let us next suppose that body *AUI* is of the same material as body *N* but has the volume of body *M*. Now by the proposition previously set forth, the speed of the [natural] motion of *AUI* and *N* will be the same. Therefore, by the first axiom, the [natural] motion of bodies *M* and *N*[23] [hence *M* and *AUI*] will be of the same speed, and these motions will take place in a definite time, as I said above.

Moreover, Aristotle, for all his keenness, had a fixed opinion that the ratio [of the speeds] of the motions through all three elements—fire, air, and water—was the same. But this view is false, and its falsity will become quite clear from the following.

Suppose, for example, that there are two bodies equal in volume and of the same shape, but of different materials and consequently of different weights. Let the bodies be *M* and *AUI*. Now suppose that body *AUI* has double the weight of body *M*, and body *M* double the weight [per unit of volume] of the medium through which those bodies move. Hence body *AUI* will have four times the weight [per unit of volume] of the medium. Hence, if these bodies move [in natural motion] for equal times, the distance traversed by body *AUI* will be three times the distance traversed by body *M*.[24] Now this is entirely clear and obvious: for these bodies are equal [in

---

taneously (*Physics*, 215b–216a). Cf. "Disputations," Ch. 1, below.

22. There has been no preparation for this proposition, that bodies of the same total weight, regardless of material or size, have equal speeds of natural motion in a void. Nor is the following paragraph, as it stands, cogent. Note that, while the 1553 *Resolutio* points in a different direction, the same proposition as this one is implied in the "Disputations" (Ch. 13, second paragraph, below). [But cf. nn. 14 and 20, above; n. 128, below. —*S. D.*]

23. Perhaps "*N*" is a slip for "*AUI*." Otherwise we have a repetition, without proof, of the proposition at the end of the preceding paragraph.

24. Thus, dist. (*AUI*) : dist. (*M*) = (4 − 1) : (2 − 1) = 3 : 1, where "dist. (*AUI*)" is the distance traversed by *AUI*, etc. (Note that in this and the next paragraph resistance of the medium is equated with buoyancy, as throughout the 1553 *Resolutio*.)

size and shape] and the resistance of the medium to *AUI* is
one part in four, and to *M* is one part in two. Or, by Archimedes, *On Floating Bodies,* [Bk. I, Prop.] 7, body *AUI* is
heavier than body *M* in that same medium in the ratio of
3 to 1.

Further, let us suppose that those bodies move in a rarer
medium with natural motion (for it is to be understood that
we are always concerned with natural motion, this being more
regular). Now suppose, for example, that body *M* is four times
as heavy [per unit of volume] as the medium. In that case
body *AUI* will be eight times as heavy [per unit of volume]
as the same medium. Therefore, according to what was said
above, if these bodies move through this medium for an equal
time, *AUI* will traverse a greater distance than *M* in the ratio
of $1\frac{4}{3}$ to 1, i.e., in the ratio of 7 to 3.[25]

Hence it is clear from this that the ratios of the [speeds
of natural] motions through different media vary. But it is
quite obvious that Aristotle held the view ascribed to him
above, since in so many passages he says that bodies maintain
the same ratios in the [speed of their natural] motions as the
ratios of their weights or lightnesses. And this was a universal
proposition with Aristotle. He held his view because he
thought that the ratio of the heaviness (or lightness) of given
bodies was constant for all media, as if the bodies were being
weighed in a void. (But from what has been said above it is
clear that this cannot be the case.) Aristotle's views may be
gathered from his words in the chapter [8] on the void in
Physics IV, where he holds that the speeds of the natural
motion of bodies moving through different elements vary
according to the ratios that these elements have to one another
in respect to lightness or heaviness.[26]

But in the case of bodies of the same material it would, of
course, be true that the ratio of the speeds of such bodies
moving [naturally] through different elements would always
be the same, provided that the bodies maintained the same
shape or underwent similar changes in shape.

From this it follows that Aristotle is in error when he says
(*Physics* VII, last chapter)[27] that if *A* is the mover, *B* the

25. Thus, dist. (AUI) : dist. $(M) = (8 - 1) : (4 - 1) = 7 : 3$.

26. E.g., *Physics,* 215b.10–12. For Aristotle, if a given body moves
naturally through two different media, the ratio of the speeds is the inverse
ratio of the densities of the media.

27. *Physics,* 249b.31–250a.4. Note that this is a case of violent motion and
really not relevant to the previous discussion, except as concerns the resistance of the medium.

moving body, C the distance traversed, and D the time of the motion, then obviously an equal force will move half of B a distance $2C$ in time D, or a distance C in half of time D, etc.

Furthermore, Aristotle says (*Physics* VII, Ch. 5)[28] that if two separate [forces] move two separate weights a certain distance in a certain time, then the two [forces] combined will move the combined weights the same distance in the same time, for the same ratio obtains. But I shall now prove that this is false.

Consider two cubes of the same material, one double the other [in volume]. Then the ratio of the [volume of the] larger to that of the smaller is equal to the cube of the ratio of the edge of the larger to the edge of the smaller (Euclid XI.37). And (by Euclid VI.18) the ratio of the [area of the] base of the larger to that of the smaller is equal to the square of the ratio [of the edge of the larger to the edge of the smaller]. Therefore the [area of the] base of the smaller is obviously greater than half that of the larger; and, consequently, the smaller body will be slower than the larger.[29]

And I say the same thing about spheres.[30] For, by common knowledge, the ratio of the surface of one sphere to the surface of a second is equal to the [square of the] ratio of the circumference of a great circle of the first sphere to that of the second. But the ratio of these circumferences is equal to the ratio of the diameters (Euclid V.16), for the ratio of the circumference to the diameter is always the same for any circle, as Archimedes proved, etc. In fact, for a long time I myself was guilty of the same error [as Aristotle].[31]

The ratio of [the speeds of] forced motions is also the same, if we first take the ratio of the motive forces and then subtract [from the respective terms of that ratio the terms repre-

28. *Physics*, 250a.25–28. Again the Aristotelian passage refers to violent motion.

29. Benedetti's point is that in the case of two cubes, one double the other in volume, the resistance offered by the medium (assumed to be proportional to the area of the cross section or base of the cube) will be greater, proportionately, in the case of the smaller cube; for its base has an area more than half that of the larger cube. This will affect the ratio of the speed of motion (both natural and violent) in a resistant medium.

30. If S and V are the surface and volume of spheres A and B, and $A < B$, then $S_A/S_B > V_A/V_B$.

31. That of taking all resistance as proportional to volume, and none to surface. In fact, in the *Resolutio* and the Vatican version of the *Demonstratio* the resistance of the medium is identified with buoyancy (weight of medium displaced) and is consequently proportional to volume. Aristotle was not himself concerned with this aspect of resistance.

senting] the ratio of the resistance of the medium.[32] That is also the case[33] when the angles [of projection] above or below the horizontal are equal, but in an order the reverse of natural motion. For forced motion is swifter at the beginning than at the end, and the contrary is true of natural motion. For natural motion is always to some small extent mingled with forced motion if the direction is horizontal or even oblique, either above or below the horizontal. In these cases nature [i.e., natural motion] acts[34] insofar as to impose some end upon the forced motion. But if the forced motion takes place perpendicularly to the horizon and toward the place to which the body naturally moves [i.e., vertically downward], then nature [i.e., natural motion] cannot prevent violence [i.e., forced motion] from always accompanying it, beginning at the point of departure.

Again, Aristotle says (*De Caelo* IV, Ch. 5) that, if void [emptiness] is the cause of lightness and plenum [fullness] the cause of heaviness, some media will move downward more swiftly, but nothing will continue to move downward forever.[35] Later[36] he concludes that a large portion of air will have to move downward more swiftly than a small amount of water, etc. But in this he does not consider the homogeneity of bodies or the ratio of void and plenum in the bodies themselves (in accordance with the intent of the ancients) with respect to the entire amount of each of the two bodies under consideration.[37]

Furthermore, Aristotle speaks again in *Physics* VII[38] of the comparison of motions, saying that a straight line is not comparable to a curved line, because some straight line would then be equal to a circular arc, or else greater or less than

32. The subtraction of one ratio from another is ambiguous. Sometimes true subtraction, at other times division, seems to be indicated. In any case, the point is that a cube or sphere having twice the volume of another cube or sphere requires less than twice the force to attain the same speed of forced motion in a resistant medium, since the resistance encountered by the smaller body is more than half that encountered by the larger.

33. For this whole paragraph, cf. *Resolutio*, p. 152, above.

34. I.e., in conjunction with the principle of forced motion, though it may later continue to act alone.

35. This hardly gives the sense of the Aristotelian passage which, in any case, is a *reductio ad absurdum*.

36. *De Caelo*, 313a.5ff.; but Aristotle's point is that such a conclusion could only arise from a false premise.

37. I.e., air and water. The meaning of this rather involved language is that Aristotle erred because he considered not relative density but total weight.

38. Cf. *Physics*, 248a.19–248b.6. Aristotle is concerned with mathematical commensurability. See Galileo's early *De motu, Opere*, I, 302–04.

that arc. (For in view of the definition given by him in *Physics* VI[39] on the speed and slowness of motion, he believed it was impossible for circular motion to be comparable to rectilinear.) Now I hold that Aristotle is mistaken in this and, particularly, in his view that it is impossible for a [straight] line to be found equal to (or even greater than or less than) a circular line. [There follows a mathematical passage dealing with the existence and constructibility of circular arcs equal to rectilinear segments, and of areas bounded by curved lines equal to areas bounded by straight lines.]

Aristotle was, to be sure, an outstanding investigator, but I cannot say, as some do (who either never read his works or, if they did read them, did not understand them), that practically every word of Aristotle is a unique idea, and that Aristotle was a god of philosophy, was never mistaken in a single word, but was divine in all his utterances. If these poor wretches knew the difference between logical proof and dependence on sensory experience, they would never have said these things. For the senses, unaided, are often deceived in such matters as these, which are not properly matters of the senses. And since we cannot be aware of the deception through the medium of the senses alone, it seems impossible to us that what appears to our senses should not be the truth.

For example, who does not believe that an image reflected in the surface of still water appears to be of the same size as that resulting from incident rays reflected from a perfectly straight [mirror] through a transparent medium?[40] But this is false (Vitellio VI.39),[41] for the surface of water is spherical, as Aristotle proves (*De Caelo* II, Ch. 4)[42] and as Archimedes proves more cogently (*On Floating Bodies*, Prop. 2).

Again, when a star appears to us [just] above the horizon, the star is not really where it appears, but is actually below the horizon. This comes about as follows. Suppose that *I* represents the star, *RAT* the horizon, *EAM* the earth with *A* its center, *E* the eye of an observer, and *EC* vapors. Then, since the visual ray passing from a rarer to a denser transparent medium undergoes a refraction[43] (Vitellio II.45), it is

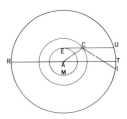

39. *Physics*, 232a.25–27.

40. I.e., as if reflected from a plane mirror.

41. There had been a recent edition: *Vitellionis . . . Opticae libri X* (Norimbergae, 1551).

42. *De Caelo*, 287a.30–287b.14.

43. [Omitting the words *a perpendiculari*. The original diagrams of Benedetti, Taisnier, and Eden differ, and none corresponds exactly with its text. A compromise diagram and textual references are adopted here. —S. D.]

clear, therefore, that star *I* is actually seen along the line *ICE*, but this line will be bent. For the pure air is rarer than the vapors, and fire rarer than air, and, finally, the matter of the heavens is rarer than fire (Vitellio X.50). Besides, the upper part of the air is rarer than the lower part. And I say the same thing, too, about [the elements] water and fire (if one may designate as fire the farthest portion of matter contiguous to the sphere of the moon), and also about the upper part of each of the elements.

Thus the star seen along line *CE* appears to be above the horizon at point *U*. Vitellio (*Optics* X, Prop. 49) shows us by instruments how stars are visible at the horizon above their true positions, because of the curvature of the rays.

Hence it follows that they do not define the horizon mathematically. They say that it is the boundary of the vision and consists of great circles on the [celestial] sphere, though, by what is proved, the circle that bounds the vision cuts the [celestial] sphere into two unequal parts, the upper part being larger than the lower. For if the horizon is the boundary of the vision and consists of great circles [of the celestial sphere], the earth does not then lie equally about the center of the universe, or the center of the universe is outside the earth. But if the earth lies equally about the center, then the boundary of vision does not consist of great circles on the [celestial] sphere, and such circles do not form the boundary of vision.

Therefore, if we see a star for the first time just above the boundary of vision, we must not for that reason suppose that it is in the twelfth station of the heaven[44] (the case must rather be understood in a logical way)—for the full strength of a star appears when it is on the great circle, whose pole is the zenith, as the star passes 90 equinoctial degrees toward the east from the intersection of this circle with the meridian. Moreover, the difference between the boundary of vision and the great circle is not a matter of merely a star's diameter; it is a matter of some degrees. And if that were not the case, we could not use Vitellio X, Prop. 49.

Therefore, they committed no small error who said that the horizon was both the boundary of vision and one of the great circles on the [celestial] sphere, and that half of the heavens is always visible to us. For in fact more than half always is visible to us because of the curvature of the rays.

44. I.e., above the horizon.

But if one wishes examples of these facts, let him consult the fourth and tenth books of Vitellio, [and] also some things in the second and fifth. Everywhere he will find evidence to show how easily we may be deceived in the sense [of sight]. And about the other senses there can be no doubt.

Therefore it is with weak and defective judgment that they speak of Aristotle as divine, a god of philosophy, and free from error. That excellent Doctor of Sacred Theology, Peter Arches, famous philosopher of the order of St. Dominic, judges the case very soundly. He confers a just measure of honor upon Aristotle; and I am not surprised at this, for Aristotle was a man of profound understanding, learned in practically all the sciences, and truly a genius in astronomy. But it is to God alone that devotion and glory are due from all.

Book of Various Mathematical and Physical Ideas

Turin, 1585

On Mechanics

Many men have written a great deal about mechanics, and they have written most ably. But nature and experience are always wont to bring to light something new or previously unknown. And it is therefore incumbent upon a high-minded and grateful individual, if he chances to discover something previously shrouded in darkness, not to begrudge it to posterity. For he himself gained a great deal from the work of others. Now it is my desire to publish a few items that will, I believe, prove not unwelcome to those who concern themselves with mechanics, items which have never before been dealt with or have not been sufficiently well explained. I may thus either show my desire to be helpful or at least give some evidence of possessing a bit of talent and industry. And perhaps in this way alone may I leave behind me proof that I ever lived at all.

Chapter 1
On the difference in the position of
the arms of a balance

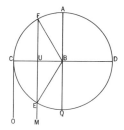

Every weight placed at the end of an arm of a balance has a greater or a lesser heaviness depending on differences in the position of the arm itself. Suppose, for example, that *B* is the center or the point that divides the arms of a balance, that *ABQ* is a vertical line or, to speak more correctly, the axis of the horizon,[1] and that *BC* is one arm of the balance. Suppose also that there is a weight at *C*, *CO* being the line of inclination, i.e., the path of *C* toward the center of the universe, and that *BC* forms a right angle with *CO* at point *C*.[2]

1. I.e., a line which, if prolonged, would pass through the center of the world and the observer's zenith.
2. In view of the assumption that all lines of gravity converge toward the center of the world, the validity of the assumption about right angles was a

166

If arm *BC* is in this position, I say that the weight at *C* will be heavier than [it would be if *BC* were] in any other position. For it will in no measure rest upon center *B*, as it would in any other position.[3]

To understand this, suppose the arm is in the position *BF*, with the same weight as before at *F*, and the line of inclination—or the path of the weight [toward the center of the universe]—is *FUM*. But the weight cannot proceed along this line unless arm *BF* is made shorter. Hence it will be clear that weight *F* exerts pressure to some extent upon center *B* by way of arm *BF*.

Now it is true that not even a weight at *C* will actually move along line *CO;* for the path of an extremity of the arm is circular, and *CO* is tangent [to this path] at one certain point. Suppose this path is *ACQ*. We must now assume that a weight at the extremity of the arm must exert pressure upon the center *B* to the extent that the line of its [downward] tendency, which we take as *FUM*, will be near center *B*. I shall prove this in the next chapter.

Suppose, for example, that *F* lies above *U*, the midpoint of line segment *BC*. Hence *UB* will be equal to *UC*. Hence it will follow that the weight will be [effectively] heavier by virtue of part *FC* than by virtue of part *AF*;[4] and that it will rest upon center *B* less by virtue of part *FC* than by virtue of part *AF*. And the more nearly horizontal a position the arm assumes in comparison with its position *BF*, the less will it rest upon center *B*, and for this reason also the heavier will the weight be.[5] And the shorter the distance to *A* from *F*, the more will the weight rest upon center *B* and hence the lighter also will the weight be.

I say that the same is true of every position of the lever arm on the lower arc *CQ*, where the weight hangs away from center *B*, pulling on that center, as in the previous case it pressed down on the center. But all these propositions will be better understood in the following chapter.

---

much debated point in medieval and Renaissance mechanics. Benedetti here treats the deviation as negligible. But see Ch. 8, below.

3. I.e., as it would if balance arm *BC* were turned about *B* so as to terminate above the horizon, e.g., in position *BF*. The case of the lowered position is stated at the end of the chapter.

4. I.e., the effective weight will increase as *F* approaches *C* and diminish as it approaches *A*.

5. I.e., effectively.

Chapter 2
On the ratio of the weight at the extremity of the arm of a balance in various positions with respect to the horizontal

The ratio of [the effect of] the weight at *C* to [the effect of] the same weight at *F* will be equal to the ratio of the whole arm *BC* to the part, *BU*, which is stituated between the center and the line of inclination *FUM*, the path of a weight freely falling from extremity *F* toward the center of the universe. In order to understand this better, let us suppose that there is a weight at the extremity *D* of the other arm of the balance *BD*, and that this weight is smaller than the weight at *C*, in proportion as part *BU* of *BC* is smaller than *BD*. It will be clearly understood from Archimedes, *On Weights*,[6] Book I, Prop. 6, that if the weight now at *C* is placed at point *U*, the balance will not move at all from its horizontal position. But weight *F* equal to *C* and placed at the extremity *F* in the position of arm *BF* is the same as if it were at point *U* in the position of the horizontal *BU*.

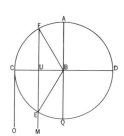

To make this clearer let us imagine a string *FU* perpendicular [to the horizontal], with the weight that had been at *F* now hanging at the extremity *U* of the string. It will be clear from this that the weight will produce the same effect as if it had been at *F*. The reason, as we have already indicated, is that the weight remaining fixed at point *U* of arm *BU* is less heavy than if it were situated at *C*, in the same proportion as *UB* is less than *BC*. And I would make the same assertion if the arm were in position *EB*. We shall easily be able to understand this if we imagine a string suspended at *U* on arm *BC*, perpendicular [to that arm] and reaching as far as *E*. Suppose there is a weight attached at this extremity [*E*], equal to the weight at *C* and hanging freely from point *E* of arm *BE*. The balance will then remain horizontal.[7] But if arm *BE* had been rigidly linked with horizontal *BD* in its position and weight *C* were hung freely by a string from *E*,[8] arm *BE* would neither rise nor fall.[9] For the weight hanging by a string from *U* is the same [in its effect] as that which had hung freely from a string at point *E* of arm *BE*. And this would be due to the fact

6. This title is always used by Benedetti for the treatise *On Plane Equilibrium*. The proposition cited is the fundamental law of the straight lever.
7. I.e., arm *BD* (with weight at *D* half of that originally at *C*) will neither rise nor fall.
8. In the previous case the string was imagined as hanging from *U*.
9. I.e., would be balanced by the weight (half of *C*) at *D*.

that in part it hung from the center *B*. And if the arm were in position *BQ*, the whole weight would be suspended from the center *B*, just as in position *BA* it would rest wholly upon that center. Hence it results that in this way a weight is [effectively] less or more heavy, according as it is more or less suspended from the center, or rests, more or less, upon the center. And this is the proximate and essential cause why it happens that one and the same weight in one and the same medium is more or less heavy.

Now I call side *BC* horizontal, supposing that it makes a right angle with *CO*, whence angle *CBQ* is less than a right angle by the size of an angle equal to that which the two lines *CO* and *BQ* make at the center of the region of the elements.[10] Yet this makes no difference, since that angle is too small to be measured.[11]

And from these considerations we may conclude that if point *U* is equally distant from center *B* and extremity *C*, weight *F* or *M* will, respectively, press down on or hang down from center *B* to the extent of half. And if *U* is nearer to *B* than to *C*, the weight will hang down from the center or press down on it to the extent of more than half; and if *U* is nearer to *C*, the weight will [hang down from or] press down [upon the center] to the extent of less than half.[12]

Chapter 3
That the [effective] magnitude of one given weight or
the magnitude of one motive force [*virtus movens*]
in comparison with another can be found by
means of perpendiculars drawn from the
center of the balance to the line of
inclination [of the weight or force]

From what we have already shown it may easily be understood that the length of *BU*, which is virtually perpendicular from center *B* to the line of inclination *FU*, is the quantity that enables us to measure the force of *F* itself in a position of this kind, i.e., a position in which line *FU* constitutes with arm *FB* the acute angle *BFU*.

To understand this better, let us imagine a balance *BOA* fixed at its center *O*, and suppose that at its extremities two

10. I.e., the center to which heavy bodies tend.
11. Literally, "of insensible magnitude."
12. **With Benedetti's account cf. Tartaglia's (i.e., Jordanus') explanation of the law of the lever and that of Guido Ubaldo.

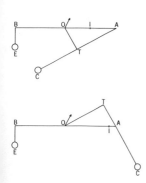

weights are attached, or two moving forces [applied], *E* and *C*, in such a way that the line of inclination of *E*, that is, *BE*, makes a right angle with *OB* at point *B*, but the line of inclination of *C*, that is, *AC*,[13] makes an acute angle or an obtuse angle with *OA* at point *A*. Let us imagine, then, a line *OT* perpendicular to the line of inclination, *CA*. Now *OT* will be shorter than *OA* (Euclid I.18).[14] Imagine, then, that *OA* is cut at point *I*, so that *OI* is equal to *OT*, and that a weight is suspended at *I*, equal to [force or weight] *C* and with a line of inclination parallel to that of weight *E*. But we assume that the weight or force *C* is greater than *E* in proportion as *BO* is greater than *OT*. Obviously, then, according to Archimedes, *On Weights*, I.6, *BOI* will not move from its position. Again, if in place of *OI* we imagine *OT* rigidly connected [in the same line] with *OB* and subjected to force *C* acting along line *TC*, the result will obviously be the same—*BOT* will not move from its position. What we have asserted is therefore true, that the [effective] magnitude of one weight in comparison with a second one is to be obtained from the [length of the] perpendiculars drawn from the center of the balance to the lines of inclination.[15] And from this it will be easily understood how much strength and effect weight or force *C* loses to the extent that it [i.e., the direction of its application] fails to make a right angle with *OA*.

And from this it will follow as a kind of corollary that the nearer the center *O* of the balance is to the center of the region of elements, the less heavy will [a given weight effectively] be.[16]

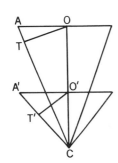

13. Note, from what follows, the sense of "line of inclination" here. The force or weight is not necessarily exerted along a plumb line but along line *AC*. The author does not show how, in the case of a weight, the line of inclination may be other than along a plumb line. But by generalizing the discussion to consider a force applied at any angle to a lever arm—and not merely the case of a lever bent at the fulcrum—he goes beyond earlier published writings. However, Duhem (*Les Origines de la statique* [Paris, 1905], I, 231) refers, in this connection, to Leonardo da Vinci ms. I. f. 30r. On the bent lever see also Moody and Clagett, *The Medieval Science of Weights* (Madison, 1953), pp. 184ff., and Clagett, *The Science of Mechanics in the Middle Ages* (Madison, 1959), pp. 81ff., 104ff.

14. Euclidean references are throughout to the *Elements*.

15. I.e., the lines along which the weights or forces exert their action.

16. If *C* is the center toward which heavy bodies tend, *O* and *O'* are the centers of the balances, and *CO* is greater than *CO'*, then perpendicular *OT* is greater than perpendicular *O'T'*, and the effect of a weight at *A* will be greater than that of an equal weight at *A'*.

Chapter 4
How all causes operating on scales and levers depend on the aforesaid causes

That the force [*vis*] in the longer arm of a balance or lever is greater than that in the shorter arm comes about from the causes which we discussed in previous chapters, namely, that a weight placed at the extremity of the longer arm may press more or less on the center or hang more or less from the center. Therefore we must first recognize the fact that balances or levers are not pure mathematical lines, but are physical, and as such exist as material bodies.[17]

Now, therefore, let us imagine that *NS* is that surface which divides the axis of a balance along its length; and let us assume that the center [fulcrum] of the balance at first is at *I*, and that the longer arm is *IU*, the shorter *IN*, and *IO* a vertical line. *IO* is equal to the thickness or depth of the balance from its upper to its lower surface. (To facilitate understanding we assume *NS* to be a parallelogram.) Now if two equal weights are placed at the ends of the arms, it is clear from experience that the weight appended at *US* will overpower the weight appended at *NX*. But we wish to investigate the cause of this effect, which cause has never, so far as I know, been assigned by anyone in the annals of literature.

We have already said that the balance or lever is material and that *NS* is its cross section, *I* assumed to be the center [fulcrum] on which the balance or lever rests. This being so, let *US* and *NX* be the lines of inclination of the weights; and let us imagine that these weights hang from points *U* and *N*, as indeed they do, even if they have been suspended from *S* and *X* – for points *U* and *N* are so joined with *S* and *X* [respectively] that what pulls on one also pulls on the other.

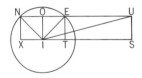

Let us also imagine two lines *IN* and *IE*, angle *OIE* being equal to angle *OIN*. Thus it will be obvious to us that if we were to suspend weight *U* (which is equal to weight *N*) from point *E*, it would, of course, have the same effect as the weight at *N* has, and would move the balance neither down nor up. For both weights would press equally on the center *I*, by means of lines *EI* and *NI*. But if the weight is placed at *U*,

17. Previous chapters developed the principles governing levers idealized as mathematical lines; this chapter shows the applicability of these principles to actual material scales and balances, whether suspended from below or above or from an intermediate point.

line *UI*, by which the body presses upon the center, becomes more horizontal than *EI*, and the line of inclination *US* is farther distant from center *I* than is line *ET*. Hence the weight in this position [suspended from *U*] is also more independent of center *I;* and it is heavier than when it was at *E*, for the reasons we gave in Chapters 1 and 2. It therefore overcomes the weight placed at *N*.

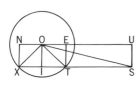

But if the center is at *O*, we shall imagine two lines *OS* and *OX*, and shall suppose that weights have been placed at *S* and *X*. Hence, since line *OS* is more [nearly] horizontal than *OX*, and the line of application [of the weight] *US* is farther distant from center *O* than line *ET*, the weight at *S* will be heavier because the extent to which it will hang from center *O* will be that much less. By the same process of reasoning as we indicated above we shall find that the same effect holds good.

In the case of the balances, *XIS* or *NOU* may rightly and properly be called horizontal, but in every other type of lever this can be said only by a kind of analogy.

The same result will be found if we assume that the center [fulcrum] lies between *O* and *I*, as each reader will be able to prove for himself without the help of anyone else.

Chapter 5
On certain facts worthy of notice

It seems to me that certain facts that are quite essential to a discussion of levers should not be omitted. What we wish to point out is essentially this: that in practice there are some levers whose fulcrum (Greek *hypomochlion*) is one of the ends of the lever and the weight to be lifted lies between these ends but near the fulcrum.

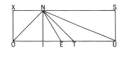

For example, if the figure *OSUX* below represents a lever with fulcrum at point *O* and weight at point *N*, it will be clear that if [weight] *N* is to be lifted, *U* must also be lifted by the operation of the hand. Now we must consider how the weight at *N* exerts a pressure on *U*. For this purpose we shall imagine straight lines *NO*, *NI*, *NE*, *NT*, and *NU*, of which *NI* is placed in the direction of the center of the universe, and *NT* makes angle *INT* equal to angle *INO*.

Now if we apply a force at *I* having an upward tendency equal to the downward tendency of *N* (apart, however, from the weight of the material of the lever), such a force will obviously sustain the whole weight of *N*. And if the weight *N* were at *X*, the whole weight would be located directly above

O, the fulcrum of the lever, and the fulcrum's resistance equal to that weight would suffice to resist it by holding it up.

But if we again place the weight at *N*, it will be clear that, if another force besides [that exerted at] the fulcrum is not exerted from below toward the upper part of the lever, the lever will necessarily be depressed on side *SU* by the force of a part of weight N^{18} — the weight of the material of the lever, as I have said, being left out of consideration. I said "a part of weight *N*" because another part of that weight presses upon fulcrum *O* itself by means of line *ON*, which does not make right angles with *OX*.

But if resistance such as to prevent the lever from being depressed is exerted from point *T*, it will, by common knowledge, be clear that the force of weight *N* will have been equally divided, one half resting upon *O* and the other upon *T*, along the two lines *NO* and *NT*, respectively. Let us now imagine that the resistance is removed from *T* and placed at *E*. It will also be clear that a greater part of weight *N* will rest upon *E*, by virtue of line *NE*, than upon *O*. For the line of direction [of the weight's application], *NI*, is nearer to *E* than to *O*, since every resistance, whether at *I* or at *E* or at *T* or at *U*, acts as a fulcrum, just as does *O*, and the one is helped by the action of the other.

And if the same resistance is placed at *U*, it will also be clear that a smaller part of weight *N* will press on *U* than on *O*, since *NI* is farther distant from fulcrum *U* than from fulcrum *O*. And the ratio of the part of weight *N* that rests on *O* to the part of weight *N* that rests on *U* will be equal, not to the ratio of angle *UNI* to angle *ONI*, but to the ratio of *UI* to *IO*.[19]

This may be clearly understood from the converse of this effect. That is to say, just as *O* and *U* are now supposed to be two fulcrums by which the weight of *N* is sustained,[20] let us imagine that *N* is a certain center from which hang two weights *O* and U^{21} having the same ratio to each other as *UI* and *IO*. Surely the scale *OS*, which we call a lever, will not, by reason of these weights, incline on either side.

Now returning to our proposition, we shall say that, since the weight of *N* rests less on *U* than it does on *O* (i.e., on *T*),

18. I.e., the part that is not thought of as pressing on the center.
19. Duhem, considering that the principle of the inclined plane was involved here, held that the ratio should be *UN* to *ON* (*Les Origines*, I, 235). But from the sequel it would seem that Benedetti is concerned here not with the inclined plane but with the straight lever.
20. The text erroneously mentions *E* as well as *N*.
21. A bent lever with rigid arms in equilibrium.

less force will be necessary at U than at T to lift up the weight of N. And consequently the farther point U is from T, the less force will be needed. Thus, when the force or resistance at U bears the same ratio to that at O as OI bears to IU, the lever will not move. But when the ratio of the resistance at U to the resistance at O is greater than the ratio of OI to IU, then the lever will be raised on the side US. But if the ratio [of the resistance at U to the resistance at O] is less than the ratio of OI to IU, then the lever will be depressed on the same side US.

. . . .

Chapter 7
On certain errors of Niccolò Tartaglia on the weights of bodies and their motions, some taken from a certain ancient writer Jordanus

Since we ought to be more devoted to the truth than to any man, as Aristotle writes, I shall at this point reveal certain errors of Niccolò Tartaglia on the weights of bodies and the velocities of locomotions.[22]

In the first place, he is in error in his [*Quesiti et*] *Inventioni diverse*, Book 8, Prop. 2, since he did not observe the effect of external resistances on the motion.[23]

The subject of the third proposition, too, is improperly demonstrated because, according to his demonstration, clearly the same thing would be true of heterogeneous bodies or of bodies differing in shape, so far as velocities are concerned.[24]

In the fourth proposition,[25] he does not properly prove what he sets out to prove. The conclusion follows more properly from what Archimedes proved in his work *On Weights*, Book I, Prop. 6.

And in the second part of the fifth proposition,[26] he fails to see that no difference in weight is produced by virtue of position in the way in which he argues. For if body B must

22. Passages of Niccolò Tartaglia, *Quesiti et inventioni diverse*, Bk. VIII, are cited in this and the next chapter by caption rather than by page.

23. In *Quesiti*, Bk. VIII, Qu. 29, Prop. 2, Tartaglia is concerned with the speeds of rise and fall of weights suspended from a balance. The principle would ideally be exemplified in a void—hence Benedetti's stricture.

24. *Quesiti*, Bk. VIII, Qu. 30, Prop. 3.

25. *Quesiti*, Bk. VIII, Qu. 31, Prop. 4.

26. *Quesiti*, Bk. VIII, Qu. 32, Prop. 5, with the two corollaries referred to below. [Diagram supplied.]

descend on arc *IL*, body *A* must ascend on arc ·*VS*, equal and
similar to arc *IL* and placed in the same way. Therefore, just
as it is easy for body *B* to descend on arc *IL*, it will be difficult
for body *A* to ascend on arc *VS*. And this fifth proposition is
the second question proposed by Jordanus.[27]

And with respect to the first corollary of this same proposition, Tartaglia states what is true, to be sure, but the cause of
this effect, as set forth earlier by Jordanus and subsequently
by Tartaglia, is not, in its nature, the true cause. For the true
cause emerges by itself from the fact that the weight hangs
down [in part] from the fulcrum of the balance, as I showed
in the first chapter of this treatise.

And from those considerations that I shall now add, it will
be clear that the second corollary is false. Let us imagine *U*
as the center of the region of elements.[28] Consider a scale
BOA, oblique with respect to *U* and consisting of equal arms,
also with equal weights at *A* and *B*. Let the lines of inclination [of the weights] be *AU* and *BU*. And let us also imagine
a line *OU* and, from the center *O* of the balance, two lines *OT*
and *OE* perpendicular to the lines of inclination [of the
weights]. Then the weight at *A* in such a position will have the
same ratio to weight *B* as line *OT* to *OE*, because of what I
proved in the third chapter of this treatise.

But line *OT* is greater than line *OE*. This I prove as follows.
Let us imagine a triangle *UAB* about which circle *UANB*,
with center *C*, is circumscribed. This center will not lie on
line *UO*, since *AOB* is assumed to be oblique with respect to
UO. Let us then imagine a line *COS* drawn from center *C* to
the circumference. This line will be perpendicular to *AB*,
according to Euclid III.3. If then we imagine the two lines
CA and *CB*, we shall have (by [Euclid] I.8) angle *ACO* equal
to angle *BCO*. Hence (by [Euclid] III.25) arc *AS* will be
equal to arc *BS*. But if we imagine *UO* produced to meet the
circumference, clearly it will cut arc *SB* at point *N*. Hence
arc *NB* will be smaller than arc *NA*, and so also angle *NUB*
will be smaller than angle *NUA* (by [Euclid] VI, last prop.).

Now let us imagine another circle whose diameter is *OU*
and whose circumference passes through the two points *E* and

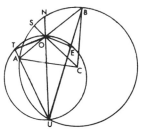

---

27. *Iordani Opusculum de ponderositate Nicolai Tartaleae studio cor-
rectum* (Venice, 1565), Quaestio 2 (p. 3b) (cited hereafter as Jordanus,
Opusculum). [Corresponding references to the English translation of Jordanus
may be found in the notes to the *Quesiti.* −*S. D.*]
28. The center toward which weights tend to move.

T, since the angles are right angles at these points, which anyone can gather by reasoning for himself if he recalls Euclid
III.30.

But since angle *OUT* is greater than angle *OUE*, arc *OT*
will be greater than arc *OE* ([Euclid,] Book VI, last prop.);
hence cord *OT* will be greater than cord *OE* (by the converse
of [Euclid] III.27). And this is what we set out to prove.

Therefore the weight of *A* in this position will be heavier
than the weight of *B*. But this directly contradicts what Tartaglia set forth in the second part of his Prop. 5; and it consequently reveals the fallacy of Corollary 2, as well as that which
lurks in Prop. 6. For since the ratio of weight *A* to weight *B*
is equal to the ratio of *OT* to *OE*, we are able to determine
that ratio, by taking account of the angles of obliquity of the
balance, i.e., angles *BOU* and *AOU:* for we must always assume some known position [for the balance].

Then if the ratio *OU* to *OB* (and to *OA*) is known, we shall
also be able to find angles [*O*]*B*[*U*] and *OAU*, and consequently *OAT* (the residue of *OAU*). Hence, knowing that
angles *E* and *T* are right angles and knowing the length of
sides *OB* and *OA*, we shall then easily determine *OT* and *OE*.

Chapter 8

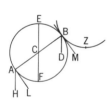

What Tartaglia writes in Prop. 6[29] and Jordanus in the
second part of Prop. 2 also contains a very serious error. For
they say that angle *HAF* does not differ from angle *DBF* otherwise than by the angle of tangency between the two circles,
as Tartaglia indicates in his figure. But this is quite false.

Thus, in the appended figure, let the balance be *BA*, with
center *C*. Let *U* be the center of the region of elements, and
AU and *BU* the lines of inclination.[30] Let us then imagine
line *BK* parallel to *AU*, which, by common knowledge, will
cut circumference *BFA* at a point *K*. We shall have angle *KBZ*
equal to angle *HAF*, that is, *UAF* (for *H*, *U*, and *D* are all
one). For according to Euclid I.29, angle *UAC* is equal to
angle *KBT* and angle *CAF* is equal to angle *TBZ*.

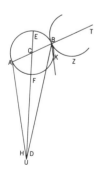

Now consider the mixtilinear angles *DBF* and *KBZ* [i.e.,
consisting of a straight and a curved line]. These two angles
have in common the mixtilinear angle *KBF*. Therefore, if the
mixtilinear angle *KBZ* is greater than the mixtilinear angle

29. *Quesiti*, Bk. VIII, Qu. 33, Prop. 6. [Diagram supplied.]
30. The respective lines along which the weights would fall toward that
center.

DBF by the angle of tangency *FBZ*, the mixtilinear angle *KBF* common to the circles will be equal to the mixtilinear angle *DBF*, that is, the part equal to the whole.

Now the whole error into which Tartaglia and Jordanus fell arose from the fact that they took the lines of inclination as parallel to each other.

Prop. 7 of Tartaglia, which is Qu. 5 of Jordanus,[31] is to be taken with a smile, I think. For since the weight of *A* is equal to that of *B*, it is [effectively] heavier than that of *B* by reason of the smaller opening of the angle of tangency at *A* than at *B*. In this the same error is committed as in the preceding, since he supposes that lines *AE* and *BD* of the figure he constructed are parallel to each other. But even if they were parallel (and angle *EAG* would then be less than angle *DBF*), the difference in the angles would still not, on that account, be a cause of the difference in the weights of *A* and *B* themselves — according to what I wrote in Chapter 4 of this treatise.

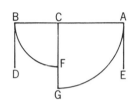

And Prop. 8,[32] which is Qu. 6 of Jordanus, is far better demonstrated by Archimedes, *On Weights*, I.6, since it was not really proved either by Jordanus or by Tartaglia. For these writers did not prove the preceding propositions, which Tartaglia cites in that eighth proposition. Nor does Tartaglia prove propositions 9, 10, 11, 12, or 13,[33] since he did not come even close to a proof of the preceding propositions.

And Prop. 14,[34] which is Qu. 10 of Jordanus, is false for two reasons.

One reason is that (on the assumption that *ADEGB* is one arm of the balance, *A* the fulcrum of the balance, *D* a weight equal to the weight at *E*, and *DK* and *EM* the lines of inclination) angles *KDE* and *MEG* are not equal to each other, since the former is internal and the latter external and opposite the aforesaid internal angle of a single triangle bounded by lines *DE*, *DK*, and *EM* produced to the center of the region of elements. Hence angle *MEG* is greater than the other (Euclid I.16). And therefore, for this reason, *E* is heavier than *D*, since it hangs down from center *A* less [than does *D*].

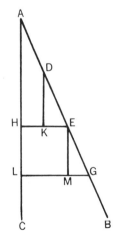

31. *Quesiti*, Bk. VIII, Qu. 34, Prop. 7. [Diagram supplied.]
32. *Quesiti*, Bk. VIII, Qu. 35, Prop. 8. [Diagram supplied.]
33. *Quesiti*, Bk. VIII, Qus. 36–40, Props. 9–13.
34. *Quesiti*, Bk. VIII, Qu. 41, Prop. 14. [Diagram supplied. The criticism insists on an impossible rigor in mechanics. — *S. D.*]

There is also another reason why E is heavier than D, namely, that E is at a greater distance from A, the fulcrum of the balance, for like reasons to those cited in Chapter 4 of this treatise.

Also without worth is Prop. 15,[35] which is Qu. 11 of Jordanus, author of the work that was brought to light from darkness by the efforts of the bookseller Troianus[36] at Venice.

Chapter 9
That the division of balances at equal intervals is
perfectly logical

It is with excellent reason that balances are divided at equal intervals, in pounds or ounces or in some other way. For suppose, for example, that there is a scale AB, with C the point of support and F the vessel that contains that which is to be weighed. Let us imagine now that the weight of arm CB on one side and the weight of arm CA, along with that of vessel F, on the other side cause balance ABC to remain horizontal. Now

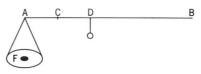

with the balance thus horizontal, imagine a weight, say, of one pound, added at point A; and at point D, at a distance from C equal to AC, imagine another weight added, also of one pound. As a matter of common knowledge, the balance will not move from its position. For since these two weights are equal, the one at D and the other at A, then if DB[37] and F were removed, clearly AD would not change its position: for DB and F, in the position in which they are found with respect to the fulcrum, are endowed with equal force [*viribus*]. Therefore, if we attach DB at D and F at A, the sum of the forces will also be equal, according to the axiom which states that if you add equals to equals the sums will also be equal.

35. *Quesiti*, Bk. VIII, Qu. 42, Prop. 15; Jordanus, *Opusculum*, p. 7.
[Benedetti's curt dismissal of this celebrated theorem may perhaps be linked with his curious criticism against the preceding proposition. —*S. D.*]
36. Publisher in 1565 of Tartaglia's edition of Jordanus' *Opusculum*.
[Curtius Troianus de Navò, the friend to whom Tartaglia willed his manuscripts. In publishing posthumous editions of Tartaglia's books, Troianus substituted his own name for that of Richard Wentworth as an interlocutor. —*S. D.*]
37. But not including the weight at D.

Now if to the weight at *A* another weight equal to it were added, we would have at *A* a weight double that at *D*. But if we want the balance to remain horizontal with the weight of *D* alone, then if the weight *D* is placed at a distance from fulcrum *C* double that of *CA*, that is, double that of *CD*, we shall attain what we wish, for the reasons indicated and with the help of Archimedes, *On Weights*, I.6. And if one were to add also another weight at *A* equal to the original weight, it would be necessary, in order for the balance always to remain horizontal, that the weight *D* be at such a distance from *C* that that distance is triple the original distance. Thus the intervals through the various degrees are kept equal.

Chapter 10
That the circumference of a circle does not have a concavity joined with a convexity, and that Aristotle was mistaken in the ratios of motions

Aristotle at the beginning of his *Questions of Mechanics* says that the line which bounds a circle seems to unite the convex with the concave.[38] But this is false. For a line of this kind has no thickness (as Aristotle himself also asserts), but is identical with the convex boundary of the circle. On the other hand, the line that bounds the surrounding surface and encloses the circle is identical with the concavity of the surface that surrounds the circle, a surface which has no convexity. And these are two lines of which one is different from the other, and not part of the other, so far as pertains to convexity and concavity.[39]

But what Aristotle writes about the twofold aspect of the motion of a single point according to a single given proportion is inadequate.

Aristotle writes as follows: Suppose that the ratio according to which the body moves is the same as that of *AB* to *AC*, and that *A* moves toward *B*, while *AB* moves down toward *MC*. And let *A* be moved toward *D* while *AB* moves toward *E*. Since the ratio of the motion is the same as *AB* has to *AC*, it follows that *AD* and *AE* have this same ratio. Therefore the

---

38. *Mechanics*, 847b.23-25.
39. From inside the circle the boundary of the circular area is concave; from outside the circle that boundary is convex. But it is not clear why the author speaks of the bounding lines as different, in view of what he has said earlier.

small quadrilateral is similar to the larger one in the ratio [of its sides]. Therefore, etc.[40]

To Aristotle I reply that the fact that point *A* moves on line *AM* from *A* toward *M* as far as *F* does not mean that it moves

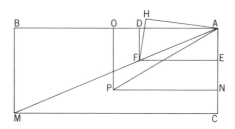

according to one definite proportion rather than some other. Thus we can suppose that the point *A* moves from *A* to *F* not only according to one ratio of uniform velocity, but also some other which is the very opposite of the first ratio — e.g., the ratio of *AC* to *AB,* it being imagined that *A* moves toward *C* and *AC* toward *BM*.

I say also that *A* moves to *F* according to the ratio of *AO* to *AN*. Thus let us imagine a line *FH* from point *F* making with *FA* an angle equal to angle *OPA;* and also a line *AH* from point *A* making with line *AF* an angle equal to angle *OAP*. Hence angle *H* will be equal to angle *O* (Euclid I.32), and triangle *AHF* will be similar to triangle *AOP*. Therefore *AH* is to *FH* as *AO* is to *OP*.

Therefore point *A* moves to *F* according to the ratio of *AO* to *OP*. Hence the discussion on this point by Aristotle is of no value.

Chapter 11
That Aristotle did not, in the first of his *Questions of Mechanics,* adduce the true cause of that which he was investigating

Aristotle, in investigating why it is that those balances which have longer arms are more exact than others, says that this comes about by reason of the greater speed of the extremities of these longer balances.[41] But this reason is not sound; for this effect is nothing but a clearer presentation, to all observers, of the obliquity of the arms to the horizontal line, and a demonstration that the balance arms in question leave this horizontal

40. *Mechanics,* 848b.15–20.
41. *Mechanics,* 849b.19–34. Cf. *Quesiti,* Bk. VII, above.

position more readily. These effects come about by themselves, neither from the speed nor from the slowness of motion, but by the principle of the lever and because of the greater
distance between the second position of the extremities in
comparison with the first.

For example, let us imagine a large balance AB in horizontal
position, with center E. Suppose the weight at B is greater than

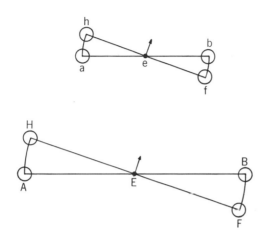

that at A. Now it is conceded that because of this the balance
will change its position. Suppose this second position is HF.
Let us also imagine a small balance aeb in horizontal position,
having weights a and b equal to the two respective weights on
the other balance. And suppose that the second position is hf,
with the angles at e equal to those at E. That is, let angle bef
be equal to BEF.

Now I say that position HF will be more exact and clearer
than position hef because the distance BF is greater than bf.
For BF is greater than bf in the same ratio as BE is greater than
be. That distance BF was traversed in a shorter or a longer
interval of time than was bf obviously makes no difference.
By the principle of the lever, then, I say that, if we consider
two balances that are alike and equal in every other respect
except in the length of their arms, the weight at B will have a
greater force [*vis*] for depressing arm EB than will the weight
at b [for depressing arm eb]. For since the balances are material and are sustained by E and e, not by a mathematical
point, but by a line or a physical surface having a material existence, some resistance arises to the motion of the arms. For
this reason, if we assume this resistance to be equal at E and at
e, it will be clear, from what I showed in Chapter 4 of this

treatise, that *B* will be heavier [effectively] than *b*, since it hangs down from *E* to a lesser extent, or presses down to a lesser extent upon *E*. And *B* will also, for this reason, move arm *EB* to a lower position with greater readiness and depress that arm much more, i.e., will make angle *BEF* greater than angle *bef*.

Chapter 12
On the true cause not perceived by Aristotle in
Questions of Mechanics 2 and 3

In investigating the cause in *Questions of Mechanics* 2, Aristotle writes as follows:[42] Why is it that, if the supporting cord is above the balance and someone removes[43] the weight that depresses the balance [on one side], the balance rights itself; but if the support is from below, the [depressed] side does not rise but remains [depressed]? Is it because, when the support is from above, more of the balance is outside the perpendicular[44] (for the supporting cord is perpendicular) and so that excess must move downward: therefore, etc.?

But the true cause why, if the support is from above and one arm of the balance is depressed and is then let free, it returns to the horizontal position, is not only the greater weight of the balance which has passed beyond the vertical line, but also the length of the raised arm found beyond the vertical line. Therefore the weight at this end is [effectively] greater in the ratio which I shall set forth in the following example.

Suppose *AB* is the balance in horizontal position and, above it, *E* the [point of attachment of the] supporting rod.[45] If we depress the arm *A* to *F*, suppose its position is *FH*. Its midpoint *G* will thus have passed the vertical *VZ* on the side of *B*, and *VZ* will cut arm *FG* at *D*, so that *DH* will be longer than *FD*. Now we must assume what is certainly true: that

43. *Mechanics*, 850a.3–7. See Duhem, *Les Origines*, I, 232f. and the references to Leonardo da Vinci; Cardano, *De subtilitate* II (pp. 369–70 in the 1663 edition of Cardano's works). Cf. *Quesiti*, Bk. VII, above.
43. Reading *abmovet* for *admovet*.
44. What is meant is that a larger part of the balance is on the side that is elevated than on the side that is depressed; hence the balance rights itself. In the second case the balance does not right itself because the larger part is on the depressed side (or, as we should say, the difference came about because in the first case the center of gravity was raised, and in the second case it was lowered, from its original position).
45. From the diagram the argument would seem sound only if a rigid rod, not a flexible cord, supported the balance. This does not seem to be what the Peripatetic author had in mind.

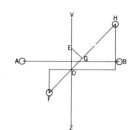

this balance in position *FH*, even if supported by point *E*, would still have the same effect as if it were supported at point *D*. Hence it follows that the weight appended at *H* is rendered [effectively] heavier than that at *F* in proportion as *DH* is greater than *DF*, for the reasons I set forth in the first chapters of this treatise. And so, even if *DH*, which is assumed to be material, had no weight at all, still that mere excess of the force [*vis*] of the weight placed at *H*, which is far greater than that of the weight situated at *F*, by virtue of the greater length of *DH*, would suffice to effect the return of the balance to the horizontal position.

Then in the second part of this problem, in which Aristotle writes that a balance remains fixed in the position in which it has been placed,[46] he is completely mistaken. For it must continue to fall until the support remains above it, with the assumption, however, that all impediment to this is removed. This proposition requires no proof, since by its own nature it is perfectly clear.

And the true cause, in the third problem,[47] is not that which Aristotle asserts. Actually this effect has its origin in that which I set forth in Chapters 4 and 5 of this treatise.

Chapter 13
That the explanation given by Aristotle in *Questions of Mechanics* 6 is not admissible

Aristotle, in seeking to give an explanation why a ship moves more swiftly when it has its yardarm raised rather than lowered, refers this effect to the principle of the lever.[48] But this does not give the true explanation. For on this kind of explanation the ship would have to move more slowly rather than more swiftly. For the higher is the sail that is struck by the force of the wind, the more is the ship's prow submerged in the water.

Actually the effect in question arises from the greater quantity of wind which the sail receives rather than from any other cause. For the wind blows more freely and more strongly in the higher region than in the lower.

---

46. *Mechanics*, 850a.20–29.
47. The explanation of the principle of the lever.
48. *Mechanics*, 851a.38ff.

Chapter 14

That the reasons devised by Aristotle in *Questions of
Mechanics* 8 are not adequate

The reasons proposed by Aristotle for getting at the truth
in *Questions of Mechanics* 8 are not sufficient. There he asks
how it happens that bodies of circular shape are easier to roll
along than others.[49] And he designates three kinds of revolv-
ing bodies: one kind is exemplified by carriage wheels; the
second by the pulley wheels with which water is drawn from
wells; the third [by the wheel] with which small vessels are
fashioned by potters.[50]

Beginning with the first, I say that there is no question that
when a round body touches a plane it touches at only one
point, as Theodosius proves [*Sphaerica*] I, Prop. 3, and Vitellio [*Optics*] I.71, and that, if a line is drawn through the center
of a sphere to the point of tangency, it will be perpendicular to
the plane touching the sphere, as is also proved in Theodosius
I, Prop. 4, Alhazen IV.25, and Vitellio I.72.[51] It is also true
that all downward tendency of such a [spherical] homogeneous
body, caused by its weight, is distributed equally about this
whole line at every part. As an example of this, we may represent on the page the circular figure appended below, *ANEU*,
tangent to line *BD* at point *A*, so that *EOA* will be perpendicular to *BD*, according to Euclid III.17, and will have as much
weight on side *AUE* as on side *ANE*.

Now then, if we imagine the center pulled toward *U* along
line *OU* parallel to *AD*, it will be clear to us that we shall pull
it without any difficulty or resistance because in such a case the
center will never change its position by moving upward from
below, i.e., will never change its position with respect to the
distance or interval which lies between it and line *AD*. Now
this center gathers unto itself the whole weight of figure *ANEU*
and commits that weight along line *EOA* to point *A* on line
BAD. And it makes no difference that point *A* moves and is
carried a greater or lesser distance toward *D* or toward *B*.
Thus, since necessarily the weight of this figure is not more
elevated at any one time than at another, but always rests in
the same way upon line *BAD*, and since it is always divided

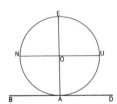

49. *Mechanics*, 851b.5ff.

50. In the first case the center moves as the wheel turns; in the second case
the center is stationary as the wheel turns in vertical position; in the third
case, the potter's wheel, the center is stationary as the wheel turns in horizontal position.

51. Reading "72" for "7" of the text.

by line *AOE* through the center, it follows axiomatically that no difficulty will arise for us in pulling this center in whatever direction we wish, as would arise in the case of any other figure which was not perfectly circular.[52]

For example, if we imagine a pentagon *KIHFL* at rest on the same line *ABK* such that the first side *IK* is completely extended on line *BK*, then, if we pull center *O* (let us assume) toward *L*, there is no doubt that center *O* would have to be raised with respect to line *AB* and would be more distant from it, moving as it does on an arc of a circle having as radius *OK*, greater than *OA* (by Euclid I.18). Thus, if from point *K* we imagine a line *KC* passing through the center of the region of elements,[53] clearly, if we wish to move center *O* from its previous position to a position on line *KC*, it will be necessary to add weight on the side of *FL* which was cut off by line *KC*, or else to take something away from the weight on the side toward the center. In whatever way this is done, it will surely be difficult to achieve. But this is not the case with a body that is perfectly circular, since the center, which is exactly in the middle of that weight, is always found to be on a line perpendicular to that plane. And note, in this connection, that even if I call it a plane, yet I do not wish it understood as a perfect plane, but as a perfectly spherical surface around the center that is the goal of [freely falling] heavy bodies.[54] For by reason of the great extent of such a surface we shall be able to picture no noteworthy difference between a perfect plane of small extent and the [infinitesimal] curvature of such a spherical surface.

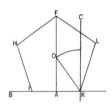

But, to return to the discussion we have undertaken about the revolution of a circular figure, it will be clear that a force no matter how small (so to speak) drawing or impelling center *O* toward *U* will cause a figure of this kind to revolve; and that half such a force would suffice to pull or push point *E*.[55]

Now let us suppose that line *NOU* were a balance situated in a perfectly circular figure *ANEU*, and that the force which was required to pull the center were divided by two, half of it being appended at extremity *U* of diameter *NOU*. It will be clear that this force would without any difficulty cause the figure to revolve over line *BAD* toward *D*, because the force or

52. E.g., an ellipse.
53. The center to which freely falling bodies tend.
54. Cf. Galileo, *De motu (Opere di Galileo Galilei*, ed. A. Favaro [Ed. Naz., 1890], vol. I), p. 301.
55. Since *E* is twice as far as *O* from fulcrum *A*.

weight here [at U] would have no counterweight on the other side of center O, the side toward N. And center O is always at rest above A in line EOA which always divides in half the whole weight of the appended figure. So much more easily, then, would the whole force, applied to the center and pushing it toward U on a line parallel to AD, cause the figure to revolve.

And if the line on which this center is pulled were not parallel to BAD, but if the line of pull were upward [or downward], i.e., above or below U, there would be some loss of force and power, and the more so the more inclined was this line of force toward AOE. And when finally it was collinear with AOE, either upward or downward, then no force whatever, not even an infinite force, would move the figure beyond the position of the original line AOE; and if it pulled upward, it would separate the figure from line BAD, but would not on that account cause center O to move away from the original line AOE.

. . . .

On the subject of this third type [of wheel—the potter's wheel] a problem may be framed: why it is that, if a wheel of this kind parallel to the horizon rests upon one point and is as evenly [balanced] as possible, and if we revolve it with all our force and then let it go, it does not rotate forever.[56]

There are four causes for this. First, such a motion is not a natural motion of the wheel. Second, even if the wheel rested over a mathematical point, it would still have to have a second point of support above to keep it horizontal, and this support would require some corporeal embodiment. Hence some friction would ensue, from which resistance would arise. Third, the contiguous air at all times constrains the wheel, and in this way resists its motion. Fourth, any portion of corporeal matter which moves by itself when an impetus has been impressed on it by any external motive force has a natural tendency to move on a rectilinear, not a curved, path.[57] And so, if some portion of the circumference were separated off from the wheel in question, no doubt the separated part would move through the

56. Cf. also Benedetti's letter to Capra (*Diversarum Speculationum*, p. 285), translated below. Also Galileo, *De motu, Opere*, I, 304–07. The problem is a classical one in medieval science, keenly discussed by John Buridan, Albert of Saxony, Nicholas of Cusa, and many others. See P. Duhem, *Le Système du monde* (Paris, 1958), VIII, 337ff.; *Études sur Léonard de Vinci* (Paris, 1955), III, 214ff.

57. An important statement, not so clearly made by any previous writer. [Were it not for the phrase "for some length of time" at the end of the following sentence, the inertial concept would be unequivocally given. —S. D.]

air in a straight line for some length of time. This we can understand by an example taken from slings used for throwing
stones. In their case the impressed impetus of motion produces a rectilinear path by a certain natural propensity, when
the stone shot out starts its rectilinear path; this path is along
the straight line tangent to that circle which the stone previously described at the point at which it was let fly. And this
accords with reason.

It is for the same reason, too, that the larger a wheel is, the
greater also is the impetus and impressed force [*impressio*]
of motion that the parts of its circumference receive. Therefore it often happens that, when we want to stop it, we do so
with great effort and difficulty. For the greater is the diameter
of a given circle, the less is the curvature of its circumference
and the more nearly the angle[58] of that circumference approaches the measure of two right rectilinear angles, that is,
the more nearly the circumference approaches the character
of a straight line. Hence the motion of those parts of the circumference is more prone to follow the tendency assigned to
them by nature, that is, a tendency to move along a straight
line.

Chapter 15
That Aristotle's explanation of *Questions of Mechanics* 9 is not to be admitted[59]

The true explanation of the ninth problem is to be sought
in the second part of Chapter 10[60] of the present treatise, and
not elsewhere.

Chapter 16
That Aristotle's explanation of *Questions of Mechanics* 10 must be rejected

The explanations given by Aristotle why empty scales move
more readily than full ones are irrelevant to the problem as set
forth. For the ratio of the moving force to the body moved
must always be considered; and Aristotle did not do this.

---

58. Reading *angulus*. The meaning is that the larger the circle, the more
nearly does each mixtilinear angle formed by the diameter and circumference
approach a right angle.
59. On the advantage of larger over smaller simple machines, such as pulleys.
60. Ch. 11 seems to be meant.

Let there be, for example, a scale *AIE* which has at the extremity of each arm merely one ounce of weight; and let there also be a scale *NIU,* exactly like the former one, which has one pound of weight on each end. Aristotle wonders that, when he adds a half-ounce weight at *E,* arm *IE* falls more rapidly than when he adds that same half-ounce at *U,* the extremity of arm *IU.* But this results from two causes.

First, there is the great difference between the ratio of one pound to half an ounce and the ratio of one ounce to half an ounce. For if the weight added to the extremity *U* were half a pound, and if the arm still moved with the same slowness, Aristotle would justifiably have cause for wonder. But this could not be, because [the addition of half a pound] would depress arm *IU* with almost the same speed with which the [added] half-ounce depressed arm *IE.* I said "almost" because some difference would remain, which arises from the second reason.

And this second reason is the resistance which comes from the supporting cord, since the greater the weight in the scale, the more pressure there is on the supporting cord at the point of support: and thus there arises more resistance in the supporting cord to the turning [of the scale] at the point in which it rests. For this [supporting cord] is a material body.

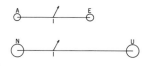

But if one wanted arm *IU* to descend with the same readiness as *IE,* it would be necessary that the addition of half a pound to the one-pound weight at *U* should exert a force sufficient to overcome the resistance of its supporting cord (at the center of arms that would have to be larger than *AI* and *IE*) in the same ratio as is the ratio of the added force of half an ounce at *E* to the resistance of *its* supporting cord.

Such reasoning, if applied to heavier and lighter wheels and to wheels impelled by any heavy bodies whatever, will reassure the doubting mind.

Chapter 17
On the true explanation of *Questions of Mechanics* 12

The true explanation why a heavy body is projected much farther from a sling than from the hand is this.[61] In the revolu-

---

61. There follows an application of the doctrine of an impressed force *(impetus).* This doctrine, which had roots in Greek thought (Hipparchus, Philoponus) and, notably through Buridan's influence, was widely held in the later medieval period, permeates Galileo's early thought (see *De motu, Opere,* I, 308ff.). For a general account and bibliography see Clagett, *Science of Mechanics,* especially Chapters 8 and 9.

tion of the sling a greater impressing of the impetus of motion in the heavy body takes place than would be the case if the hand alone were used. For when the projectile is released from the sling, it takes its path, with nature as its guide, from the point whence it has quit [the sling] on a line tangent to the circle which it last made. And there is no doubt that a greater impetus of motion can be impressed on the body by the sling, since from repeated revolutions an ever greater impetus comes to the body in question.

Now the hand (and I say this despite Aristotle)[62] is not the center of the motion of that projectile while it is being revolved, nor is the sling the radius. Indeed, the hand is itself moved so far as possible in a circle; and this circular motion causes the body itself to be carried around in a circle. Now that body, once it receives a small impetus, would, by a certain natural tendency, seek to proceed in a straight line, as is clear in the appended figure. Here E indicates the hand, A the body, and AB the straight line tangent to circle $AAAA$ when the body is free.

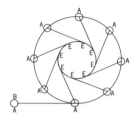

Now it is true that the impressed impetus gradually and continuously decreases.[63] Hence the downward tendency of that body, caused by its weight, enters at once and, mingling itself with the impressed force, does not permit line AB to remain straight for long, but causes it quickly to become curved. For body A is moved by two forces, one impressed by force, the other by nature—contrary to the view of Tartaglia who says that a body cannot at one and the same time be moved by forced and natural motion.[64]

Nor should we pass over in silence a certain effect worthy of notice in this connection, namely, that the greater the increase in the impetus in body A caused by increase in the speed of the revolution of E, the more must the hand feel itself drawn by body A through the medium of the string. For, the greater is the impetus of motion impressed on body A, the more does that body tend to traverse a rectilinear path: and hence the greater also is the force with which it pulls in order to enter on a rectilinear path.

62. Cf. *Mechanics*, 852b.6ff.

63. I.e., after the body leaves the sling. [Again, this vitiates the inertial concept that is repeatedly adumbrated by Benedetti. —S. D.]

64. Cf. Tartaglia, *Nova Scientia*, Bk. I, Prop. 5, and Benedetti's reference in his letter to Muzio Groto (*Diversarum Speculationum*, p. 365), translated below. See also Galileo, *De motu, Opere*, I, 300, marginal note. The passage is discussed by Duhem, *Études*, III, 211.

Chapter 18
On *Questions of Mechanics* 13

The explanation of Question 13[65] should be based entirely on the lever. We must imagine the axis of the cylindrical beam as the fulcrum. For the rest, the whole explanation depends on the fourth and fifth chapters of the present treatise. But there is one difference between this machine and the lever, namely, that the beam produces a certain resistance by reason of the pressure it exerts at the points [of support] as it re-volves—a resistance greater than the fulcrum causes in the case of the lever.

Chapter 19
On *Questions of Mechanics* 14

The principles of Question 14[66] depend on those of the lever. For example, suppose stick *ABCD* is to be broken in half by a pressing with the knee at point *O*. We shall then see most clearly that, by holding the hands far from the center at *A* and *C*, we shall break the stick more easily and with less effort than if we were to place our hands near the middle of the stick, at *E* and *I*. The reasons for this are the same as were set forth in the first chapters of this treatise.

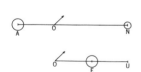

Let us imagine straight lines drawn from point *O* to points *A, E, I,* and *C*. We shall, then, clearly perceive, on the basis of the principles we have already set forth, that points *E* and *I*, by the medium of lines *EO* and *IO*, press more upon center *O* than do points *A* and *C* by virtue of the two lines *AO* and *CO*. Hence [the hands] will also exert greater force at *A* and *C* than at *E* and *I*.

Chapter 20
On the true explanation of *Questions of Mechanics* 17

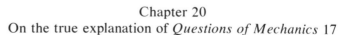

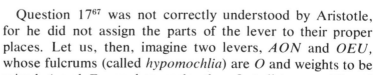

Question 17[67] was not correctly understood by Aristotle, for he did not assign the parts of the lever to their proper places. Let us, then, imagine two levers, *AON* and *OEU*, whose fulcrums (called *hypomochlia*) are *O* and weights to be raised *A* and *E*, equal to each other. Let distances *AO* and *EO* be equal to each other, but let *ON* be equal to *OU*. It will

---

65. On the capstan and the windlass.
66. The rest of the paragraph indicates what the problem is.
67. The explanation of the action of a wedge.

be clear that to raise *A* it will be necessary to depress *N*, and to raise *E* it will be necessary to lift *U*. And because all are assumed equal, it will also be clear, by common knowledge, that the force at *N* required to raise *A* will also be sufficient at *U* to raise *E*. For in their own case *E* and *U* press upon their center *O* at angles equal to those at which the two forces *A* and *N* press upon their own center *O*. And all the principles cited for lever *AON* in Chapters 4 and 5 of our treatise may apply to lever *OEU*, as I abundantly showed in that fifth chapter.

Now suppose a piece of wood *DEFG* is to be split along its grain, and let *ABC* be the wedge which has penetrated as far as *TX* under the force of a hammer *P*. It will be clear from this that opening *IMR* in the wood, after the wedge has been driven in along the grain, will be longer than *XBT*, the part of the wedge which has entered. We must now imagine two levers like the one *UEO*, mentioned above, in this way—that points *I* and *R* of the wood correspond to *U* at the end of the lever; that *TX* corresponds to the force applied at *U;* that the resistance at point *M* corresponds to weight *E* of the lever *OEU;* and that part *K* (the part just beyond *M* toward *FE*, the other end of the log) corresponds to fulcrum *O*. Thus, the longer are lines *IMK* and *RMK*, the more easily will forces *TX* drive *I* and *R* apart.

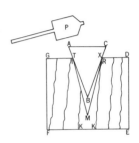

Chapter 21
On the true and intrinsic explanation of compound pulleys

To understand the true and intrinsic explanation why the multiplication of wheels in compound pulleys enables a small force to move upward or raise large weights,[68] let us imagine the two blocks of pulleys illustrated below and spread out as follows. That is, let *AB* be a small fixed beam parallel to the horizon, and let wheels be affixed to the under side of the beam. And directly opposite this beam let there be another beam *CD*, which can be moved up and down and to which are attached the same number of wheels and pins[69] [as are attached to the upper beam]. When later a rope has been attached to a fixed point *B*, and the rope has been made to pass over the wheels, both those above and those below, and a weight *E* has been attached to the small movable beam *CD*, and the

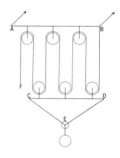

68. See above, "On Mechanics," Ch. 18. Cf. Duhem (*Les Origines*, I, 233–35).
69. I.e., the device which attaches the wheel to the beam.

[free] end *F* of the rope that passes over the wheels is then pulled, the same effect will clearly take place as always takes place when the wheels are united [in blocks]. And an explanation of this effect will more readily come to our understanding from the accompanying figure.

Let us consider, separately from the preceding figure, a balance *GH* with fulcrum *K* so situated that arm *GK* is double arm *KH*. Now if we assume a weight or moving force of one pound at point *G*, and of two pounds at *H*, clearly these two forces at these distances from the center will be equal to each other for the reasons already set forth in previous chapters, and the scale will remain horizontal. Clearly, then, the addition of any force, however small, to *G* will move the balance from its horizontal position. Now let the force [*virtus*] of *H* be applied to a point *I* midway between *G* and *K*, arm *KH* being no longer considered, and let the force at *I* be applied in the same direction as when it was at *H*, but let the force at *G* be applied in the opposite direction, opposite to that in which it was previously applied. It will be clear, by common understanding and for the reasons mentioned in Chapter 5 of this treatise, that *GH* will always remain in the same position without motion, and such a balance we shall call mobile[70] and primary.

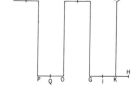

Let us now imagine a cord *EK* that descends from fixed point *E* and supports point *K*, the end of diameter *GK*, which I take to represent the diameter of one of the lower wheels of the compound pulley. And let *NLM* represent the diameter of one of the upper wheels of the small fixed beam, on the side where *G* exerts its force and parallel to diameter *GK*. In diameter *NLM* let *L* be the fixed center and let point *G* be joined to point *M* by the cord. And let this cord be perpendicular to the first diameter, *GIK*, as well as to the second, *NM*, so that angles *NMG* and *MGK* are right angles.

Let us also imagine the force of *G* applied to extremity *N*, but with opposite direction, i.e., downward. Now this force, as a matter of common understanding, will possess the same power to sustain diameter *GIK* unmoved as it had when it acted at *G* with upward direction. And so also diameter *NLM* will not incline in one direction any more than in another. Since a certain force is found at *N* equal to half of the force at *I*, which force at *I* has an effective force tending to depress

70. I.e., ready to move in response to a minimal force. [Or, perhaps, movable vertically without disturbance of equilibrium. − *S. D.*]

G, i.e., *M*, equal to half of itself, it follows that *NM* must remain unmoved. . . .[71]

.   .   .   .

Chapter 23
On the essential explanation of *Questions of Mechanics* 24

The true explanation of the effect described in Question 24[72] has not yet, so far as I know, been noticed by anyone, though it is not very difficult or obscure. Let us consider two concentric circles *CF* and *BG* so joined together that, if one of them revolves in a circle, the other does so too, as the wheels of a wagon revolve.[73]

And first let us imagine that the larger circle revolves over line *FI*. When that circle is at *L*, line *FI* is tangent to the circumference of that circle at point *C*. Hence line *GM* will be

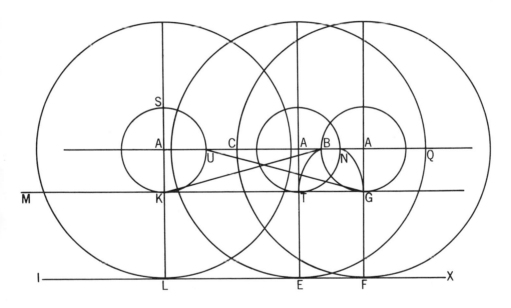

71. Benedetti goes on to consider the condition for equilibrium as more pulleys are added, and arrives at the formulation for the force required to maintain equilibrium in terms of the number of pulleys.

72. The so-called problem of Aristotle's wheel. On the history of the paradox see *Osiris*, 9 (1950), 162–98, and C. de Waard (ed.), *Correspondance du P. Marin Mersenne*, I, 13.

73. I.e., without slipping (where successive points of the circle touch a single point of the base line) or skidding (where a single point of the circle touches successive points of the base line).

tangent to the circumference of the smaller circle at point *B*, now in position *K*. *KG* will be equal to *FL* (Euclid I.34) because (Euclid III.17) angles *F* and *G* are equal. Hence *FL* and *GK* are parallel (Euclid I.28), and so also, by the same proposition, will *KL* and *FG* be parallel.

Now the reason why arc *GB* traverses line *GK* —greater than itself—is that, by virtue of the revolution of *FC*, each point of arc *GB* is, in the course of its motion, carried over line *GM* farther toward *K* than if it were moved solely by the revolution of *GB* on line *GM*. Thus, for example, when, by reason of the revolution of the greater circle, center *A* is found situated on line *LK*, point *G* will have completed path *GU*, and point *B* path *BK*, and also all the other points between *G* and *B* will have traversed long paths, having already been carried along by the large circle.

Next, let us imagine that these circles have been moved by reason of the revolution of the smaller circle, and that part *GT* of straight line *GM* was traversed by arc *GB*. When, therefore, *B* is at *T*, path *BT* will have been traversed by *B*, and path *GN* by *G*. Now these paths are much shorter than those in the previous case, because points *B* and *G* are now revolved on shorter radii.[74] I say the same of all the other points between *G* and *B*. And in this case point *F* will be at *Q* and point *C* at *E*. Therefore all points of tangency between *F* and *C*[75] not only will not have been carried forward,[76] but rather will have been carried backward from their initial position.

Thus there is no reason for such wonder if, while the larger circle revolves,[77] arc *GB* of the smaller circle is seen to traverse the whole line *GK*; and while the smaller circle revolves, arc *FC* is seen to traverse a path no longer than from *F* to *E*. For, when the larger circle revolves, every point of arc *GB* partakes of two motions in the same direction. For example, point *B* moves toward *M* not only because it is carried around center *A* —since that center moves toward *M* —but because, in addition to this, it is carried by the larger circle to line *KL* in the direction of *M*.

74. The controlling radius of revolution is now that of the smaller circle; previously it was that of the larger circle.

75. I.e., all points on the arc as they successively come into tangency with line *IX*.

76. I.e., in the first quarter-turn of the circles.

77. I.e., in the first case, the case in which the larger circle, in one revolution, traverses on its base line a path equal to its circumference.

But when the smaller circle is revolved, each point of arc *FC* has two contrary motions, of which one is toward *I* by virtue of the revolution of the smaller circle and the other away from *I* because the greater circle turns about center *A*.[78] Hence every point of contact of the greater circle with the straight line *FI* is forced backward toward *X*.

Chapter 24
On the true explanation of *Questions of Mechanics* 30

When a man is seated (but not in the fashion of the Turks)[79] and wishes to get up on his feet, he draws back his heels, [so that the lower legs] make an acute angle with the thighs and hips from below, and bends his abdomen forward, so as also to produce an acute angle above. The true explanation why he does this is in order that the weight of the whole body may evenly, that is, from opposite sides, encompass the straight line which passes through the place where his feet are at rest toward the center of the universe; that is, so that there may result an equilibrium of the weight of the body about that line which serves as a support underfoot. And so, then, by the opening of the two angles about that line, the body can be straightened up without any difficulty and without any danger of falling in either direction.

Chapter 25
On the explanation of the thirty-fifth and last of the *Questions of Mechanics*

The true explanation why objects found in whirlpools always come together toward the middle of these whirlpools arises from the fact that the centers of the whirlpools are always more depressed. Hence, the motion of those objects toward the center is merely the downward motion of those objects by reason of their weight and heaviness. For whirlpools have an almost conical shape, concave with [vertex] angle below and upper base circular. And the true explanation of the phenomenon in question is that which I have given, not that which

---

78. Benedetti means that in this case, where the smaller circle, in one revolution, traverses on its base line a path equal to its circumference, *F* reaches the (horizontal) level of *A* to the right, not to the left, of the starting position of *A*.

79. E.g., on a chair (not on the ground, with legs crossed under him—the Turkish fashion).

Aristotle asserts. But in the case of all those other problems that I have omitted, Aristotle's explanations are correct.

Disputations on Certain Opinions of Aristotle

Such is the influence and authority of Aristotle that it is difficult and hazardous to write anything contrary to his teachings, and especially so for me, since I have always had the highest admiration for his wisdom. But impelled by the pursuit of truth — and if he were alive, he would himself be motivated by this love for truth — I have not hesitated to set forth certain matters in which I have been constrained to disagree with him. And I base my disagreement on the unshakable foundation of mathematical philosophy on which I always take my stand.

Chapter 1
Aristotle on the velocity of natural locomotion: how and where his view differs from ours

Wishing to prove that in the nature of things there is no void, Aristotle says (*Physics* IV, Ch. 8)[80] that if the same body moves through different media, e.g., through air and water, the ratio of the speed of the body's motion through air to the speed of its motion through water will be one and the same as the ratio of the rareness of air to the rareness of water. And in the last part of the same chapter he writes:[81] "For since we observe that objects having more heaviness or lightness, if they are of the same shape, cover an equal distance with greater speed in the ratio that the bodies bear to each other, surely the same would be the case in the void."[82] The philosopher also sets forth another reason, writing (*Physics* VI, Ch. 2) that the magnitude is divided in the same ratio as the time.[83] And in *De Caelo* I, Ch. 6,[84] he writes that the times are in inverse proportion to the weights, so that if half a certain weight moves [a given distance in natural motion] in the time of one hour,

80. *Physics,* 215a.31–215b.12.
81. *Physics,* 216a.13–16. I.e., for natural downward motion, speed is to speed as weight is to weight; and, for natural upward motion, speed is to speed as lightness is to lightness.
82. From the difficulties inherent in this conclusion, Aristotle goes on to argue the nonexistence of the void.
83. Cf. *Physics,* 233b.6–7, 22–30. Note that, in natural downward motion, Aristotle holds that weight is to weight inversely as time is to time (for the traversing of a fixed distance).
84. *De Caelo,* 273b.30–274a.2.

the whole weight would move [that distance] in half an hour. In *De Caelo* III, Ch. 2,[85] he plainly declares in two passages that the speed [of natural motion] of the smaller body is to the speed of the greater one as the bodies are to each other. In Ch. 5 of the same book he affirms the same proposition with an example drawn from fire.[86] And also from many other passages we can gather that Aristotle believed that two bodies of the same material and shape maintain the same ratio in the speeds of their [natural] motions as the ratio of their sizes.

Many others, too, have held the same view, most recently Niccolò Tartaglia in [*Quesiti*] Bk. VIII, Qu. 29, Prop. 2, where he professes to demonstrate that this is a true proposition. But he does not see how great is the difference in the resistances which can arise from the difference in shapes as well as from difference in sizes. Indeed he does not even consider these differences [in shape].[87]

Chapter 2
Certain preliminaries to clarify why we reject the views of Aristotle on the speed of natural locomotion

Since we have undertaken the task of showing that Aristotle was mistaken on the subject of locomotion, certain preliminary statements should be made that are completely true and known by themselves as objects of understanding. To begin with, any two bodies, heavy or light, equal in volume,[88] of similar shape, and similarly situated,[89] but made of different material, will maintain the same ratio between the speeds of their natural locomotions as the ratio between their weights or lightnesses[90] in one and the same medium. And this is obvious from their nature if we consider that greater speed or slowness can arise from no other source than from four causes (so long as the medium is uniform and quiet), i.e., from greater or lesser

85. Perhaps the references are to Ch. 2 of Bk. IV: cf. *De Caelo* 308b.18-21, 309b.12-15. But the next reference is to Bk. III.

86. *De Caelo,* 304b.15-18.

87. But Tartaglia was discussing the balance, not free fall. [Moreover, he did point out the effect of shape in an earlier book; cf. *Nova Scientia* (18). —*S. D.*]

88. The Latin term here is *area,* but Benedetti uses the word elsewhere to denote volume. In any case, equality of area would, in conjunction with the next phrase, here imply equality of volume.

89. Explained in the next sentence.

90. "Weights" in the case of natural downward motion in a medium lighter than either of the bodies compared; "lightnesses" in the case of the natural upward motion in a medium heavier than either of the bodies compared. The measure of weight and lightness will, as the sequel indicates, be the difference between the density of a body and that of the medium.

weight or lightness, from differences of shape, or, in the case of bodies of the same shape, from differences in the position of the body with respect to the line of inclination (which is a straight line drawn[91] between the center of the universe and the circumference thereof), and, finally, from unequal size. Hence it will be clear that, in the absence of variation in the form of the body, either in size or in kind, or in the position of the body having that form, the motion will be in proportion to the moving force [*virtus*], i.e., to the weight or lightness. And what I say about the kind, the size, and the position of that form, I say with respect to the resistance of the medium itself. For the differences and inequalities of shapes and variation in position have no small effect on the motions of these bodies. For a small figure more easily divides the continuity of the medium than does a large one,[92] and so also an acute figure divides the medium more swiftly than a blunt figure, and one with an angle moves more swiftly if the angle is in front than otherwise.

Therefore, whenever two bodies are subjected to or receive one and the same resistance to [the motion of] their surfaces, [the speed of] their motions will turn out to be to each other in precisely the same proportion as their motive forces. And, conversely, whenever two bodies have one and the same heaviness or lightness,[93] but are subject to differing resistances, [the speeds of] their motions will have the same ratio to each other as the inverse ratio of the resistances: and these resistances will have the same ratio to each other as their surfaces, with respect to shape alone, or size alone, or position alone,[94] or any or all of these factors. But this is predicated on the basis of what was set forth above, namely, that the body which, when compared to the other, is of equal weight or lightness, but is subject to smaller resistance, moves [in natural motion] more swiftly than the other in the same proportion as its surface is subject to a smaller resistance than is that of the other body, by reason of its being able more readily to divide the continuity of the air or water.

91. I.e., through the center of gravity of the body. If the body is so placed, e.g., on an inclined plane, that it cannot directly move along its line of inclination, its speed will be correspondingly affected.

92. From what follows Benedetti seems to refer to the cross section or to the total surface.

93. Total weight (or lightness) rather than specific weight (or lightness) is referred to.

94. The two bodies may differ in the shape, area, or position of their surfaces, or in more than one of these respects. But there is no discussion of how, for example, differences in position are to be quantified. (Cf. note 92.)

For example, if the surface of the larger body to that of the smaller body were in the ratio of 4 to 3, the speed of the larger body to that of the smaller would be in the ratio of 3 to 4. That is, the speed of the smaller body would be greater than that of the larger in the ratio of 4 to 3.[95]

Another proposition must also be presupposed, i.e., that the speeds of the natural motion of a heavy body in different media are proportional to the weight of the body in those media. For example, if the total weight[96] of a heavy body is indicated by *AI*, then when the body is placed in a medium less dense than itself (for if it were placed in a medium denser than itself it would not be heavy, but light, as Archimedes showed), that medium subtracts part *EI*, so that part *AE* of that weight remains free. And if the same body were then placed in some other medium denser [than the first] but still less dense than the body itself is, this medium would subtract part *UI* of the weight, so that part *AU* of that weight would remain.

I say that the ratio of the speed of the body through the less dense medium to its speed through the denser medium will be as *AE* to *AU*. And this is more reasonable than if we were to say that the ratio of the speeds was as *UI* to *EI*: for the speeds attain their ratio solely from the motive forces (since the form of the body remains one and the same in quality,[97] size, and position). What I have just now said is quite in keeping with what I wrote above. For to say that the ratio of the speeds of [natural motion of] two heterogeneous bodies of the same shape and size and in the same medium is equal to the ratio of their weights is the same as saying that the ratio of the speeds of a single body moving [naturally] through different media is equal to the ratio of the body's weights in those media.

Chapter 3
That there may be a body such that the ratio of its speeds
in different media is equal to the inverse
ratio of the densities of the media

It is possible in the nature of things that a body may be found having such a density that the speed of its natural mo-

---

95. The bodies are assumed to have the same total weight and the same shape. The resistance of the medium is taken as proportional to the surface area (or to the area of the cross section) of the body; and the ratio of the speeds is taken as equal to the inverse ratio of the resistances. (Note that the method used by Stevin and others to refute the proportionality of resistance to volume might also be applied to refuting the proportionality to surface: see *Principal Works of Simon Stevin* [Amsterdam, 1955], I, 513.)

96. As weighed in a void?

97. I.e., shape.

tion through air is to the speed of its natural motion through water as the density of water is to the density of air. For example, let the density of water be denoted by UI, the density of air by EI, the weight of a body in air by EA and the weight of the same body in water by UA, but in such a way that the ratio EA to UA is equal to the ratio UI to EI. Hence by the last assumption of the preceding chapter, the speed of the body through air will be to the speed[98] of the same body through water as EA to UA, and therefore (by Euclid V.11) as UI to EI.[99]

Chapter 4
On an oversight in Aristotle's proof in
Physics IV, Ch. 8

From the above it is clear that what Aristotle writes (*Physics* IV, Ch. 8) is in general not true, namely, that the velocities of [natural] motion of a body through different media are to each other [inversely] as the densities of these media. Thus, suppose the ratio UI to EI is the ratio of the density of water to that of air, and the ratio EA to UA is the ratio of the weight of a body in air to its weight in water, the ratio EA to UA being, however, greater or less than the ratio UI to EI.[100]

'It will follow that, since speed through air is to the speed through water as EA to AU, it will not be as UI is to EI. Hence, because of this it is quite unreasonable to suppose that the ratio of the speeds [of natural motion] of all[101] heavy bodies through air is the same as the ratio of the speeds of the same bodies through water, as Aristotle held.[102]

98. Reading *velocitati* for *proportioni* (in the next to the last line of the chapter).

99. The point of the chapter is to show that the Aristotelian formulation happens to be correct only in a special case. The special case turns out to be that in which the density of the body is equal to the sum of the densities of the media. For if $EA/UA = UI/EI$ (the Aristotelian formulation), then $AU = EI$ and $AU + AE = AI$.

100. But not equal (the case of Ch. 3).

101. But note that the proposition is true (Ch. 8) of bodies of the same material (and shape).

102. According to Aristotle, the ratio of the speeds of natural motions of a given body through air and water is equal to the ratio of the densities of water and air. It follows that (for Aristotle) the ratio of the speeds of two bodies in air is the same as the ratio of the speeds of these bodies in water (or in any given medium).

Chapter 5
Illustrations of our propositions

Let us assume, for example, that the density of water is twice that of air,[103] and that a heavy body has a density twice that of water, and hence four times that of air. Therefore it would lose half its total weight when placed in water and a fourth part of its total weight when placed in air,[104] according to Archimedes *On Floating Bodies,* [Bk. I, Prop.] 7. It would therefore move in water by virtue of half its weight and in air by virtue of three-fourths of it. Hence the ratio of the moving force of the body in air to the moving force of the same body in water will be as 3 to 2.

Suppose such a body is *A.* And suppose there is another body, *B,* of the same shape and size as body *A,* but with density three-halves the density of water. Thus its density will be three times that of air. Hence body *A* will be heavier than body *B* in air in the ratio of 3 to 2, and therefore will be swifter than body *B* [in its natural motion] in air in the same ratio. But body *B* will have a weight in air twice its weight in water, since two thirds of its weight[105] remain to it in air and only one third in water. And so I agree with Aristotle that the [natural] motion of body *B* in air will be swifter than in water in the same ratio as water is denser than air, according to Euclid V.11.[106]

But if, besides all this, body *A* were also swifter in air than in water in the same proportion [i.e., 2 to 1], it would follow, from Euclid V.16, that the ratio of the velocity of *A* in water to the velocity of *B* in water would also be 3 to 2. But since the density of body *A* is twice that of water, and the density of body *B* is three-halves that of water, it will follow that the ratio of weight *A* in water to weight *B* in water is 2 to 1. Therefore, according to the first assumption of Ch. 2, the ratio of the speed of *A* in water to the speed of *B* in water will be 2 to 1, not 3 to 2. If, then, the ratio of the speed of *A* in water to that of *B* in water is 2 to 1, and if the ratio of the speed of *B* in air to the speed of *B* in water is also 2 to 1 (whence, by Euclid V.9, the speed of *A* through water will be equal to the speed of *B*

103. An unrealistic assumption, of course, but the example given is purely hypothetical to show the contradiction inherent in Aristotle's formulation.
104. I.e., if the body had originally been weighed in a void.
105. I.e., of its weight in a void.
106. This would be the special case of Ch. 3 (see note 99, above).

through air), and if the ratio of the speed of *A* through air is to that of *B* through air as 3 to 2, then the ratio of the speed of *A* through air to the speed of *A* through water will be 3 to 2 (by Euclid V.7), and not 2 to 1.

By these arguments we attain a corroboration of the truth of the last assumption of Ch. 2, namely, that the ratio of the speed of natural motion in different media of a body that has weight in those media is equal to the ratio of the weights of that body in those media. But it is understood that we refer to those media that are not greasy or fat, such as oil, milk, or other such substances, which are altered by any very small amount of cold or heat and become impermeable.

Chapter 6

That the ratio of the weights of the same body in different media is not the same as the [inverse] ratio
of the densities of those media. Hence unequal
ratios of speeds necessarily result[107]

In the case of every heavy body the ratios of its weights in various media vary, and therefore the ratios of the speeds are also unequal. For example, suppose there is a body *A* whose

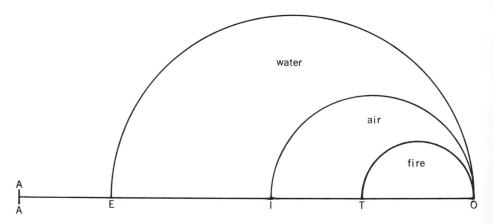

total weight is *OA*, which in water is diminished to the extent of *EO*, so that only weight *AE* is left to it. And suppose that its loss of weight in air is *IO*, so that only weight *AI* remains to it. Now let us suppose another medium less dense than air

107. I.e., the ratio of the weights of the same body in different media (which is the same as the ratio of the speeds) varies, depending on the media, and is not the same as the [inverse] ratio of the densities of the media.

in the same ratio as air is less dense than water, a medium in which body *A* loses part *TO* of its weight. Then according to Archimedes, *On Floating Bodies,* [Bk. I, Prop.] 7, the ratio of *EO* to *IO* will be equal to the ratio of *IO* to *TO*.

Suppose also that the ratio of *AI* to *AE* is equal to the ratio of *EO* to *IO*. Then I say that the ratio of *TA* to *AI* will not be equal to the ratio of *IO* to *TO*. Therefore, since the ratio of *AI* to *AE* is equal to the ratio of *EO* to *IO*, by disjunction the ratio of *EI* to *EA* will be equal to the ratio of *EI* to *IO*. Therefore, according to Euclid V.9, *AE* will be equal to *IO*. But since *EO* is to *IO* as *IO* is to *TO*, then (by Euclid V.11) the ratio of *AI* to *EA* will be equal to the ratio of *IO* to *TO*.

But since, as we have seen, *AE* is equal to *IO*, the ratio of *TA* to *IA* cannot be equal to the ratio of *OI* to *TO*. For if it were, then, by disjunction, the ratio of *IT* to *IA* would be equal to the ratio of *IT* to *TO*, and (by Euclid V.9, cited above) *AI* would be equal to *TO*. But it would be completely contradictory for *TO*, which is smaller than *OI*, that is, smaller than *AE*, to be equal to *AI*, which is greater than *AE*.

Yet the same result could obviously be proved as follows. If *IO* is equal to *AE*, *IO* will also be smaller than *AI*, since *AE* is a part of *AI*. But, by the same reasoning, *OT* is smaller than *OI*. A fortiori, therefore, *TO* will be smaller than *IA*. Therefore, according to Euclid V.8, the ratio of *IT* to *TO* will be greater than the ratio of *IT* to *IA*: and, according to Prop. 28 of the same book, the ratio of *IO* to *TO* will be greater than the ratio of *TA* to *IA*. According to Euclid V.12, therefore, the ratio of *IA* to *EA* will be greater than the ratio of *TA* to *IA*. And this will also be true of the ratio of the speeds.

Chapter 7
If heavy or light bodies of the same shape and material but of unequal size move naturally in the same medium, the ratio of their speeds is quite different from what Aristotle thought

I must now show that, if two bodies of unequal size, but of the same shape and material, move in natural motion in one and the same medium, the ratio [of the speeds] will be different from that which Aristotle taught.

Suppose, then, that bodies *A* and *O* are unequal [in size] but of the same shape and material; and suppose *A* is the larger body and consequently (as all will agree) heavier than *O* in the same ratio as it is larger than *O*.

Now Aristotle writes that the ratio of the speed of body *A* to that of body *O* (when each of them moves naturally) will be equal to the ratio of the size (or of the weight) of body *A* to that of body *O*.

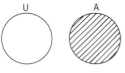

And so let us imagine a body *U* having the same size and shape as body *A* but the same weight as body *O*, whatever may be its material. Hence, from the first assumption of Ch. 2, we are assured that the ratio of the speed of body *A* to that of body *U* will be equal to the ratio of the weight of body *A* to that of body *U*.

Therefore, according to Euclid V.9, Aristotle must admit that the speed of body *O* manifests itself as one and the same as that of body *U*. But this proposition would clearly contradict the second[108] assumption of Ch. 2 of our book. Hence the opinion of Aristotle is false.[109]

The same could be proved with the help of a body, *I*, of the same size and shape as body *O*, but, as far as weight is concerned, equal to body *A*. Hence, according to the first assumption of Ch. 2 of this book, the speed [of natural motion] of body *I* would be greater than that of *O* in proportion as body *I* is heavier than body *O*. Therefore, according to Euclid V.9, Aristotle is forced to assert that body *A* is as swift as body *I*. Thus the same contradiction results [as before], according to the second assumption of Ch. 2 of this book.

Chapter 8
That two unequal bodies of the same material will maintain the same ratio of speed in different media[110]

Any two unequal bodies of the same shape and of the same material, moving naturally through various media, always maintain one and the same ratio of speeds.

Suppose the two unequal bodies are *A* and *B*, *A* being larger than *B*, but having the same shape and material as *B*. Let the total weight of *A* be *XO* and of *B* be *US*.[111] Again, let us sup-

---

108. The text reads "first." But cf. p. 169, lines 21ff., of the original text of Ch. 2.

109. If we use *v* for speed and *w* for weight, Benedetti's argument is as follows: $v_A/v_O = w_A/w_O$ (Aristotle). But $v_A/v_U = w_A/w_U$ (Ch. 2, first assumption). Therefore $v_O = v_U$. But this contradicts Ch. 2, second assumption. (The alternative argument of the next paragraph is entirely analogous.) Note that Benedetti seeks to refute Aristotle on the basis of assumptions which are themselves inconsistent with the Aristotelian formulation.

110. I.e., the ratio of the speeds of the natural motion of two bodies of the same material and shape, but of different size, is the same in any medium.

111. As weighed in a void.

pose that body *A*, when placed in water, loses part *OE* of its
weight *OX*, and that *B*, when placed in water, loses *CS;* also
that, when the bodies are placed in air, *A* loses part *IO* and
B part *TS*. Now the body of water to which *EO* corresponds
is equal [in volume] to *A*, and the body of water to which *CS*
corresponds is equal [in volume] to *B*, as is proved by Archimedes.[112] It follows, therefore, by an axiom of our science,
that *OX* is to *EO* as *US* is to *CS*. And for the same reasons *XO*
is to *IO* as *US* is to *TS*, and *OX* is to *SU* as *EO* is to *CS* and
as *OI* is to *ST*. Hence, according to Euclid V.19, *XI* is to *UT*
as *XO* is to *US*, and *XE* is to *UC* as *XO* is to *US*. Thus, according to Euclid V.11, *XI* is to *UT* as *XE* is to *UC*.[113]

Now, if from these ratios are subtracted the ratios of the
resistances that the bodies meet from without, the ratios that
remain will be equal to each other, according to the third axiom
at the beginning of Euclid, Book I.[114] And these are the ratios
that the velocities of these bodies will maintain.

Chapter 9
Did Aristotle deal correctly with the ratios of motions
in the void?

When Aristotle notes at the end of *Physics* IV, Ch. 8,[115]
that the ratio [of the speeds] of such bodies[116] moving in a void
would be the same as in a plenum, this statement is,[117] with all
due respect to Aristotle, completely false. For in a plenum the
ratio of the external resistances in the case of these bodies is
subtracted from the ratio of the weights, so that the ratio of
the speeds remains. And this last ratio would be annulled if the
ratio of these resistances were equal to the ratio of the weights.

112. *On Floating Bodies*, Bk. I, Prop. 7. Since the bodies are of the same
material, the weights both in and out of any medium are in the same ratio.

113. Up to this point, Benedetti has shown that the ratio of the weights in
air *(XI : UT)* and in water *(XE : UC)* are equal. But in this system of Benedetti (as distinguished from that employed in the 1553 *Resolutio* and the
Vatican version of the *Demonstratio*), the medium, apart from producing an
effect of buoyancy, offers resistance to a body's natural motion within it. The
ratio of the said resistance encountered by the two bodies is the same in any
medium and (Ch. 2) equals the ratio of the surfaces of the bodies.

Whether the equal ratios *(XI : UT)* and *(XE : UC)* are *diminished* by, or
(Ch. 18) *divided* by, the same quantity (ratio of resistances), equality will
be maintained. See also note 147, below.

114. If equals are subtracted from equals, the remainders are equal.

115. Cf *Physics*, 216a.2–7, 13–16. But Aristotle is here seeking to show
the impossibility of a void by a *reductio ad absurdum*.

116. I.e., bodies of the same shape and material but of unequal size.

117. Omitting *sit*.

For this reason the ratio of the speeds of the bodies in a void would be different from what it is in a plenum.[118]

Chapter 10
That bodies of the same material would move with equal speed in a void[119]

I assert for the following reason that these bodies [i.e., bodies of the same material] would move with equal speed in natural motion in a void.

For suppose that there are two bodies of the same material, *O* and *G*, and that *G* is half the size of *O*. And suppose further that there are two other bodies, *A* and *E*, of the same material

as the first two, with each of them equal [in size] to *G*. Now let us conceive of both these latter bodies as placed at the ends of a line whose midpoint is *I*. Clearly, point *I* will have as much weight as the center of *O*. And *I*, by virtue of bodies *A* and *E*, would move in a void with the same speed as the center of *O*. And the fact that those bodies *A* and *E* are separated by the length of the line does not in any way change their speed [of natural motion]. Each would have the same speed as *G*. Therefore *G* would have the same speed as *O*.

118. [Benedetti's argument appears to me to be that Aristotle is right only when weights and resistances are proportional, and that this is not generally true. Note 116, above, limits Aristotle's position sharply; I read "such bodies" as meaning "heavy bodies."—S. D.]

119. As early as 1553 Benedetti had enunciated the proposition for the general case of the natural motion of bodies of the same material and shape, but of different size, *in any medium* (see p. 151, above). This is also the doctrine of the Vatican version of the 1554 *Demonstratio*, J. Taisnier's plagiarism thereof in *Opusculum perpetua memoria dignissimum* (1562), pp. 16ff., and G. Cardano, *Opus novum de proportionibus* (1570), V, Prop. 110. Note Stevin's criticism (*The Principal Works*, I, 511). Cf. also Galileo, *De motu*, *Opere*, I, 263ff.

But in the Padua version of the 1554 publication (see p. 158) and also in the present work, Benedetti asserts that the speeds are equal *only in the void*, on the ground that in a plenum there would be an additional frictional resistance that would disturb this equality (unless this resistance itself were proportional to the weights of the bodies: Ch. 11).

Chapter 11
Bodies of the same material and shape but of unequal size, if they are subject to resistances proportional to their weights, will move with equal speed

By the same reasoning as we set forth in the preceding chapter, it could be shown that, if the two bodies, O and G, are subject to resistances proportional to their weights, the bodies will move with equal speed in a plenum—as I mentioned in passing at the end of Ch. 9.[120] For point I would have the same speed as the center of O, since I is moved by a weight equal to that by which the center of O is moved. And since, by hypothesis, both bodies A and E [together] would encounter as much resistance as O alone would encounter, bodies A and E, though separated, would have the same speed as if joined together. Therefore, the speed of G would equal that of O.[121]

Chapter 12
Here it is proved that the ratio of the weight of a denser to that of a less dense body in denser media is greater than the ratio of the same bodies in a less dense medium; and that the ratio of the weights of the bodies is not equal to the ratio of the densities of the media

If we have two bodies, A and B, of equal size, with A denser than B, I shall prove that the ratio of the weight of A to the weight of B will be greater in a denser medium than in a less dense medium.

Suppose that PG is the total weight[122] of body A and QK the total weight of body B, so that PG is greater than QK. Suppose OG is the weight which the denser medium subtracts from weight PG, and NK that which the same medium subtracts from weight QK. Also suppose that FG is the weight which the less dense medium subtracts from PG, and IK the weight which the same medium subtracts from QK. Hence OG

---

120. In Ch. 9 the ratio p/q is "annulled" by a ratio p/q as a subtrahend. Here the ratio p/q is conceived as being reduced to equality (1:1), ostensibly by p/q being taken as divisor. A ratio 1:1 is also spoken of as a "nullity"—with arithmetic sense of $1 - 1$. That is, the ratio p/q may be conceived arithmetically as $p - q$ (or $q - p$) and may be thought of as "annulled" by the same ratio $p - q$ (or $q - p$).

121. Cf. *Galileo Galilei On Motion and On Mechanics* (Madison, 1960), pp. 29–30.

122. As weighed in the void or in a medium whose density is considered negligible (e.g., air).

will be equal to *NK* and *FG* will be equal to *IK*. For the bodies are assumed to be equal in volume.[123]

Therefore the ratio of *PF* to *QI* will be greater than that of *OF* to *NI*, by an axiom of our science. For if one divided *PF* at point *C* so that *CF* were equal to *QI*, the interval[124] [arithmetic difference] *CF* to *QI* would be equal to that of *OF* to *NI*, namely, zero—but the ratio *PF* to *QI* would be greater than the ratio *CF* to *QI*, according to Euclid V.8; and hence, according to Euclid V.12, the ratio *PF* to *QI* will be greater than the ratio *OF* to *NI*. Hence, by Euclid V.33, the ratio *PO* to *QN* will be greater than the ratio of *PF* to *QI*. And the speeds will also have this ratio [*PO* : *QN*] to each other, and this is what concerns us.

Now since the ratio *PO* to *QN* is greater than the ratio *PF* to *QI*, therefore, by conversion, the ratio *PO* to *PF* is greater than the ratio *QN* to *QI*, and, by inversion, the ratio *QI* to *QN* is greater than the ratio *PF* to *PO*. Hence, if the ratio *PF* to *PO* were equal to the ratio *OG* to *FG*, then the ratio *QI* to *QN* would not be equal to the ratio *OG* to *FG*, or (what is the same thing) to the ratio *NK* to *IK*. In Ch. 5 I illustrated these principles with examples.

And since the velocities depend on the weights, it follows that the ratio of the speeds of two heterogeneous bodies is not the same in different media. And this is contrary to what would follow if we accepted the view of Aristotle in *Physics* IV, Ch. 8.[125]

Chapter 13
That the truth is quite different from what Aristotle teaches at the end of *Physics* VII

To assign the ratio of the speeds of two bodies moving naturally is not as easy as Aristotle thought it was (last chapter of *Physics* VII).[126]

Thus, suppose there are two bodies, *B* and *D*, of different materials and sizes but equal in weight and alike in shape. And suppose that the ratio of the resistances which they encounter from the medium in their [natural] motion is as *AE*

123. The term *area* in the text refers to that which the surface encloses.
124. The Latin term is *proportio*, which may be rendered "ratio" (relation), so that the ratio *a* to *b* sometimes means *a/b* (called "geometric") and sometimes *a* − *b* (called "arithmetic").
125. This does not justly represent Aristotle's view, which involves a comparison of plenum with a void, not of one plenum with another.
126. VII.5. But that chapter concerns violent, rather than natural, motion.

to OI.[127] And let AU and OC denote the total speeds unreduced by any resistance. These will be equal to each other, as is obvious from our assumptions.[128]

Now suppose that there are two other bodies, V and M, which, like the first pair, B and D, are in the same medium, but are made [each] of a different material from that of [the corresponding member of] the first pair [B and D], though of the same size and shape [as the corresponding member of that pair].[129] Also let their resistances be designated by NR and TS, respectively, and their speeds, when undiminished by any of the resistances, by NX and TG. Hence NR is equal to AE, TS is equal to OI, and NX is equal to TG. But NX and TG will not be equal to AU and OC.

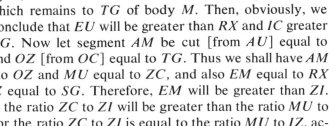

Let us, for example, assume that they [NX and TG] are smaller [than AU and OC]. Now let us suppose that EU is the speed which remains to body B when resistance AE is applied to it, the resistance diminishing the whole of AU by AE. And suppose that IC is the speed which remains to OC of body D, RX that which remains to NX of body V, and SG that which remains to TG of body M. Then, obviously, we shall conclude that EU will be greater than RX and IC greater than SG. Now let segment AM be cut [from AU] equal to NX, and OZ [from OC] equal to TG. Thus we shall have AM equal to OZ and MU equal to ZC, and also EM equal to RX and IZ equal to SG. Therefore, EM will be greater than ZI. Hence the ratio ZC to ZI will be greater than the ratio MU to ME (for the ratio ZC to ZI is equal to the ratio MU to IZ, according to Euclid V.7). But the ratio MU to IZ will be greater than the ratio MU to ME (Euclid V.8). Hence the ratio ZC to ZI will be greater than the ratio MU to ME (Euclid V.12). Therefore the ratio CI to ZI will be greater than the ratio MU[130] to ME (Euclid V.28). And the ratio CI to UE will be greater than the ratio ZI to EM, that is, greater than the ratio

127. The text here erroneously has "OI to AE."

128. The assumption, apparently, is that bodies of the same total weight would move with equal speed in natural motion in a void. This is not made explicit in the present treatise. Since bodies of the same material, whatever their size, also move naturally with equal speed in a void (Ch. 10), the assumption is equivalent to asserting the equal velocity of natural motion of all bodies in a void. See the 1554 *Demonstratio*, above.

129. The assumption, not clearly stated, is that V and M are equal to each other in weight (so that NX and TG are equal), but of a weight different from that of B and D. V has the same shape and size as B (but different material and weight); M has the same shape and size as D (but different material and weight).

130. Reading MU for U.

SG to *RX* (Euclid V.27).[131] But this did not occur to Aristotle.

The same can also be proved in other ways. Thus, one may say that the ratio *EM* to *MU* is greater than the ratio *IZ* to *ZC*. (For the ratio *EM* to *MU* is equal to the ratio *EM* to *ZC* [Euclid V.7]. But the ratio *EM* to *ZC* is greater than the ratio *IZ* to *ZC* [Euclid V.8]. Hence the ratio *EM* to *MU* will be greater than the ratio *IZ* to *ZC* [Euclid V.12]). Hence, *componendo*, the ratio *EU* to *MU* will be greater than the ratio *IC* to *ZC*, and *permutando*, the ratio[132] *EU* to *IC* will be greater than the ratio *MU* to *ZC*. And, by Euclid V.33, the ratio *EM* to *IZ* will be greater than the ratio *EU* to *IC*.

Chapter 14

What follows from the above

From the preceding chapter it may clearly be seen that the view of Aristotle expressed in the first part of the last chapter of *Physics* VII[133] is, in general, not true. For there, assuming that body *B* (of our last chapter) is half of body *D* in volume,[134] though [*B* and *D* are] equal to each other in weight, Aristotle says that *B* will have twice the speed of *D*. But in the last chapter I took *EU* as the residual speed of body *B*, after subtracting the part, *EA*, which resistance[135] takes from the speed. And I took *IC* as the residual speed of body *D*, *RX* for that of body *V*, and *SG* for that of body *M*.

Now let Aristotle tell us which of these two ratios [*EU* : *IC* or *RX* : *SG*] will be as 2 to 1. For if this is true of either of them, it cannot possibly be true of the other, as I have shown above, even if the two bodies *V* and *M* have the same conditions as *B* and *D*.[136] And the reason that induced Aristotle to his view could only have been that he thought the resistances proportionate to the volumes [*magnitudines*] of the bodies. That is, since *B* had half the volume of *D*, it would encounter half the resistance that *D* would encounter. But

131. I.e., the ratio of the speeds of *B* and *D* will not be the same as the ratio of the speeds of *V* and *M*.

132. Omitting *quam*.

133. *Physics*, VII.5. But, as indicated above, that chapter of Aristotle concerned violent, not natural, motion. Benedetti might have cited, e.g., *Physics*, IV.8 (216a.13–16).

134. Latin *area corporea*.

135. I.e., buoyancy. But for Aristotle the primary resistance to violent motion is the weight of the body that is being moved. He speaks of the resistance of the medium only in connection with natural motion.

136. See the beginning of Ch. 13.

even if this were true, it would not necessarily follow that for bodies in general the ratio of the speeds will be the same as the [inverse] ratio of the resistances, as we showed in the previous chapter.

Chapter 15
Was the Philosopher correct in holding that resistances are proportional to [the volumes of] moving bodies?

In holding that resistances are proportional to [the volumes of] bodies, Aristotle erred. For if the surfaces were proportional to [the volumes of] bodies, doubtless the resistances too would be proportional to [the volumes of] these bodies.[137] This is on the assumption that these surfaces are similarly placed while the bodies in question are in motion. But the ratio between the surfaces is not the same as the ratio between the [volumes of the] bodies. Thus Aristotle was mistaken on this point.

I say that the ratio between the surfaces is not the same as the ratio between the [volumes of] bodies. For if we first consider spherical figures, we shall find that the ratio between [the volumes of] two spheres is always the cube of the ratio between their diameters, according to the last proposition of Euclid XII. But the ratio between the surfaces of two spheres is equal to the ratio between [the areas of] great circles of these spheres (Euclid V.16), since (according to Archimedes, *Sphere and Cylinder* I.31) every spherical surface is four times the area of a great circle of the sphere. But the ratio between the two circles is equal to the square of the ratio between their diameters (Euclid XII.2). Hence the ratio between the [volumes of the] bodies will not be equal, as Aristotle believed, to the ratio between the surfaces, but to the three-halves power thereof.

And I say the same thing of bodies of similar shape bounded by plane surfaces. I base my argument on Euclid XI.36 and VI.18, from which we shall find that the ratio of the [volumes of] bodies is equal to the cube of the ratio of the sides, and that the ratio of the surfaces is equal to the square of the ratio of the sides. Hence the ratio of the [volumes of] bodies will be equal to three-halves power of the ratio of their surfaces.[138]

---

137. Benedetti's assumption is that there are resistances proportional to surfaces. See Ch. 2, third and second paragraphs before the end.
138. Cf. Galileo, *Discorsi, Opere,* VIII, 135.

Thus, if the ratio of the speeds were equal to that of the surfaces, then the ratio of the speed of body *B* to that of body *C*
would have to be equal not to the ratio of [the volumes of] the
bodies, but to the two-thirds power of the ratio of the volumes
of the bodies.

Chapter 16
Alternative proof of the same proposition

It can also be proved in another way that what Aristotle
wrote in the first part of the last chapter of *Physics* VII is
not, in general, true. What Aristotle wrote[139] is as follows: If
A is the mover, *B* the body moved, *C* the distance which it is
moved, and *D* the time required, obviously a force [*potentia*]
equal to *A* will move half of *B* in the same time [*D*] over a
distance double *C*, or over the distance *C* in half of time *D*.
For a like proportion will thus be maintained.

Suppose, then, that body *O* (of Ch. 7) is equal in weight to
body *U* (of the same chapter), but in volume[140] smaller than
U by half, the shapes of the two bodies being the same. Now

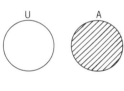

let us imagine a third body, *I*, homogeneous with *U*, but the
same as *O* in size and shape. Body *I* will be smaller than *U* by
half, and for this reason *U* will be twice as heavy as *I*, and,
consequently, *O* will also be twice as heavy as *I* (Euclid V.7).

Therefore, the speed of body *O* will be twice that of *I*, by the
first assumption of Ch. 2 of this book. Now since, according
to Aristotle, *O* would also have a speed double that of *U*, it
would follow (by Euclid V.9) that the speed of *I* would be
equal to that of *U*.[141] And this I showed, in Ch. 7 of the present
book, to be false.[142]

Chapter 17
On another error of Aristotle

In the last chapter of *Physics* VII[143] Aristotle writes as
follows: "If two [forces] separately moved two separate
weights through a given distance in a given time, then the

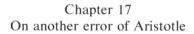

139. *Physics*, 249b.30–250a.4. Note again that the Aristotelian passage
refers to violent motion. [Chs. 13–17 seem to treat motion in general.—*S.D.*]
140. Latin *area corporea*.
141. The argument is sophistical, for Aristotle was concerned here with
violent motion, while Benedetti's assumptions (Ch. 2) have to do with natural
motion.
142. The falsity was not shown directly in Ch. 7. Rather it was shown
(Ch. 10) that only in a void could the proposition be true, but that in a plenum,
unless resistances were proportional to volume (Ch. 11), it was not true.
143. *Physics*, 250a. 25–28.

combined force will move a body formed by combining both weights the same distance in the same time. For thus the same ratio is maintained."

But this cannot in general be true in a plenum, since I have already proved in Ch. 14 that the same ratio does not hold between the surfaces of bodies and the [volumes of the] bodies themselves.[144]

Chapter 18
How to determine the ratio of the speeds[145] of two bodies of like shape and material but unequal size

Even though it is not easy to determine the ratio of the [speeds] of natural motion of two bodies alike in shape and material but unequal in size, I shall nevertheless show how we may accomplish this.

Let us suppose, for example, that there are two bodies A and O, both spherical and of the same material, but of unequal size, A being the larger. Now we wish to find the ratio of the speeds of their natural motion. Consider a spherical body I, made of different material [from that of A and O] but equal in weight to body O, and of a size such that the ratio of its surface to that of body A is equal to the ratio of its weight to that of body A.[146] This done, let us find the ratio between the surfaces of bodies I and O. This ratio will always be equal to the square of the ratio of the diameters, as I have pointed out in Ch. 15. Let the ratio of the spherical surfaces of O and I be subtracted from equality.[147] Therefore, what will remain will be the ratio of the speeds between the two bodies O and I, that is, between O and A.

For example, if the ratio of the surface of O to that of I were 4 to 3, that ratio being subtracted from equality, there would result the ratio of 3 to 4.[148] Hence, the speed of the larger body (which in the present case is assumed to be O) would be in the

---

144. Since one resistance is not proportional to volume (or weight) but to surface, the ratio of the speeds will not be maintained. But no distinction is made between the cases of natural and violent motion.

145. I.e., of "natural" (i.e., "free") motion.

146. So that (taking v as the speed of free fall in a plenum, w as weight, s as surface) we have $v_I = v_A$ (see Ch. 11). For, from the way I is constructed, $v_A/v_I = (w_A/w_I) \div (s_A/s_I) = 1$. Or, alternatively, if we use the arithmetic instead of the geometric sense of *proportio*, $v_A - v_I = (w_I - w_A) - (s_I - s_A) = 0$.

147. Thus, $v_I/v_O = (w_I/w_O) \div (s_I/s_O) = s_O/s_I$. Hence $v_A/v_O = s_O/s_I$.

"Subtracting" the *"proportio"* a to b from equality means, geometrically, $(1/1) \div (a/b) = b/a$, or, arithmetically, $(1 - 1) - (a - b) = b - a$, the result in either case being the corresponding *"proportio"* b to a.

148. See the preceding note.

ratio of 3 to 4 to the speed of the smaller body *I*. Or we may say that *I* would be swifter than *O* in the ratio of 4 to 3, according to the second assumption of the second chapter of this book. But *I* has the same speed as *A* (Ch. 11).[149] Therefore the ratio of the speed of *A* to that of *O* is as 4 to 3.[150]

Chapter 19

How fruitless is Aristotle's attempt to prove that the void does not exist

From what we have demonstrated above we may readily see that the argument which Aristotle constructed in *Physics* IV, Ch. 8, in order to disprove the possibility of a void is erroneous. In order to show this more easily, let us imagine corporeal media of indefinite extent,[151] one rarer than another, in whatever proportion we please, beginning with any one of the media. Let us also imagine a body *Q* denser than the first medium.[152] Suppose that the total weight of this body is *AB*,[153] and that, when placed in this [first] medium, it loses part *EB* of that weight, that in the second medium it loses part *IB*, and so on through the various degrees. Hence it will be clear to us that in none of the media will body *Q* retain its total weight *AB*.

Now if Aristotle should ask me the ratio of the speed of body *Q* through a void to its speed through a plenum, I should set down the ratio of *AB* to *AE*, for example, saying that [the natural motion of] body *Q* will be swifter in a void than in that [first] plenum in the same ratio as the ratio in which *AB* is greater than *AE*.[154] And we shall call the density of that plenum *EB*. But at this point Aristotle will say that some other medium may be assumed rarer than *EB* in the same proportion as *AE* is less than *AB*. Suppose that it is *IB* in which Aristotle believes that body *Q* will be as swift as in a void.[155] But that is

---

149. Since *I* was so constructed that the ratio of the resistances (i.e., the ratio of the surfaces) of *I* and *A* is the same as the ratio of their weights.

150. By the method outlined in this chapter the ratio of the speeds of spheres of the same material but different sizes falling freely in a medium is equal to the ratio of their radii (i.e., ratio of total weights divided by ratio of surfaces).

151. Or, perhaps, of indefinite number.

152. I.e., denser than the densest of the media.

153. If this is the weight of the body in a void, then Benedetti, by assuming *AB* as finite, is begging the whole question.

154. Note that there is no mention here of the factor of resistance, apart from the buoyancy of the medium. But, since only one body is involved, the factor of resistance (taken as a divisor) would not affect the ratio of the speeds.

155. Aristotle's hypothetical assumption of a finite speed in the void (e.g., *Physics*, 215b.21ff.) was designed to force a contradiction by showing that

where he errs, for the ratio of the speed of body Q in medium IB to its speed in medium EB will be equal to the ratio of IA to EA, according to the last assumption of Ch. 2 of this book. And this ratio would be smaller than the ratio of AB to AE (Euclid V.8).

·   ·   ·   ·   ·   ·   ·   ·   ·   ·   ·

Chapter 23
That rectilinear motion is continuous, despite Aristotle's dissent

Aristotle says (*Physics* VIII, Ch. 8) that it is impossible for a body to move in a straight line, first in one direction and then in the opposite direction, i.e., going and returning on the same line, without [an interval of][156] rest at the extremities. I hold, on the contrary, that it is possible.[157]

To consider this question, let us imagine a circumference UAN moving with continuous motion about its center O in either direction, to the left or to the right, and let us imagine a point B, wherever we wish, outside this circle. Let two lines BU and BN be drawn from B tangent to the circle at points U and N. And between [i.e., joining] these two line segments let us imagine another line drawn anywhere such as UN or CD or EF or GH. Take any point A on the circumference of the circle, and consider that point connected with B by a line BA fixed at B but movable according as point A moves. Hence sometimes this line will coincide with BU, sometimes with BN; sometimes it will move from BU toward BN, and sometimes from BN toward BU. This is what happens in the case of the line of progression and retrogradation of the planets: thus circle UAN will represent the epicycle [of the planet] and B the center of the earth.

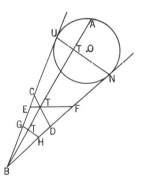

---

a body would then have the same speed in a certain plenum and in the void. Benedetti seeks to counter this argument by showing that the speed through the void would be greater than that through *any* plenum. The key is Benedetti's assumption that the speed varies as the weight of the body in the plenum (i.e., weight in void *minus* weight of plenum displaced by the body). Aristotle, however, had assumed that the speed varied inversely as the density of the medium.

156. The distinction between an instantaneous velocity of zero and a non-instantaneous interval of rest is the key to the problem—a distinction that was not clearly made before the seventeenth century. But Benedetti seems to have understood Aristotle in the sense I have indicated.

157. On the subject of the so-called *quies media* in the Middle Ages, see the references in Duhem, *Études*, II, 442. See also Duhem, *Le Système*, VIII, 255, 272; Clagett, *Science of Mechanics*, p. 514; de Waard (ed.), *Correspondance du P. Marin Mersenne*, II, 400. Cf. also Galileo's early *De motu, Opere*, I, 323–28.

It will now be clear that, when line *BA* coincides with *BU* or with *BN,* it will not be at rest: for it will return instantaneously, since *BU* and *BN* touch the circle in a point, and *BA* will always intersect any of the lines *UN* or *CD* or *EF* or *GH,* the point of intersection being *T.* Let us now imagine that an object moves along point *T* on any of these lines. It will be clear that such an object will never be at rest [for an interval] even if it is at one of the extremities. Hence Aristotle's opinion is not sound.

Chapter 24

Did the same eminent authority hold a correct view of
forced and natural motions of bodies?

Aristotle at the end of *Physics* VIII[158] expresses the opinion that a body moved by force and separated from its original mover is or has been moved over a period of time by air or water, which follows it. But this cannot be so, since the air which enters the space left by the body, in order to eliminate the void, not only does not drive the body forward but actually hinders its motion. For air is forcibly drawn by the body, and though this air is separate from the air in front of the body, it similarly exerts resistance. To the extent that the air in the [forward] part is condensed, the air in the rear is rarefied. Hence, since the air is rarefied under compulsion [*per vim*], it prevents the body from continuing its flight with the same speed with which it started, since every agent suffers in acting.[159] Therefore, while the air is forcibly acted on by the body, the body too is forcibly acted on by the air. Now this rarefying of the air is not natural, but forced. For this reason the air resists and tries to draw [the body] to itself, but since nature does not permit a void to exist between one and another of the bodies,[160] these bodies are always in contact. And since the moving body cannot leave the air behind, its speed is reduced.

Thus, in the case of such a body, speed originates independently of an original mover, from something naturally impressed, from an impetus received by the moving body. This impression

158. *Physics,* 266b.27–267a.20.
159. The body exerts itself in bringing about the condensation and rarefaction of the air and thus suffers, i.e., loses force, and is impeded in its motion by that air.
160. E.g., between the projectile and the air.

or impetus continuously increases in rectilinear natural motions, since the body always has within itself the moving cause, i.e., the tendency to go toward the place assigned to it by nature.[161]

Now Aristotle should not have said (*De Caelo* I, Ch. 8)[162] that the nearer a body approaches its terminal goal the swifter it is, but rather that the farther distant it is from its starting point the swifter it is.[163] For the impression[164] is always greater, the more the body moves in natural motion. Thus the body continually receives new impetus since it contains within itself the cause of motion, which is the tendency to go toward its own proper place, outside of which it remains only by force.

Aristotle also was in error when he wrote (*Physics* VIII, Ch. 9; *De Caelo* I, Ch. 2)[165] that there is a motion compounded from rectilinear and circular motion. For this is quite impossible.

Chapter 25
That rectilinear motion is not also natural motion primarily and essentially, whatever Aristotle may have thought

The upward or downward rectilinear motion of bodies in nature is not primarily and essentially natural motion. For natural motion is perpetual or, more properly, unceasing, and cannot be other than circular. No portion [of matter], when joined with its whole, can have any kind of natural motion other than that which characterizes the whole. But if the part is severed and separated from its whole and moves freely, it proceeds on its own and by the shortest possible path to the place fixed by nature for its whole. And such motion of this body is not primarily and essentially natural, since it is generated by a cause contrary to its nature, i.e., from the fact that

161. On the question of Benedetti's originality here see Duhem, *Etudes,* III, 212f. [Another difficulty with the assumption that Galileo consulted Benedetti's work is his failure to develop a theory of acceleration in the way that follows, rather than his awkward hypothesis in *De motu* of a lingering contrary impetus. −*S. D.*]

162. *De Caelo,* 277a.27–29.

163. For similar expressions in the thirteenth and fourteenth centuries (Richard of Middleton, Walter of Burley, and John Buridan), see Duhem, *Le Système,* VIII, 252; Clagett, *Science of Mechanics,* pp. 550, 560. [Tartaglia adopted the same view. −*S. D.*]

164. I.e., the body's impetus.

165. Cf. *Physics,* 265a.15; *De Caelo,* 268b.18.

it is outside its own place and is found there contrary to its nature. Thus motion of this kind is partially, but not entirely, natural. On the other hand, that motion is essential and natural which preserves the essence of the body in question. But this rectilinear motion does not do that, for it destroys that nature.[166] Therefore such motion is not primarily and essentially natural.

Chapter 26
It is not sound to say that every body is heavy in its proper place, as Aristotle held

Aristotle writes as follows (*De Caelo* IV, Ch. 4): "All bodies except fire have heaviness in their own place, and an indication of this is that an inflated bladder has more weight than an uninflated one, etc."[167]

In this passage he clearly shows that he does not know the cause of the heaviness or lightness of natural bodies, which is that the density (or rareness) of the heavy (or light) body is greater than the density (or rareness) of the fluid medium in which it is placed.

The example of the inflated bladder, which Aristotle himself gave, ought at least to have opened his eyes to a view of the truth, which shines through clearly. It is quite true that the inflated bladder has more weight than an empty one or one in which air is not enclosed under pressure. But the reason for this is that, when the bladder is inflated, the quantity of air pumped into it by force occupies a smaller space than if the same quantity of air were permitted to wander freely. Thus the air [in the inflated bladder] is in a sense forcibly condensed. Moreover, a dense body always sinks in a less dense [medium] (and a less dense body rises in a denser [medium]). That is the reason why the inflated bladder descends, since it is filled with a body more dense than the medium which surrounds it. The reason is not that air is heavy in air or water in water.

· · · · · · · · · · · ·

166. Presumably because its nature is to be at rest in its proper place. It would seem from this that the rectilinear motion of a terrestrial element, even toward its proper place, is only in a secondary sense natural; the circular motion of the celestial element is the only essentially natural motion. [This chapter curiously contradicts Benedetti's promising start on an inertial concept in "On Mechanics," Ch. 17. —S. D.]

167. *De Caelo*, 311b. 8–10. The result of the experiment with the bladder was debated even in antiquity. See Simplicius, *Commentary on Aristotle, De Caelo*, ed. Heiberg, pp. 710f. Cf. also Galileo, *De motu, Opere*, I, 285–89.

Chapter 28
That the cause of the twinkling of the stars was unknown
to the great Stagirite

When Aristotle says that the twinkling of stars takes place by reason of our viewing them at so great a distance, he commits a serious error. He also errs when he supposes that vision takes place by the emission of something [from the organ of vision]. His assertion is contrary to what he said elsewhere, indeed, is contrary to the truth.

Thus the twinkling of the stars comes about not because of our vision or because of some change in those stars, but because of the inequality of the motion of transparent media. This fact can be clearly seen. For if there is some smoke rising between an object and us, the object will appear to us as if it is trembling. And this will be the more likely as the object is more distant from the smoke. There will therefore be no occasion to wonder if we observe that the fixed stars twinkle more than the planets. The light of a star, as it comes to our eye, continuously penetrates the various transparencies by means of the continuous motions of the intermediate bodies that serve as media. Hence the light [of the fixed stars] varies, and more in the case of farther distant than of nearer stars, as anyone can see from the example of smoke, given above, and from glasses that have a surface that is not plane but irregular.

Chapter 29
That there may be a continuous motion without end on
a finite straight line[168]

All writers have up to now considered it impossible to conceive of a continuous motion without end taking place on a finite straight line. But in this they were deceived.

Let us imagine two parallel lines *BR* and *TX*, of which *BR* is prolonged without end in a given direction; and let us suppose that a point moves continuously on that line in the direction in question. And let us imagine a fixed point *C* on line *TX*, and between *C* and *A* consider a straight line *CA* and also a line *RX* fixed between the two given parallels. Suppose *CA* moves away from *B*, cutting *RX* in point *I*. This point of inter-

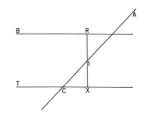

---

168. Cf. the example given by Galileo, *De motu, Opere*, I, 331. [Diagram and point references have been modified here to clarify the passage. —*S. D.*]

section will move continuously from *R* toward *X* over an end-less time and yet will never coincide with point *X*.

.   .   .   .

Chapter 35
That, despite Aristotle's contrary opinion, a rectilinear motion may be compared to a curved one[169]

But to return to Aristotle, he is mistaken when he says (*Physics* VII, Ch. 4)[170] that a rectilinear motion cannot be compared to a curved motion. And he also errs there in say-ing that no straight line can be found equal to the circumfer-ence of a given circle. For Archimedes proved in his book on the *Quadrature of the Circle* that the right triangle, one of whose sides about the right angle is equal to the radius of a circle and the other to its circumference, will be equal [in area] to that circle. Therefore that triangle, which will be equal to a given circle and will have one of its two sides about the right angle equal to the radius of that circle, will also have the other side enclosing the right angle equal, of necessity, to the circumference of that circle. Hence there can be one cer-tain straight line equal to a circular line. And this is contrary to the view of Aristotle, who did not recall what he himself wrote in a section on related things[171] when he said that the quadrature of a circle could exist even if at that time there was no knowledge of it. If, then, this quadrature can exist, there can also exist, for the reasons already given, a straight line equal to the circumference of that circle.

But if Aristotle had said that the circular motion of the heavenly bodies was not comparable to the rectilinear motion of the [four] elementary bodies, he would have been right — not because one of these motions is circular and the other rectilinear, but because the celestial motion is regular, not sometimes slow and sometimes fast, but always maintaining one and the same speed, whereas the contrary is true of the motion of the [four] elementary bodies. And a further reason [he might have given] is that there never has been nor will there ever be any of these natural rectilinear motions, as they are called, as swift as the motion of the heaven. For if we wish to consider the diurnal motion of 24 hours, according to the

169. Cf. the 1554 *Demonstratio* (12), above; Galileo, *De motu, Opere,* I, 302–4.
170. *Physics,* 248a.19–248b.6.
171. *Categories,* 7b.27–33.

general view,[172] we shall find by calculation that the moon in quadratures with the sun, when it is at the equator, moves 500 Italian miles or thereabouts per minute, and that in conjunctions and oppositions with the sun it moves 1000 miles or thereabouts per minute: also that the sun at the time of the equinoxes moves 18,000 miles per minute, that Saturn near its position at the equator moves 260,000 miles and more per minute. And as for the speed of the fixed stars situated near the equator, one may make one's own estimate, and, in fact, this will seem very difficult to some. But this difficulty does not occur in the beautiful system of Aristarchus of Samos that has been so divinely expounded by Nicholas Copernicus. And it is clear that neither the reasons adduced by Aristotle nor those set forth by Ptolemy have any validity against this system. Indeed, in its own proper motion the sun moves about 48 miles per minute, the moon, when it is in conjunction with or in opposition to the sun, 36 miles, and at quadratures, 18, Saturn 24, Jupiter 40, Mars 100, Venus 26, and Mercury 5. But in the rapid [diurnal] motion [required by the geostatic theory] Saturn moves 260,000 miles per minute, as we have said, Jupiter about 170,000, Mars 75,000, Venus 10,000, and Mercury 2000.

But one of the four [terrestrial] elements, even if it moved rectilinearly and more swiftly than a heavenly body, but without preserving uniformity, as the heavenly body does, could not in any way be compared to the latter. For so-called natural rectilinear motion always increases its speed because of the continuous impress which it receives from a cause continuously linked with the body, namely, the natural propensity for proceeding to its own place by a shorter path. And so, even if that elementary body, as it passed from a slower to a swifter motion, could exceed the [speed of] motion of some celestial body, those two motions would intersect at only one point, which could not be divided or separated into parts. That is to say, only at one instant of time would the motions be equal, so to speak. And I speak not only of circular celestial motion as compared with the rectilinear motion of the [terrestrial] elements, but of any kind of motion whatever, whether both are rectilinear or both curved, when any one of them is irregular.

172. I.e., the view that the earth is at rest at the center. The tremendous velocities of the heavenly bodies, especially of the fixed stars, required by this assumption had often been invoked as an objection to it (e.g., by Copernicus, *De revolutionibus* I.8; Oresme, *Le Livre du ciel et du monde* II.25).

Chapter 36
That the view of those who hold that many worlds exist was not adequately refuted by Aristotle[173]

The chief argument with which Aristotle seeks to refute the view of those who assert the existence of many worlds consists of this: he holds that the particles of [the element] earth which might be assigned to *other* worlds have their inclination [i.e., the direction of their natural motion] toward the center of *this* world, and, similarly, that the fire of *those* worlds will have a propensity [for motion] toward the circumference of *this* world.

But surely this argument is so weak that it falls by itself. For Aristotle does not consider that, if they could be spoken of as worlds, each of them would have its own center and its own circumference, and the various earths and fires would have tendencies to the centers and circumferences, respectively, of their own respective worlds; and a particle of earth [from one world] would not seek the center of another world. For example, if the system of the learned Aristarchus is correct, it will be perfectly logical for that which takes place in the case of the moon to take place also in the case of any of the five other planets. Thus, just as the moon with the help of its epicycles revolves around the earth, as if on the circumference of a certain other epicycle in which the earth is like a natural center (i.e., is in the middle), carried around the sun by the sphere of annual motion, so also Saturn, Jupiter, Mars, Venus, and Mercury may revolve about some body situated in the center of their major epicycle. And this body, also having some motion about its axis, may be opaque, possessing conditions like those of the earth, with conditions on the epicycle in question similar to those on the lunar epicycles described.

Chapter 37
Was Aristotle right on the subject of the passage of light through a void?

In *De Anima* II[174] Aristotle expressed the view that light proceeding from a luminous body cannot pass through a void.

173. For the medieval debate on this ancient problem, see Duhem, *Etudes,* II, 57–96, 408–23.
174. *De Anima,* 419a.12–21.

But this view does not seem plausible. For the less dense a body is, the more apt it is to be diaphanous. And the less dense that body is, the smaller the quantity of matter it contains. And again, the more diaphanous it is, since it consists of very rare matter, the freer, obviously, will be the passage of light through it. Hence the smaller is the quantity of matter in the given space, the more clearly will light pass through that space. It follows, therefore, that where no matter existed, light would pass freely and in its entirety. The blue color which we see in the depths of water and air[175] is the color of water and air—the color marking the resistance made by the air and the water to [the passage of] light. But if there were no body present, then this light would not be deflected at all, but instead would pass in a straight line without any hindrance.

.　.　.　.　.　.　.　.　.　.　.　.　.　.

Chapter 39
Examination of the validity of Aristotle's view on the unalterability of the heavens

Aristotle in *De Caelo* I.22[176] writes as follows: The evidence of the senses is sufficient to make us believe this, at least with human belief, and in all past time according to tradition no change has been observed in the outermost heaven as a whole, nor in any proper part of it.

But in this passage Aristotle did not consider that the same thing could be said about the earth if it were viewed from such a distance. Indeed, it may be asserted without any doubt that if the earth were endowed with the sun's light and someone wished to observe it from the eighth sphere,[177] he would not be able to see it at all: for the so-called stars of the first magnitude, which are thought to be more than a hundred times as large as the earth, are not seen except as points.

End of Disputations on Certain Opinions
of Aristotle

175. I.e., the blue of sea and sky.
176. *De Caelo*, 270b.11–16.
177. I.e., the sphere of fixed stars.

Letters

On the force of impact from a cannon, according to
different elevations. And on certain errors
of Niccolò Tartaglia on this subject

To the distinguished Lord Joseph Cambiano of the Lords
of Ruffia, noble knight and prefect of the artillery
of war of the Most Serene Duke of Savoy[178]

I studied certain questions, while I had the leisure because
of the absence of the Most Serene Duke, and I have decided
to write of these matters to you. If you approve of my ideas,
I would not be loath to publish them sometime, while if you
disapprove of them, I would quickly suppress them. This is the
kind of question: Why is it that a cannon-shot has more striking force when the shot is aimed upward than when horizontally, as Tartaglia writes (*Quesiti* I.2)? The problem has
not been [correctly] treated as yet by anyone, so far as I know.

The arguments of Tartaglia are of no weight. For if they
were valid, it would follow that, if the cannon were so inclined
that its angle of depression below the horizon were equal to its
angle of elevation above the horizon [in a previous test], the
force of the shot from either of these positions would be the
same. And if some difference did arise by reason of the weight
of the ball shot from the cannon, what would happen is that
the ball would be swifter in downward rather than in upward
motion, since [in the former case] the weight is not so much
opposed to the motion [as in the latter case].

But this does not take place. What, then, is the true reason
why it happens that the shot strikes with greater force when
the cannon is more rather than less elevated? In general, it is
virtually the same reason as the reason why a body of denser
material, but similar [in form] and equal [in size][179] to a body
of less dense material, moves more swiftly than the less dense

178. With the whole letter cf. Galileo, *De motu, Opere*, I, 337–40. See also
R. Caverni, *Storia del metodo sperimentale in Italia* (Florence, 1895), IV,
513, and Duhem, *Études*, III, 213, 221. Benedetti's whole discussion is
based on the false assumption that the initial part of the trajectory is rectilinear.
This assumption was the one prevailing at that time; it was also made for
practical purposes by Tartaglia himself. Tartaglia, however, had suggested
in 1537 (see *Nova Scientia*, above) that in fact the entire trajectory was
continuously, though at first insensibly, curved.

179. I so interpret, rather than "in weight," since the reference is not to
the usual principles of dynamics but to what, on first glance, seem to be exceptional cases. Cf. Galileo, *De motu, Opere*, I, 313. Actually, the question

body, when both have been propelled by one and the same or by an equal force.

And, in particular, it is the same as the reason why a greater explosion is produced by gunpowder which is put in an underground place, tightly sealed in iron containers.[180]

It is also like the reason why in ball-playing the ball can be batted farther with the wooden bat when it is hit opposite the direction of its [previous] motion rather than along the same path as that motion.[181] Now this happens because the motive force [*virtus movens*] strikes the body in such a case with greater and more intense force [*vis*]. For the more the body that is to be moved in such a situation resists the motive force (within definite limits, however), the more force does it gather in that brief interval of time: and this motive force then moves the body with so much more impetus, and clings to it so much more strongly as it drives it on, that it produces a greater effect than if the body had readily surrendered itself to the motion.

And this is what happens when the cannon's angle of fire is raised. For the weight of the cannonball—and this weight is what resists the motive force—gives the ball the opportunity of gathering that force in much greater measure than would be the case if the ball were shot at a lesser angle of elevation.[182]

Now since the increase in this kind of motive force bears no ratio to the weight of the cannonball, I would infer that, in the gathering of such a great amount of motive force, much more is collected than would suffice to shoot the ball, as evidenced by the great increase in speed.[183] For the more time is available for converting the gunpowder into fire, the greater quantity of fire is produced. Thus it happens that so much more space is required,[184] and hence the firing of the cannon is so much more powerful.

---

has to do with the conditions for increasing the length of the straight line segment that was assumed to be the first part of the trajectory. (Benedetti later criticizes the assumption that the force of the shot is greater during any part of this supposed initial straight line segment than it is in the curved portion before the natural downward motion. But obviously at the very top of the trajectory of a missile projected vertically upward the force is zero.)

180. Cf. Galileo, *De motu, Opere,* I, 339.

181. Cf. Galileo, *De motu, Opere,* I, 338.

182. This is the first of Galileo's two explanations (*De motu, Opere,* I, 338 *fin.*).

183. Reading *magni* (for *magnae*) and understanding the passage to mean that the speed of increase of motive force is greatly increased in the short additional time.

184. For the expansion of the powder into fire.

But, as I have said, the increase [in motive force] takes place with such speed that this motive force far exceeds the resistance exerted by the weight of the cannonball. This is the reason why the effect we are discussing takes place, an effect that is familiar in experience.

But that argument, by which Tartaglia in *Quesiti* I.3 thinks that he can reduce the Spanish Ambassador to an impossible position on the subject of the trajectory of the cannonball, is without foundation. For it must not always be said that the more swiftly a cannonball moves, the straighter is the path of its motion. For it could be said in answer to Tartaglia that, for a certain limit of speed, the cannonball is apt to move in a perfectly straight path over a certain distance, but if it moved more swiftly, it would not move in a straighter path over the same distance, but would continue its motion in a straight line over a longer distance. Thus he would have no reply to make except that he assumes that which he denies in *Quesiti* I.18. There he says that, when the cannon ball is near the muzzle opening, it is not moving as swiftly as when it is some distance past that opening, and that this fact is due to the resistance of the aerial cylinder.[185]

But the fact that the cannonball moves in a straight line [whose length is determined] according to whether the angle of elevation of the cannon is greater or smaller is due to the following reason. The line of natural [downward] motion [of the cannonball] does not form a right angle with the line of violent motion [i.e., the line of initial projection]. And to the extent that the angle they form differs from a right angle, whether it is acute or obtuse, to that extent [the natural downward tendency] has lesser force.[186] And this clearly happens in virtually the same way as I described in Ch. 3 of my treatise *On Mechanics*. For in shots aimed [obliquely] upward, the path of the violent motion of the ball toward the end[187] [of that part of the trajectory], beginning from the place [of firing] of the ball, makes an obtuse angle with the path of natural down-

185. [The view here imputed to Tartaglia was not his. Concerning this and the "aerial cylinder," see note 189, below. —*S. D.*]

186. I.e., the natural downward tendency is less able to effect a change in what is presumed to be the initial straight line of projection of the ball. This is another way of saying that the shot is stronger (in the sense of maintaining a longer initial straight line trajectory) the more it departs from the horizontal. Cf. Galileo's second explanation, *De motu, Opere*, I, 339f.

187. I.e., the end of the straight line segment that is presumed to be the initial portion of the trajectory.

ward motion; and in shots aimed [obliquely] downward it makes an acute angle.[188]

And I shall not at this point overlook an error worth noting which Tartaglia commits in the same discussion. He supposes, without qualification, that a body exerts an impulsive or percussive force with greater impetus when the body is moving in a straight line. But if that were so, it would follow that a heavy body, if projected perpendicularly upward, would have, at any stage of its trajectory, a stronger striking force than it would have at any stage of another trajectory resulting from any upward oblique projection. But I leave to you to contemplate how false this is.

Also false is the argument Tartaglia adduces in *Quesiti* [I.]4. For the air does not persist in motion as much as he thinks: indeed, such violent agitation ends quickly, and even more quickly than if it had acted upon a bag full of feathers, quite apart from any cannon.

And the argument which he adduces in *Quesiti* [I.] 18 on the passage of the cannonball through the cylindrical aerial body is clearly false. For the air which had previously been enclosed in the cannon immediately bursts out of it, gives way, and is divided by the ball, as if it had never had that [cylindrical] form; and the surrounding air does not resist it.

But that the ball is swifter at a certain distance [from the muzzle] than at the beginning [of its flight][189] would, if true,

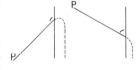

188. The obtuse and acute angles seem to be those indicated in the accompanying diagram, *P* being the position of the cannon in each case.

[In the next paragraph, Benedetti again misrepresents Tartaglia's view. *Quesiti,* Bk. I, Qu. 3, follows immediately on Tartaglia's denial that any portion of the trajectory can be mathematically straight; see *Quesiti* (10r–11v), above. Thus Benedetti's phrase "without qualification" is entirely unjustified. —S. D.]

189. Whether a projectile reaches its maximum speed sometime after projection was a question that had been widely debated from the time of Aristotle (*De Caelo,* 288a. 22). Tartaglia upheld the negative (*Nova Scientia,* Bk. I, Prop. 3). [In *Quesiti,* Bk. I, Qu. 18, Tartaglia replied to a gunner who had declared that cannonballs are more effective at some distance from the muzzle. Tartaglia said (1) that a cylinder of air, compressed by the ball, precedes it for some distance, and (2) that any obstacle close to the muzzle reflects that compressed air back against the ball, cushioning the shock. The question was not one of swifter motion, and Tartaglia's only reference to speed was to say that the ball moves faster than the "aerial cylinder." —S. D.]

depend on another cause. And that cause would in part be like that which causes bodies in natural motion to move more swiftly, the farther distant they are from the beginning of their natural motion: for such a body would move over some distance as if moved by natural motion.

. . . .

On the motion of a millstone and of a top [etc.][190]

To the Illustrious Joannes Paul Capra of Novara, Master
of the Hospice of the Duke of Savoy and a man
of surpassing intellect and purity of character
renowned also for his noble lineage

If the doctrine of transmigration of souls, which Pythagoras, the father of Italian wisdom, invented, were true, I should imagine that your soul and mine had once belonged to hunting dogs.

You ask me this question in your letter. Suppose a millstone rested on a virtually mathematical point and was set in circular motion; could that circular motion continue without end, it being assumed that the millstone is perfectly round and smooth?[191]

I answer that this kind of motion will certainly not be perpetual and will not even last long. For apart from the fact that the wheel is constrained by the air which surrounds it and offers resistance to it, there is also resistance from the parts of the moving body itself. When these parts are in motion, they have by nature a tendency [*impetus*] to move along a straight path.[192] Hence, since all the parts are joined, and any one of them is continuous with another, they suffer constraint in moving circularly and they remain joined together in such motion only under compulsion. For the more they move, the more there grows in them the natural tendency to move in a straight line, and therefore the more contrary to their nature is their circular motion. And so they come to rest naturally: for, since

190. The other topics in this letter do not concern mechanics.
191. Benedetti also discusses this problem in his "Mechanics," Ch. 14. For an earlier treatment, cf. J. Buridan, *Quaestiones super libris quattuor de caelo et mundo*, II, Qu. 12 (Clagett, *Science of Mechanics*, p. 561). Cf. also Duhem, *Études*, I, 120, III, 214–27, and Galileo, *De motu, Opere*, I, 304–07.
192. Benedetti's denial of circular impetus is an important step toward the formulation of the laws of motion in classical mechanics. [Rather, this pas-

it is natural to them, when they are in motion, to move in a straight line, it follows that, the more they rotate under compulsion, the more does one part resist the next one and, so to speak, hold back the one in front of it.

As a result of this tendency to straight-line motion possessed by the parts of any rotating body, a top while spinning around with great force remains for some time quite erect over its iron point, not inclining toward the center of the universe on one side any more than on another. For the inclination of each of its parts during such motion is hardly at all toward the center of the universe, but is directed far more transversely, at right angles to the line of direction,[193] i.e., at right angles to the vertical or the axis of the horizon. Thus such a body must necessarily remain erect.

Now in saying that the line of inclination of those parts is "hardly at all" toward the center of the universe, my reason is that those parts are never completely deprived of this tendency, a tendency which makes the spinning top rest on its point. However, it is true that, the more swiftly it spins, the less does it press upon that point; indeed, the body becomes that much lighter. This is obvious if we consider the example of the missile shot by a bow or any other instrument or machine for hurling missiles. The swifter the missile is in its forced motion, the stronger is the tendency it has to traverse a straighter path and, therefore, the weaker is its tendency toward the center of the universe, and, for this reason, the lighter does it become.

But if you wish to see this truth more clearly, imagine that while the body, i.e., the top, is spinning around very rapidly, it is cut up or divided into many parts. You will observe not that those parts immediately fall toward the center of the universe, but that they move in a straight line and, so to speak, horizontally. No one, so far as I know, has previously made this observation on the subject of the top.

From such motion of the top or of a body of this kind it may be clearly seen how mistaken are the Peripatetics on the subject of the forced motion of a body. They hold that the body is driven forward by the air which enters [behind it] to occupy the space left by the body. But actually the opposite

---

sage appears to me a denial of conservation of angular momentum; the theory offered is ingenious, but quite wrong. — *S. D.*]

193. I.e., of free fall of heavy bodies.

effect [that is to say, resistance] is produced by the air.[194]

.   .   .   .

On the revolution of the wheel used at the well and other problems[195]

To Capra

The rope to which the bucket is attached can be turned far more easily about the axle if a wheel is fitted to the axle; and it can be turned even more easily, to the extent that the wheel is larger and the axle narrower. But it is turned most easily if the extreme circumference of the wheel consists of a thin layer of denser, and therefore heavier, material. There are several reasons for this.

First, every heavy body, when moved either naturally or by force, receives on itself an impression and impetus of motion, so that, even if separated from the motive force, it moves by itself for some length of time. (Indeed, if it is set in *natural* motion, it will always increase its velocity: for then the impetus and impression [of motion] are always being increased, since the motive force is always joined to the body.) Thus, if we move the wheel with our hand and then remove the hand from it, the wheel will not immediately come to rest but will turn for some length of time.

The second reason is that every heavy body, when moved either naturally or by force, tends, by its nature, to move in a straight line.[196] This we can clearly observe if we shoot stones from a sling after whirling our arm around. For the more swiftly the sling is rotated and the motion increased, the more weight do the ropes sustain, and the more pull do they exert on the hand. And this results from the natural tendency inherent in the stone to proceed along a straight line. So it is that the weight of the circumference of the wheel is more easily rotated and moves by itself for a longer time, to the extent that the circumference is farther distant from the center, since its

194. Benedetti's point is that, in the case of the spinning top, the top does not leave the space it occupies, so that there is no empty space for the air to enter. Buridan had used the argument in *Quaestiones super octo phisicorum libros Aristotelis,* VIII, Qu. 12 (Clagett, *Science of Mechanics,* p. 532). On the motion of the top, cf. B. Baldi in Duhem, *Études,* I, 118, and the section on Baldi in the Introduction (above).

195. Cf. Duhem, *Études,* III, 212, 217.

196. See the preceding letter.

path is that much less curved. And consequently, for that reason, the larger the wheel and the more its weight is concentrated near the circumference, the longer lasting will be the impetus of motion impressed upon the wheel.

The third reason is that while the rope turns around, it does so nearer the mathematical axis of revolution than does the heavy material at the circumference of the wheel. And so, by the principle of the lever, when the wheel is in motion, its impetus is not subjected to an equal resistance from the counterpoised weight of water in the bucket.

. . . .

On the device which drives forward and raises water

To Capra

Why do I hold that, in the case of a fountain, the vessel or tank [pump cylinder] in which the piston that drives the water forward is inserted need not have a diameter greater than that of the pipe through which the water must rise?[197] The reason is that, if the diameter [of the tank] were greater, the piston needed to drive the water forward would have to be much heavier than a whole mass of water such as would fill a pipe as tall as the fountain and yet as broad as the tank.

Suppose, for example, that the whole pipe or tube through which the water rises is F, and the tank is AU, which is as tall as F, while F is narrower than AU. Now when these two vessels are filled, it will be clear that the water in F will be sufficient to resist all the water in AU and the water in AU will resist that in F, though the water in AU is of greater quantity and weight than that in F. Now this comes about from the fact that the water in AU does not press on the water in F with all its weight, because that weight is divided proportionately over the base of the tank.

For example, suppose there is a vessel $BDNM$ in the shape of a hollow cone or a truncated cone, and full of water. Let

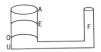

---

197. The discussion is on the difference in the force necessary for the piston to move the water, depending on the size of the opening which the piston fills. We have here the basis for (if not a precise formulation of) the principle of Pascal and an important step in the development of the hydraulic press. But note that in Benedetti's discussion the motive force is merely the weight of the piston, there being no additional force exerted upon it. Cf. also Stevin's discussion published in the following year. The entire development from Leonardo da Vinci to Galileo is dealt with by Duhem, *Études*, I, 198-220.

the diameter of its opening be *BD*, many times larger than the diameter *MN* of its base below. And let us consider *BD* divided into parts each equal to *MN*, and let us at these divisions

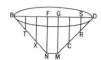

imagine vertical lines drawn toward the center of the universe to points *R, C, M,* and *T, X, N,* as one can see in the figure drawn below. Let us think of these lines as representing so many curved and conical surfaces, and let us imagine that there is water in the parts between the lines. This water will, with its weight, be at rest over a greater surface than a surface [everywhere] equally distant from the center of the world, or than the surface of base *MN*. For example, consider the water between *GM* and *SR* whose weight is distributed along width *MR*, which is greater than *GS*. Suppose then that *MC* is equal to *GS*. It will be clear that *MC* will not sustain the whole weight of the water found between *GM* and *SR*. The reason is that the whole quantity of water tends [by its weight] toward the center of the world. Hence the bottom or base does not sustain any other weight but that of water *FM*.

But if someone were to call this into question, saying that the water surrounding the position of the body of water *FM* presses laterally on that water, the answer must be that the body of water *FM* presses equally on the water surrounding it: for they are homogeneous bodies, and equal parts of homogeneous bodies have equal strength.

But to return to vessels *AU* and *F*, I say that just as water *F* suffices to resist water *AU*, so any weight equal to *F*, of any

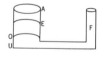

material, placed in tube *F*, will also suffice, provided that that body [e.g., a piston] so fits the concavity of tube *F* that it does not permit the passage of any air or water between its outer boundary and the inner boundary of tube *F*. This is self-evident.

But in the case of vessel *AU*, since it is, by hypothesis, wider than *F*, no other body will suffice to resist the water in *F* unless it is as heavy as all the water in *AU*, *AU* being as long as *F*. Hence, if the water in *F* weighed no more than a single pound, and vessel *AU* were ten times as great [in area of cross section] as *F*, it would then be necessary to place in *AU* a body completely filling out its concavity and having a weight of ten pounds, in order to suffice to sustain the water of *F*. And to drive that water up would require a weight of more than ten pounds.

Let us now assume that that body were so much more dense than water that it would occupy no more space than *OE*. Then that body *OE*, but no lesser body, would suffice to push forward water *F*.

On the reason for the tension in the cord of a certain
balance, and on a certain comparison of the
circle and the ellipse

To Angelo Ferrario, expert surveyor of the Most Serene Duke
of Savoy

You will recall that we were together a few days ago, at
the splendid villa where our Most Serene Duke was sojourn-
ing, while you did the leveling for the garden. And I asked you
whether you knew how it happens that, when a balance stands
at right angles to its base [i.e., stands level], the cord which
joins the ends of the balance with the base in the form of an
isosceles triangle is more tightly stretched than when the balance stands oblique to the base, the cord in that case forming
a scalene triangle with the balance.

For example, suppose that line *DBC* is the balance, *BEU* its
base, and *DEC* the cord, which sometimes makes an isosceles
triangle with the balance and sometimes a scalene triangle.
I asked you then whether you knew the reason why cord *DEC*
was tight in Figure 1 and loose in Figure 2, as we saw. You
made some answer, which I do not now remember, but, since
I promised that I would send you my explanation, that is the
reason why I am now doing so.

Observe, then, that the explanation is merely that in Figure 1
the two lines *CE* and *DE,* when joined together in one straight
line, are longer than they are in Figure 2. And therefore the
cord, since it is one and the same cord in both Figures 1 and 2,
is loose in Figure 2, not taut, as it is in Figure 1.

That you may have a perfect understanding of this truth,
consider the circle *FEI* drawn below, with radius equal to
BE, and diameter *FI,* on which you may consider balance
DBC placed, along with Figures 1 and 2. I shall prove that
lines *DE* and *EC* of Figure 1 are together longer than *DE* and
EC of Figure 2.

Let us suppose, then, that *BE* is the semi-minor axis of an
ellipse whose figure is traced with *D* and *C* as centers [i.e., foci].
The circumference of the ellipse will, of course, fall outside
the given circle, and will be tangent to that circle at only one
point, namely, point *E* of Figure 1, which is distinct from point
E of Figure 2.[198] Then if line *DE* of Figure 2 is extended to
meet the circumference of the ellipse at *G,* and from *G* to *C*
line *GC* is drawn, it will be clear that the two lines *DE* and *EC*

198. In what follows it is important to note which of the two points *E* is
referred to.

of Figure 1, added together, will be equal to the sum of *DG* and *GC*. (This is easily seen from Apollonius, *Conics* III.52.) But we already know (Euclid I.21) that lines *DG* and *GC* are together longer than the sum of lines *DE* and *EC* of Figure 2. Therefore, *DE* and *EC* of Figure 1 are together longer than *DE* and *EC* of Figure 2. Which was to be proved.

[The rest of the letter contains a comparison of certain mathematical properties of the ellipse and the circle.]

. . . On certain errors of Tartaglia

To Muzio Groto

·   ·   ·   ·   ·   ·   ·   ·   ·   ·   ·   ·   ·

But when I wrote you at the end of that letter that Tartaglia had erred in his *New Science,* I, Prop. 5,[199] I had good reason to write that. For Tartaglia says that no uniformly heavy[200] body can, during a given interval of time, move with natural and forced motion mingled together at the same time.[201]

In this Tartaglia is mistaken because he does not observe that there is an increase in the velocity of one motion together with a decrease in the velocity of the other motion at the very same time. This is clearly seen in the trajectory of the body's motion in the example given by Tartaglia himself. There the speed of motion in the interval *CD* increases insofar as it is natural and decreases insofar as it is forced. For the speed increases as the body approaches the horizon, and decreases as it moves away from line *AB*. But if the motion from point *C* to point *D* were purely forced motion, as Tartaglia thinks, the body would not fall at all, for the motive force at *A* cannot produce such an effect. Hence the descent of the body proceeds not from any force but from nature, by reason of the weight the body has in that medium, the air. But if Tartaglia had said that the motion was purely natural, this would have been false, for the reason that the purely natural motion of an unhindered body outside its natural place takes place on a straight line, and not on a curve, as one can see [that it does] in the motion between *C* and *D*.

·   ·   ·   ·   ·   ·   ·   ·   ·   ·   ·   ·   ·

199. [Benedetti's previous letter to Groto is lost. −*S. D.*]

200. A "uniformly heavy" body is defined by Tartaglia (Bk. I, Def. 1, *Nova Scientia*) as one that, by reason of the weight and shape of its material, will not encounter any noticeable resistance on the part of the air in any of its movements.

201. On "compound impetus," see Duhem, *Études,* III, 189ff., 222ff.

[From a letter to Franchino Trivulzio]

I now turn to the question you ask about the motion of a stone falling toward the center of the world,[202] it being assumed able to reach and pass that point. I hold that Niccolò Tartaglia and Francisco Maurolico were right and Alessandro Piccolomini wrong, and that the best illustration was that given by Maurolico.[203] But if you cannot understand this illustration, at least have confidence in the authority of such men who so far outshine Alessandro Piccolomini in these sciences as the sun outshines the other stars.

Now the stone would move past the center and then return, but with a diminution in its impressed force, in just about the way those thoughtful scholars describe, until after many such movements up and down [i.e., oscillations away from and toward the center], it would finally come to rest at the center of the world.

But for a clearer understanding consider a cord (as in the illustration adduced by those scholars) to [one end of] which a weight is appended. Let this cord be equal to the axis of the horizon, i.e., let its fixed end be located in the *primum mobile*[204] at the very zenith of your horizon. Then the arc through which the stone [i.e., the weight] would oscillate over a path equal to the diameter of the earth would not differ appreciably from a straight line. Now the stone, which is an earth's radius distant from the center of the world, would, as you know, move back and forth. It would therefore act in the same way if the cord were longer by that earth's radius and thus reached the center. For the difference of an earth's radius is practically nil in comparison with the radius of the *primum mobile*.

Some notes on Archimedes

To the most Learned and Reverend Lord Vincenzio Mercato

What I told you elsewhere is true, that the mind cannot rest altogether satisfied with the two propositions of Archimedes

---

202. I.e., the center to which weights tend to move, identical, in the present context, with the center of the earth.

203. See Tartaglia, *Nova Scientia.* Bk. I, Prop. 1; Maurolico, *Cosmographia,* Dialogus I, pp. 15–16 (Duhem, *Études,* III, 195f.). For earlier discussions by Albert of Saxony and Oresme (who gives the same example of a pendulum), see Clagett, *Science of Mechanics,* pp. 566, 570. See also R. Dugas, *A History of Mechanics,* (Neuchâtel, 1955), pp. 59, 90 (where Erasmus' treatment of the problem is cited).

204. The sphere forming the outermost boundary of the universe.

that appear under the numbers 4 and 5 in Tartaglia's translation and under the numbers 6 and 7 in the Basel edition.[205]
There he deals with the centers of the scale or balance. Note,
then, that in Prop. 4 all parts of the weights are considered to
be [evenly] distributed along the entire length of *LK*. Now I
want you to imagine drawn from points *E* and *D* on line *LK*
two lines, *EO* and *DU*, equal to each other and virtually perpendicular to *LK*, i.e., directed toward the center of the universe. Imagine also *OU*, parallel to *LK* and divided at point *I*
above *G*. Thus no one can doubt, since *G* was the center of the
whole weight suspended from *LK*, that *I* similarly will be the
center, since it is situated directly above *G*, that is, in the same
line of direction. And this needs no demonstration, since it is
quite clear by itself. Hence *O* will obviously be the center of
the weight applied to *LH* and *U* the center of the weight applied to *HK*.

We know, then, that *I* is the center of the two parts, *LH* and
HK considered as joined into the whole *LK*. Now, therefore,
if we consider *LK* to be divided or disjoined at point *H*, we
shall nevertheless find that *I* is still the center of the weights.
What this means is that *LK* is really continuous, though divided
at point *H*. Thus point *I* will not, because of that division, be
any more or less the center of the two weights *LH* and *HK* if
one of them is entirely suspended from *O* and the other entirely
from *U*. Thus in the matter of the length *OU*, divided as indicated, we shall have what we set out to prove.

I leave the rest of the proposition to you.

The proposition about which I told you that Archimedes
was silent deals with the subject of two weights in equilibrium
at the ends of a balance at certain predetermined distances
from the fulcrum. I say that, if one of these weights remains
stationary and the other is moved farther from the fulcrum,
that second weight will fall; while if that weight is appended
nearer the fulcrum, it will rise. Now instances of this proposition are observed all the time and everywhere, and I suppose
that Archimedes omitted it because it is so simple, since it
practically follows from the preceding proposition.

Suppose, for example, that there is a scale *AU*, with fulcrum
I and weights *A* and *U* appended, the ratio of *A* to *U* being
equal to the ratio of *IU* to *IA*. Now I say that, if weight *U*
is placed nearer the fulcrum, say, at *O*, while weight *A* remains

205. *On Floating Bodies*, Bk. I, Props. 6 and 7. The Prop. 4 here discussed
is Bk. I, Prop. 6, of the standard edition.

fixed, arm *IOU* will rise, and, conversely, if weight *U* is placed farther from the fulcrum, arm *IOU* will fall.

Let weight *U* be moved to *O*, as was said, nearer the fulcrum. Then arm *IO* will be shorter than arm *IU*. Hence the ratio of *IO* to *IA* will be less than that of *IU* to *IA*, and consequently less than that of weight *A* (which is *NE*) to weight *U*. Thus, if from weight *NE* is taken a part of it, *E*, such that the remainder *N* has the same ratio to weight *O* as *IO* has to *IA*, the balance will then not move. But if part *E* is replaced, we may obviously conclude that *A* will fall, and thus *O* will rise. And you can prove the converse by yourself by similar reasoning.

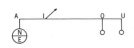

[There follow other discussions on mathematical details in Archimedes.]

Guido Ubaldo
Marquis del Monte

Mechanicorum liber

Pesaro, 1577

With the commentaries of
Filippo Pigafetta
from his Italian translation

Le Mechaniche

Venice, 1581

Abridged translation from the Italian
with notes by
Stillman Drake

Preface of Guido Ubaldo Marquis del Monte

To Francisco Maria II Illustrious Duke of Urbino

There are two qualities, Illustrious Prince, that are usually very effective in adding to men's power, namely, utility and nobility. It seems to me that these join in making the subject of mechanics attractive and in rendering it desirable in comparison with all others. For if we measure the nobility of something by its origin (as most people now do), the origin of mechanics is, on one side, geometry and, on the other side, physics. From the union of these two comes the most noble of the arts, mechanics. For if we hold that nobility is related both to the underlying subject matter and to the logical necessity of the arguments (as Aristotle on occasion asserts), we shall doubtless consider [mechanics][1] the noblest of all. It not only crowns and perfects geometry (as Pappus attests) but also holds control of the realm of nature. For whatever helps manual workers, builders, carriers, farmers, sailors, and many others (in [apparent] opposition to the laws of nature) — all this is the province of mechanics. And mechanics, since it operates against nature or rather in rivalry with the laws of nature, surely deserves our highest admiration.

Now it is certainly true, and freely admitted by anyone who learned it previously from Aristotle, that all mechanical problems and all mechanical theorems are reducible to the wheel and depend therefore on its principle, which is apprehended no less by the senses than by reason. The wheel is the device that is best adapted to movement, and the more so the larger it is.

In addition to this nobility, [mechanics possesses] the highest utility in matters pertaining to life. And this utility exceeds

1. Words in square brackets are interpolations by the translators, Italian or English. The author's preface, omitted by Pigafetta, was translated by I. E. Drabkin. Like Pigafetta's dedication and preface, it was printed without pagination. The Italian edition of 1581 was used for pagination of the text as shown in the running heads.

all other forms of utility obtained from the various arts for the reason that other branches unfolded their usefulness after a long interval of time had elapsed from the birth of the world; but mechanical usefulness was so necessary to men even from the very beginning of the world that, had it been removed, the sun itself would have been removed from the world. For under whatever constraints the life of Adam was passed, though he may have had to ward off the onslaughts from the sky in huts thatched with straw and in narrow hovels and cottages, and by clothing his body, and though he himself may have been concerned only with warding off the rain, snow, wind, sun, and cold — whatever was the case, the situation was all mechanical.

And the case with this subject [mechanics] is not the same [historically] as that of winds which are strongest in the place where they arise, but arrive at a distant point broken and weakened. But rather the [historical] situation is the same as generally occurs with great rivers, which, while they are small where they rise, are continually increased and have a broader bed the farther they are from their source. Thus with the passing of time mechanical power began to distribute equally the labor of plowing by the use of the yoke and to have the plow drawn around the field.

And then mechanical power showed men how, with teams of two and four animals, to move produce, merchandise, and all kinds of cargo — to export from our country to neighboring lands and, again, to import from those lands to our own. Moreover, when things came to be measured not merely by necessity but by their beauty and usefulness, it was thanks to mechanical ingenuity that we could move ships by oars; that with a small rudder at the extremity of the stern we could steer huge triremes; that often with the hands of one individual, rather than with the hands of many workmen, we could now lift heavy stones and beams for our construction workers and architects; and now with a kind of swing-beam we could draw water from wells for gardeners.

By mechanical power, too, are wines, oils, and unguents pressed out by presses used for liquids and forced to give up to the owner whatever liquid they contain. Hence, too, by means of two forces pulling in opposite directions we have divided stout trees and great masses of marble. Hence also in war, in the building of ramparts, in fighting at close quarters, in attacking and defending places, there are almost infinite uses [of mechanics].

With the help of mechanics, too, those who work with wood, stone and marble, wines, oils and unguents, iron, gold, and other metals, as well as surgeons, barbers, bakers, tailors, and all workers in the useful arts, make many important contributions to human life.

And as for certain recent manipulators of words who deprecate mechanics, let them go and wipe away their shame, if they have any, and stop falsely charging [mechanics with] lack of nobility and lack of usefulness. If they still do not wish to do so, let us leave them, I say, in their ignorance; and let us rather follow Aristotle, the leader of the philosophers, whose burning love for mechanics is sufficiently proved by the acute *Questions of Mechanics* which he gave to posterity. In this achievement he greatly surpassed Plato. For, when Archytas and Eudoxus were keenly exploring the usefulness of mechanics, Plato (as Plutarch tells us) discouraged them from this course, on the ground that they were revealing to the masses and making public the noblest possession of philosophers and betraying, as it were, the secret mysteries of philosophy. But surely, at least in my judgment, such a view [that is, Plato's] is to be completely rejected, unless perhaps we wish to praise the detached contemplation of so noble a subject yet to impugn the fruits, the usefulness, and the goal of the art.

But in comparison with all other mathematicians Archimedes alone is to be praised most eloquently, for God willed that in mechanics he should be a unique ideal which all students of that subject might keep before them as a model for imitation. For he made a model of the universe all enclosed in a quite small and fragile glass sphere, with stars that imitated the actual work of nature and so accurately exhibited the laws of the heavens by their precise motions that the hand that rivaled nature deserved the following encomium: "So does his hand imitate nature that nature herself is thought to have imitated his hand." Archimedes, with the help of a block and tackle, pulled a load of 5000 pecks with one hand. Alone with his machines he pulled a heavily loaded ship onto the shore and then pulled it toward himself as if it were being moved in the sea by oars or sails. And then he pulled it from the shore back into the sea (something that all the [human] strength of Sicily could not have accomplished). His too are those engines of war with which Syracuse was so defended against Marcellus that the operator of those engines was always called a hundred-handed Briareus by the Romans.[2] Finally, relying on this art, he made so bold as to give utterance to a statement in such [apparent] conflict with the laws of nature: "Give me a place

2. Many of the feats of Archimedes are recounted in Plutarch's life of Marcellus.

to stand, and I shall move the earth." And not only do we show
in the present book that this could have been done with a lever,
but, in fact, all antiquity seems to me to have been completely
convinced of this (though possibly this will appear remarkable
to many). For antiquity attributed to Neptune a trident, like
a lever; and by virtue of it he is everywhere called Earth-
shaker by the poets. Indeed, it is with this in mind that our most
celebrated poet introduces Neptune raising the shoals with that
device so that they may be visible to the Trojans, "with his
trident he raises and opens up the vast shoals." [3]

Other mechanicians were Hero, Ctesibius, and Pappus. And
though they did not perhaps reach the pinnacle of mechanics,
as did Archimedes, still they had remarkable understanding of
the subject of mechanics and were all great men. Indeed this
is especially true of Pappus, so that no one could, I believe,
blame me for following him as my leader. I have more readily
done so for the reason that Pappus does not depart even a
nail's breadth from the principles of Archimedes. For I have
always wished in this branch of science to follow in the footsteps of Archimedes. And though his thoughts on the subject
of mechanics have for some years been widely sought by
scholars, still his very learned book *On [Plane] Equilibrium*
is extant; in that book I believe that practically all the teachings of mechanics are gathered together, as in an abundant
store.[4] Surely, if the mathematicians of our time had a better
knowledge of this book, they would have found that many
ideas, which they themselves now declare valid and correct,
are there very acutely and properly shaken and overturned.
But let them see for themselves.

I return to Pappus, who, deeply devoted to a richer application of mathematics and to increasing the profits to be derived
from such application, made a thorough and brilliant investigation of the five primary machines, that is, the lever, pulley,
wheel and axle, wedge, and screw. And he proved that, in the
case of machines, everything that could properly be considered
as sharply defined or definitely established was reducible to
those machines which [potentially] are capable of unlimited
force. I wish that the ravages of time had not caused any loss
in the writings of so great a man. For such a thick mist of ignorance would not have covered almost all the earth, nor would
there have been such ignorance of the subject of mechanics

3. [A misreading of Vergil, *Aeneid*, I, 145; Neptune raised the ships of the
Trojans, not the shoals. — *I. E. D.*]
4. Guido Ubaldo published a paraphrase of this book of Archimedes in 1588.

that men are thought of as leading mathematicians who, by their inept distinctions, remove some difficulties, but not those that are very arduous or obscure.

Thus, there are found some keen mathematicians of our time who assert that mechanics may be considered either mathematically, removed [from physical considerations], or else physically.[5] As if, at any time, mechanics could be considered apart from either geometrical demonstrations or actual motion! Surely when that distinction is made, it seems to me (to deal gently with them) that all they accomplish by putting themselves forth alternately as physicists and as mathematicians is simply that they fall between two stools, as the saying goes. For mechanics can no longer be called mechanics when it is abstracted and separated from machines.

Yet in the midst of that darkness (though there were also some other famous names), Federico Commandino shone like the sun; he, by his many learned studies, not only restored the lost heritage of mathematics, but actually increased and enriched it. For that great man was so endowed with all mathematical talents that Archytas, Eudoxus, Hero, Euclid, Theon, Aristarchus, Diophantus, Theodosius, Ptolemy, Apollonius, Serenus, Pappus, and even Archimedes himself (for his commentaries on Archimedes smell of Archimedes' own lamp) seem to have lived again in him. And, lo, just as he had been suddenly thrust from the darkness and prison of the body (as we believe) into the light and liberty of mathematics, so at the most inopportune time he left mathematics bereft of its fine and noble father and left us so prostrate that we scarcely seem able even by a long discourse to console ourselves for his loss. And yet in his endless concern with the elucidation of other parts of mathematics, he either left mechanics completely untreated or touched on it just casually.[6]

Therefore I began to devote myself more eagerly to this study, and, in making my way through every branch of mathematics, I never lost sight of my course to find whatever could be appropriated and derived from each of these branches, so that I may be better equipped to perfect and embellish mechanics. But now I think that, while I have not completed the treatment of everything that pertains to mechanics, still I have advanced to such a point that I can bring some help to those who learned from Pappus, Vitruvius, and others, what a lever

5. Cf. Tartaglia, *Quesiti et inventioni diverse*, Bk. VII.
6. Commandino's only contribution to mechanics, not counting translations and commentaries, was his *Liber de centro gravitatis* (Bologna, 1565).

is, a pulley, wheel and axle, wedge, and screw, and how they should be arranged so that weights may be moved, and the many properties present in those machines by virtue of the lever, properties connecting force and weight, which they are eager to learn about. For that reason I thought it was the proper time for me to emerge and give some example of the work I did on this subject.

Now, in order that my whole work might be more easily built up from its foundation to its very top, certain properties of the balance had to be treated, particularly the case when one arm of the balance is depressed by a single weight. On this subject it is strange what disastrous errors were made by Jordanus (who has enjoyed the greatest authority among recent writers) and others who proposed to discuss this subject. Surely it was a difficult task that we undertook, one perhaps beyond our powers. Still, aspiring to notable results, as we do, we deserve to have our efforts and industry meet with the everlasting approval and applause of all good people, since we have devoted all our powers to a study so noble, so magnificent, and so praiseworthy.

This study, such as it is, we have decided to dedicate to you, illustrious Prince, and, surely, there are many obvious reasons for this plan and decision of ours. First there are your hereditary services to our family, by which you have laid us under such an obligation that we readily recognize that we ought to be prepared to give up our blood and even our lives in keeping with your worth. In addition, there is the fact, of no small import, that from boyhood you were so inflamed with a passion not only for all studies, but especially for mathematical studies, that you would consider your life bitter and unhappy unless you had embraced them. And then, occupied in the study [of mathematics], you passed the first part of your life in gaining an understanding of the subject, and often raised your voice, as was worthy of a prince, to say that you were especially fond of mathematics for the reason that mathematics in particular can emerge from that domestic and private kind of life into the sun and dust, as they say. And, indeed, in clear proof of these [public] interests would be the ardent desire for military skill that you manifested from early youth; but it would only reveal my limited mind, were I to try to set forth the things that could be expected from you. For when you were quite a young man you went on to the early accomplishment of many outstanding things. Thus, when the foundations had been laid by his Holiness Pope Pius V for the sound union of Christian princes, you, eagerly essaying to overcome the enemies of Christ, won for yourself true and solid glory. And every time there was deliberation on high policy, you uttered sentiments which showed the highest prudence joined with the greatest elevation of spirit. I shall omit your many other outstanding and heroic acts performed in those times, lest I seem to you to be announcing things which are already known to all. Though all these achievements are great and outstanding, men still await from you achievements far greater and more outstanding. Farewell, then, noblest adornment of the world, and, if at any time you have some leisure, do not disdain to examine these products of my study.

Dedicatory Letter of Filippo Pigafetta

To the Illustrious Signor Giulio Savorgnano Count of Belgrade, etc.

My Revered Lord: — Inasmuch as the science of mechanics is highly useful to many and important actions in our lives, there is good reason that philosophers and ancient kings gave it more than a little study and that princes favored excellent engineers and enriched them. Certainly this science is of the highest theoretical value and of subtlest structure, for it deals with that part of philosophy which treats of the elements in general, and of the motion and rest of bodies according to their positions; thus we assign the cause of their natural movements, and thus by machines we force bodies to leave their natural places, carrying them upward and in every direction, contrary to their nature.

Both these goals are carried out by propositions arising from matter itself and artificial structures and instruments. And thus it is necessary to consider this subject in two manners: one that regards theory and the application of reason to things that must be done, making use of arithmetic and geometry, astrology, and natural philosophy; the other that is carried out in practice and requiring activity and manual labor, utilizing architecture, painting, design, the arts of the builder and carpenter and mason and related crafts, and in such a way that these things become intertwined and are a mixture of natural philosophy, mathematics, and the practical arts. So that whoever is instructed by clever men and has learned from childhood the previously mentioned sciences and can also design and work with his hands may become a skilled mechanic, inventor, and maker of marvelous works.

This knowledge includes infinite things of use to men in war and peace; it is useful to cities, farms, and commerce; medicine takes from it the devices for restoring broken and dislocated bones to their places. Thus Oribasius in his book on machines includes many instruments taken from mechanics and converted to medical uses, as the tripaston of Archimedes. The art of navigation receives other aids, such as the rudder, which, placed behind or beside the vessel, moves and guides it easily, though it is very small with respect to the whole vessel; oars, which like levers drive it forward; and masts and sails. Windmills, watermills, mills turned by living beings, wagons, plows, and other farm devices are reducible to mechanics. So are the weighing of things with balances, the drawing of water from wells by pulleys or by cranes, called in Latin *tollenones,* which are like huge balances. The manner of conducting water and raising it from deep valleys to heights is similarly derived. The ancients called those persons mechanics also who produced miraculous effects by means of wind, water, or ropes — such as various sounds, or songs of angels, and even the expression of words as by human voices; and those who made clocks which were run by wheels or by water or which measured time by means of the sun and distinguished the hours. Mechanics are those who make celestial spheres showing the various heavens and the movements of the planets and other heavenly bodies like a miniature universe, by the equal movement in rotation given by water power, as we are told was done by Archimedes of Syracuse, the first master. But the moving of very great weights with small force by means of diverse instruments and devices is the chief function of mechanics; thus do balances, steelyards, levers, pulleys, wedges, mills, gears and smooth wheels, all sorts of screws, mangles, windlasses, augers, and many other things involving these. According to Aristotle, all these things reduce to the lever, the circle, and the round framework, which moves the more rapidly the larger it is.

The art of fortification of palaces and places, and of defending them, [an art] which may be called military architecture, is a mechanical profession, for with bastions and barricades and other defenses a man with few soldiers essays to repel many by means of machines and instruments and to maintain his advantage. Furthermore, the fabrication and operation of warlike instruments is the special province of this science, as, for example, catapults and slings and the like, which hurl fire and stones and masses of iron weighing 250 pounds to a great distance (as much as 300 paces, according to Silius and Vitruvius) with ruinous force; arrows and bolts as large as beams; things that strike with damage nearby, such as rams, crowbars, and sledges; maritime devices such as grappling hooks and rope ladders, bridges, floating towers, and similar ancient devices. These in turn have been replaced by artillery, which, causing frightful damage by means of a small amount of incendiary material, is itself governed by mechanical considerations.

This science, which embraces innumerable other uses, both pleasing and necessary to men, has in other times existed in various states of development in the hands of its practitioners. To begin with, in distant ages past, before the Trojan War, Dedalus of Athens was a great master mechanic; he was first to invent the saw and axe, the plumb line, the auger, the mast, sailyard, and sail, and other things; he designed the intricate Cretan labyrinth, and ultimately he tried to make himself and his son two pairs of wings, by which they might traverse the air like angels, as the poets sing.

One may believe that, in erecting the temple of Solomon, the greatest as to size, mastery of architecture, and ornamentation of any that has ever been made, and in constructing the pyramids and many other edifices of ancient times that have filled the world with amazement, excellent mechanics took part, raising on high immense stones and [doing] other things that such men attempt. Later came Eudoxus and Archytas, both worthy engineers. Of Archytas, we read that he made a wooden dove so masterfully adjusted and inflated that it flew through the air like a real dove. These men in turn were followed by Aristotle the philosopher, who propounded in writing a few very elegant questions of mechanics. Then came Demetrius the king, called the taker or destroyer of cities, because he made machines and devices by which he could descend on them from above and capture them. Perhaps these were similar to the machine called the Trojan Horse, which enabled the Greeks to take Troy, for Pausanias in his *Attica* says that he considers it madness to believe that this was really a horse, and not a machine designed to assault city walls and capture the city. It was this king who first raised mechanics to a degree of honor.

Archimedes, who was the best of all craftsmen up to his time in this profession and who is like a light that has since illuminated the whole world, brought the reputation of mechanics to a peak from the poor and vile art it had been (as Plutarch tells us in his life of Marcellus) and caused it to be numbered with the most noble and prized military arts. For when Marcellus attacked Syracuse by sea and by land with great Roman armies, Archimedes, by various machines and ingenuities, always repelled the forces, to their great shame and damage, as Livy, Plutarch, and others have narrated at length, naming his devices. And when Marcellus sailed near the walls to attack them with rams, good Archimedes with his derricks and iron claws lifted the ships into the air, releasing them to drop into the sea and sink, treating many ships thus, until the navies held back from approaching the walls. Nor did he cease to plague the enemy with this, but, as Galen notes in the third book of his *Temperaments,* and Giovanni Zonara and Tzetzes confirm, citing Diodorus and Dionysus, he built certain large concave mirrors in the right proportion for the distances of those ships from the walls, and, exposing these to the direct rays of the sun, he set the [ships] afire, as if miraculously. And when armies attacked by land, he struck them with other devices, so that neither at sea nor on land could the foe shield himself from the cleverness of that excellent mechanic, new defenses and terrible offenses forever appearing. Pappus of Alexandria cites the fortieth device of Archimedes, to show that his inventions numbered at least forty. Marcellus, seeing that nothing was to be gained by his attempted assaults and that his men were being exposed to danger simply by the existence of that valorous old man, came to share the opinion of his whole army that the defense of Syracuse was governed by divine power. Hence he changed the course of his warfare, turning it into a siege and strictly preventing any foodstuffs from entering the city.

These, then, were the reasons for which mechanics rose to such glory that the Romans later honored it in their armies. Caesar took prisoner the chief of the smiths of Pompeii, called Magio Cremona; Vitruvius was made captain of catapults by Augustus Caesar, which would be equivalent to captain of artillery in our armies. This glory was afterward maintained by many eminent writers and masters of mechanics, such as Ctesibius of Alexandria, Hero of Alexandria, another Hero, Athenaeus, Bion, and Pappus of Alexandria (who cites Carpus of Antioch), and by Heliodorus, Oribasius, and other Greeks who flourished at various periods. These men taught the theory, construction, and use not only of warlike machines but of all the others that belong to mechanics. Among the ancient Latins, Varro wrote of architecture and thus had to mention mechanics; Vitruvius, Vegetius, and some others spoke of the manufacture of military machines and machines to move weights, helping to sustain among men the reputation of mechanics.

But with the fall of the Roman Empire and the appearance

of the barbarians in Italy, Greece, Egypt, and [places] where arts and letters had prevailed, nearly all the sciences declined miserably and were lost. Mechanics in particular was for a long time neglected. In war only slings, ballistae, crossbows, cranes, and a few such instruments were used until artillery, which had little by little fallen into disuse, came back. And as to the part of mechanics which deals with the moving of weights, very little understanding remained. Indeed, it appears that for a certain time the noblest arts and teachings, such as literature, philosophy, medicine, astrology, arithmetic, music, geometry, architecture, sculpture, painting, and above all mechanics, fell into dark shadow and lay entombed, and were later restored to light. As to mechanics, Jordanus Nemorarius, who wrote of the science of weights, began to resuscitate it somewhat, and then Leon Battista Alberti in his architecture; Tartaglia opened the road to many mechanical theorems; Victor Fausto in the Venetian arsenal showed himself a fine mechanic; the Reverend Monsignor Barbaro, Elector of Aquileia, in his commentary on the tenth book of Vitruvius, named the instruments used to move weights; Georgius Agricola in the sixth book of his *Metals* collected many machines for the raising of weights; and some others. Finally the author of the present work, proceeding in a very different manner from the others named, has taught in an admirable order and with true and certain reasoning, best among Latin writers, the whole science of moving weights.

Now just as the modern writers I have mentioned, and above all the author of the present book, have elaborated and elevated mechanics by their words and their books, so your Excellency has celebrated it and praised it in discourses and by means of actual operations, making familiar and domestic various machines, constructed on the most profound theory, and performing experiments in the moving of the greatest weights that can possibly be of use to man. So it may be truly affirmed that you, on the one hand, and the authors of these treatises, on the other, have restored to mechanics its pristine honor, which had been lost to it from ancient times to our own.

It is now about forty years since your Excellency proposed for diversion some forty problems in mechanics to Niccolò Tartaglia, a man much admired at his time in that profession; he delighted to solve subtle questions of mechanics and mathematics. In his dialogues he introduced many great persons as speakers, though sometimes he made them say things of which they were ashamed. Your problems were mostly difficult ones;

some he attempted to solve, but from others he excused himself, saying that each of them would require a volume by itself, as one may read in his books on the new science.[7]

Now it is no wonder that you penetrated so deeply into this matter, could work in mechanics so well, and were a master of the whole art of fortifying places and every other military matter, for you were raised by your distinguished father in the company of men learned in science and other affairs. Among these was Constantine Lascari, a noble and learned Greek, by whom you were instructed successfully in letters, arithmetic, geometry, astrology, and geography and were taught to design and work with your hands in various ways — to ride, to handle arms, to fire the arquebus and artillery pieces, to make gunpowder, and to practice the excellent art of the bombardier — to live soberly, and in work to tolerate heat, cold, and every discomfort. All of these are things that guide the spirit and harden the body to military enterprises.

At the age of sixteen, you were sent with twelve horsemen, mostly Turkish, and with sufficient funds to serve in the entire war in Italy from the capture of Francis I of France to the general peace which ensued in the year 1529. This war involved practically every known military movement, through the large armies which confronted one another, the quality and number of undertakings, and a thousand other important events and stratagems that took place; and above all because in one field or another and at all seasons the foremost soldiers of the world were fighting in great numbers with prudence, astuteness, and bravery, for the honor of conquering and being the victors . . .

The Christian princes having returned in peace, you dedicated yourself to the service of your Serene Lords, where in the most important and greatest charges and in two wars you have added fifty years of splendid service to the two hundred of your Savorgnan predecessors; and during this time you have made some fifty great catapults in different provinces of their states, well thought out and masterfully made, with great economy of public funds.

Returning now to mechanics, I may mention that some years

7. Savorgnano proposed one of the problems discussed in the *Quesiti* (Bk II. Qu. 9), and Tartaglia dedicated to him his second *Ragionamento* published in 1551. In that dedication he said that Savorgnano had sent him 29 questions of great ingenuity, on which a book might be written for each, and that he would attempt to deal only with two. Those two concerned the weights required to submerge a container filled with air; they were relevant to Tartaglia's discussion of the diving bell.

ago I visited your fortress at Osopo. There I was delighted to see your warehouse of arms neatly arranged, a magazine of warlike machines and machines to move weights, of which you have through your industry fabricated perhaps a dozen different sorts, some to drag weights, some to raise great weights with little force. One has but a single toothed wheel, yet it draws up steeply five of your cannons by the strength of Gradasso, your dwarf. Another, with but an ounce of force placed on the handle, sets in motion 114,000 pounds of weight; and since a man usually has 50 pounds of force in his arm, if this were used on the same handle, it is evident that the said machine would have the incredible power of moving more than eight million pounds. These machines can be carried by a mule, and some even by a man; they are necessary for various affairs, especially the handling and transporting of great artillery pieces. . . .

Now in peace it has pleased your Excellency to investigate for your amusement many and various sorts of arrangements for the moving of weights, to utilize these in the construction of stone dams to hold back the force of the Tagliamento, that it might not damage the fields of Osopo, and to make use of them in time of war. Thus did Archimedes, according to Plutarch; for in time of peace, at King Hieron's request, he invented many machines for sport and as a geometrical pastime, and when war came, he was able to use these against the Romans. Various authors attest that he, seated at a certain machine (called tripaston according to Oribasius) which was operated by three ropes, drew from the sea to the land a great ship of the king's. With the power of his left hand he moved by means of this instrument a load of 5000 bushels. Computed at 45 pounds weight per bushel, the total came to 225,000 pounds; and he boasted he could move the earth if he had a place to rest his lever. As to that machine described by Pappus in the eighth book of his *Collections,* which had five wheels on axles and a worm gear with its handle, I am sure that you could design instruments to do as much.[8]

Having seen and tested these various devices at Osopo and having been shown by you for the first time this book, which

___

8. Pappus flourished about the end of the fourth century of our era. He left extensive commentaries on the work of his predecessors, including Hero of Alexandria, who described the machine here referred to. Hero's chief work on mechanics, however, remained unknown until the nineteenth century, when an Arabic translation was found.

you highly commended, I formed the idea that it would be useful to translate it into our native language, so that those might understand it and profit from it who have no knowledge of Latin. The work finished and printed, I send it to your Excellency, for you are master of this subject and encourage literature, which comes to naught unless favored (after God) by great gentlemen. If to some degree I shall by my labors bring something useful to lovers of mechanics, let them know that it is to you they owe this work to a large degree.

From your affectionate servant,
Filippo Pigafetta[9]

Venice, 28th June 1581

9. Pigafetta was the author of several books, including one on the moving of the obelisk at Rome.

Preface of Filippo Pigafetta

To the Reader:—The present book contains six treatises, the first on the balance and steelyard, the second on the lever, the third on pulleys, the fourth on the windlass, the fifth on the wedge, and the last on the screw, all which are mechanical instruments. It is entitled *Mechanics*. But this word "mechanics" is perhaps not understood by everyone in its true sense, and some are even found who consider it an insulting word; for in many parts of Italy a man is called a mechanic in scorn and degradation, and in some places people are offended to be called even "engineer." Hence it will perhaps not be out of place to mention that "mechanic" is a most honorable term according to Plutarch, meaning business pertinent to military affairs, and is appropriate to a man of high position who knows how with his hands and his heart to carry out marvelous works of singular utility and delight to mankind.

To name some among the many philosophers and princes of past centuries, Archytas of Tarentum and Eudoxus the companion of Plato, whom Plutarch mentions in his life of Marcellus, were excellent engineers and mechanics; King Demetrius was a clever inventor of war machines and worked with his hands also; and among the Sicilian Greeks the most famous mechanic and engineer was Archimedes of Syracuse, who was of noble lineage and a relative of King Hieron of Sicily.

Although in the same work Plutarch affirms that Archimedes disparaged mechanics as base and vile and material and did not deign to write of it, and that he employed himself on machines not as a principal work but merely for amusement and as a geometrical game, requested by the king, yet we read in other authors that he wrote a book on the measurement and proportions of every kind of vessel, devising the shape of the great ship of Hieron, in which nothing was lacking. Pappus of Alexandria quotes from Archimedes' book on the balance, which is entirely mechanical; also in the eighth book of his *Mathematical Collections* he shows an instrument for the moving of weights, the fortieth invention of Archimedes, of which he said: "Give me a place to stand, and I shall move the earth." The mechanic Carpus wrote that Archimedes composed a book on the making of spheres, which is a mechanical task. Moreover, this same Archimedes himself more than once cites [mechanics] in his book on the *Quadrature of the Parabola,* with these words: "Since it is demonstrated in the *Mechanics . . .* " referring to some propositions of his book on equiponderance, which is entirely mechanical. Also a part of his book on the *Quadrature of the Parabola* and the second book of his work on *Bodies in Water* are mechanical. From this it is seen that Archimedes not only performed mechanical works, but also wrote many treatises of it. Plutarch admits that Archimedes rose in reputation more from his mechanical undertakings than from any other teaching and, indeed, by means of these gained the fame not of human science but of divine wisdom. Hence one may ask why Plutarch allowed himself to say that Archimedes disparaged mechanics? Surely he would have been wrong to show little esteem for that which gained him much greater fame than any other science he possessed.

Among the Romans, Vitruvius was a good mechanic and served as captain of catapults and other war machines under Octavius Caesar; he wrote a book on architecture and made a fortune from it.

Hence to be a mechanic and an engineer after the example of these great men is not unworthy of a gentleman. Mechanics is a Greek word, meaning a thing made artificially to move, as by a miracle and beyond human power, when great weights are moved with small force; and in general it includes every structure, machine, instrument, windlass, mangle, or masterfully discovered device constructed for such effects, and many others in any science, art, and practice. I mention these concretely to put the matter in a form suited to the taste of most men, leaving accurate definition to a more appropriate time.

It should be added that under this general title the author has contented himself at present to teach (and he is the first Latin writer to do so)[10] by means of easy and plain demonstrations merely the method of understanding and operating the six mechanical instruments, to which all others may be reduced. For these are basic and fundamental, and there may be compounded in various ways combinations of two, three, or more; thus the windlass may be combined with the pulley, the screw with the windlass or the lever, and so on. This may be done at

10. Although this seems an exaggeration, such earlier Latin works as those of Vitruvius and Agricola were of quite a different kind.

will by anyone who can proceed with good judgment in various works, as the author notes at the end of this volume.

Now although the author has reasoned of these machines in good method and admirable order, and nothing inherently obscure must be mastered, yet it requires a man's whole mind, and the demonstrations should be read attentively more than once with concentration.

The Book of Mechanics
of Guido Ubaldo

Composed in Latin in 1577
Translated into Italian in 1581

With Commentaries by Filippo Pigafetta

DEFINITIONS

The *center of gravity* of any body is a certain point within it, from which, if it is imagined to be suspended and carried, it remains stable and maintains the position which it had at the beginning, and is not set to rotating by that motion.

This definition of the center of gravity is taught by Pappus of Alexandria in the eighth book of his *Collections*. But Federico Commandino in his book *On Centers of Gravity of Solid Bodies* explains this center as follows:

The center of gravity of any solid shape is that point within it around which are disposed on all sides parts of equal moments, so that if a plane be passed through this point cutting the said shape, it will always be divided into parts of equal weight.[11]

AXIOMS [*Communes Notiones*]

1. If, from things of equal weight, other things of equal weight be taken, the remainders will be of equal weight.
2. If, to things of equal weight, other things of equal weight be added, the wholes will be of equal weight.
3. Things equal in weight to the same thing are equal in weight to each other.

POSTULATES [*Suppositiones*]

1. Every body has but a single center of gravity.
2. The center of gravity of any body is always in the same place with respect to that body.
3. A heavy body descends according to its center of gravity.[12]

---

11. In the book mentioned in note 6, above, Commandino gave first the definition used by Pappus and then his own alternate definition, adopted by Guido Ubaldo here.

12. That is, the measure of descent is given by the descent of the center of gravity.

On the Balance

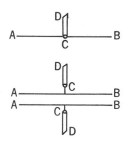

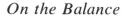

Before a discussion of the [actual] balance, to make matters clear, let the straight line *AB* be a balance, with its support [*trutina*] *CD*, which in accordance with common practice is kept always perpendicular to the horizon. The stationary point *C*, about which the balance turns, is (though improperly) called the center of the balance, even if it is above or below the balance, and *CA* and *CB* [in the first diagram, or their equivalents in the others] are called the arms or distances of the balance. And if the center of the balance is above or below *AB*, let there be drawn from it a line at right angles [to *AB*] which sustains the balance and will be called the *perpendicular*;[13] and however the balance may move, this will always remain perpendicular to it.

LEMMA

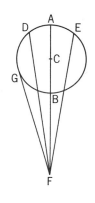

Let the line *AB* be perpendicular to the horizon, and describe the circle *AEBD* having the diameter *AB* and the center *C*. I say that the point *B* is the lowest place on the circumference *AEBD*, and the point *A* is the highest; and that any other points such as *D* and *E*, which are equidistant from *A*, are situated equally below it; and that those points which are closer to *A* are higher than those which are more distant from *A*. [The proof is omitted.[14] *F* represents the center of the world.]

PROPOSITION I

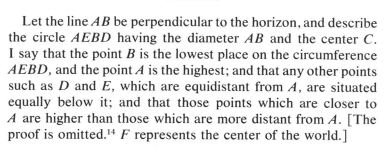

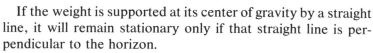

If the weight is supported at its center of gravity by a straight line, it will remain stationary only if that straight line is perpendicular to the horizon.

[The proof, omitted here, rests on Postulate 3.]

From this it may be deduced that a weight supported at any point in any manner will never remain at rest unless the line drawn from the center of gravity to the point of support is perpendicular to the horizon.

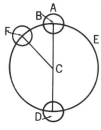

Thus, let the weight be supported by the lines *CG* and *CH*. I say that the line *BC* being perpendicular to the horizon, the weight will remain at rest; but the line *CF* being not perpendicular to the horizon, the point *F* will move downward to *D*,

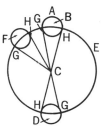

13. When used in this sense hereinafter, the word is italicized.

14. Many proofs of a classical or an obvious character are omitted in the present translation. Proofs included are those typical of the period or illustrative of the particular concerns of Guido Ubaldo.

where it will rest, and the line *CD* will be perpendicular to the horizon. All which may be shown by the foregoing reasons.

PROPOSITION II

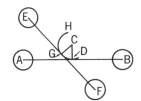

A balance parallel to the horizon, with its center above [and] having equal weights at its extremities which are equidistant from the *perpendicular* [*CD*], when moved from this position and released, will return and rest in it.

[The proof, omitted, is based on Archimedes, *On Plane Equilibrium*, I.4.]

PROPOSITION III

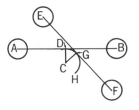

A balance parallel to the horizon, with its center below [and] having equal weights at its extremities which are equidistant from the *perpendicular* [*CD*], will be at rest; but if moved and left tilted, it will move toward the lower side.

[The proof, omitted, is based on Archimedes.]

PROPOSITION IV

A balance parallel to the horizon, having its center within the balance and with equal weights at its extremities, equally distant from the center of the balance, will remain stable in any position to which it is moved.[15]

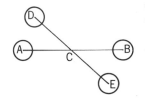

Let the balance be the straight line *AB*, parallel to the horizon, with its center *C* in the line *AB* and the distance *CA* equal to the distance *CB;* and let the weights *A* and *B* be equal, and have their centers of gravity in the points *A* and *B*. Let the balance be moved to *DE* and left there. I say, first, that the balance *DE* will not move and will remain in that position. Now since the weights *A* and *B* are equal, the center of gravity of the combination of the two weights *A* and *B* will be at *C*. Hence the same point *C* will be the center of gravity of the balance and of the whole weight. And since the center of gravity of the balance, *C*, remains motionless while the balance *AB* together with the weights moves to *DE*, the center of gravity is not moved. Therefore the balance *DE*, being hung on this, will not move, by the definition of the center of gravity. The same likewise happens with the balance *AB* parallel to the horizon or in any other position. Hence the balance will remain where it is left; which was to be demonstrated.

---

15. On p. 26v, Pigafetta states that Guido Ubaldo actually succeeded in the construction of such a balance.

Although we have considered in the foregoing only the weights of the bodies which are at the ends of the balance, without that of the balance itself, yet, since the arms of the balance are equal, the balance will behave the same whether we consider its weight together with those of the bodies or without them, for the same center of gravity without weights will be that of the balance alone. Likewise if the weights are attached to the ends of the balance, in the usual manner, it will be the same, provided that the lines drawn from where the weights are attached toward the center of heavy things (the balance being moved in any manner) go to meet in the center of the world, since, when the weights are attached in this manner, they bear down as if they had their centers of gravity in those same points. Whence we may consider the results in just the same way.

But with regard to this last conclusion, many things are said by men who believe otherwise.[16] Hence it will be well to dwell further on this; and according to my ability I shall endeavor to defend not only my own opinion but Archimedes too, who seems to have been of the same opinion.

Things being as before, let there be drawn the line *FG* plumb to *AB* and to the horizon; and with the center *C* at the distance *CA* describe the circle *ADFBEG*. The points *ABDE* will be on the circumference, because the arms of the balance are equal. Now these authors are of the opinion that the balance *DE* does not move to *FG*, nor remain at *DE*, but returns to the line *AB* parallel to the horizon; I shall show that their opinion cannot stand. For if what they say is true, this result will occur because either the weight *D* is heavier than the weight *E* or the weights are equal but the distances at which they are placed are not equal; that is, *CD* does not equal *CE*, but is greater. But it is clear from our assumptions that the weights at *D* and *E* are equal, and that the distance *CD* is equal to *CE*. Now since they say that the weight placed at *D* is heavier in that position than is the weight placed at *E* in its lower position, then, when the weights are at *D* and *E*, the point *C* will no longer be their center of gravity, inasmuch as they would not be stable if suspended from *C*. But that center will be on the line *CD*, by Archimedes, *On Plane Equilibrium*, I.3. It will not be on *CE*, the weight *D* being heavier than the weight *E;* let it therefore be at *H*, from which, if they were suspended, the weights would remain stationary. And since

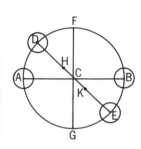

16. A marginal note here names Jordanus, Cardano, and Tartaglia.

the center of gravity of the weights joined by *AB* is at the point
C, but that of those placed at *D* and *E* is the point *H*, when
the weights *A* and *B* are moved to *DE*, the center of gravity *C*
would be moved toward *D* and would approach closer to *D*,
which is impossible. For the weights remain the same distance
apart, and the center of gravity of any body stays always in the
same place with respect to that body. And though the point *C*
is the center of gravity of two bodies *A* and *B*, yet these being
joined together by the balance, so that they are always the
same, the point *C* will be their center of gravity as if they were
a single body, since the balance together with the weights
makes a single continuous body whose center of gravity remains always at the center. Therefore the weight placed at *D*
is not heavier than the weight at *E*. If they should say the
center of gravity was not in the line *CD*, but must be in *CE*,
the same error would follow.

To this error if the weight *D* will move down, the weight *E* will
move up. Therefore a weight heavier than *E*, put in the same
place, will weigh the same as the weight *D*, and it will happen
that unequal heavy bodies, placed at equal distances, will
weigh equally. Add, then, to the weight *E* some heavy thing
so made as to counterpoise *D* if suspended from *C*. But it
having been shown above that point *C* is the center of gravity
of equal weights placed at *D* and *E*, then if the weight *E* shall
be heavier than *D*, the center of gravity will still be in the line
CE; let this center be *K*. But by the definition of the center of
gravity, if the weights were suspended from *K*, they would be
stable. Hence if they were hung from *C*, they would not be
stable, which goes against the assumption, for the weight *E*
will move upward. For if they weighed equally when hung
from *C*, it would come about that there would be two centers
of gravity for one body, which is impossible [by the first
axiom]. Therefore the weight placed at *E*, heavier than that
at *D*, would not weigh the same as *D* being hung from *C*. The
equal weights placed at *D* and *E*, therefore, hung from their
center of gravity, will weigh equally and will stay motionless;
which was to be proved.

To this last contradiction these authors reply that it is impossible to add to *E* so small a weight that, if indeed they [*E*
and *D*] were suspended from *C*, the weight *E* would not continue downward to *G*.[17] But we have assumed this to be possible, and believe it can be done. For, the excess of weight *D*

17. A marginal note here cites Tartaglia [*Quesiti,* Bk.] VIII, 6 [Qu. 33].

over weight *E* having some ratio and quantitative part, we imagined it to be not only minimal but also capable of infinite division. They seek in the following manner to prove that no such weight can be found, since it is not just minimal, but still less.

Things being taken as before, and from the points *D* and *E* the lines *DH* and *EK* being drawn perpendicular to the horizon, let there be taken another equal circle *LDM*, with center *N*, which is tangent to the circle *FDG* at the point *D*. *NC* will be a straight line, and, since the angle *KEC* is equal to the angle *HDN*, and the angle *CEG* is likewise equal to the angle *NDM*, being contained within equal radii and circumferences, the remaining mixed angle *KEG* equals that of *HDM*. Accordingly, they assume that the smaller the angle contained between the vertical line and the circumference, the heavier the weight will be in that position. So that, as the angle *HDG* contained between *HD* and the circumference *DG* is less than the angle *KEG* (that is, than the angle *HDM*) in this same proportion will the weight at *D* be heavier than it would be at *E*. But the ratio of angle *MDH* to *HDG* is smaller than any other ratio that exists between greater and smaller quantities; therefore the proportion of the weights at *D* and *E* will be the smallest of all possible ratios, or, rather, will not be a ratio at all. That the ratio of *MDH* to *HDG* is the least of all, they demonstrate by this necessary reason: that *MDH* exceeds *HDG* by a curvilinear angle *MDG*, which angle is less than any angle made by straight lines; and since no smaller angle can be found than *MDG*, the ratio of *MDH* to *HDG* will be the least of all ratios.[18]

This reasoning seems very frivolous; for, though the angle *MDG* is less than any angle made by straight lines, it does not follow that it is the least of all possible angles; inasmuch as, if we draw the line *DO* from *D* perpendicular to *NC*, it will touch both the circumferences *LDM* and *FDG* at the point *D*. But since the circumferences are equal, the mixed angle *MDO* is equal to the mixed angle *ODG*. Hence one of the angles, that is, *ODG*, will be less than *MDG*; that is, less than the minimum. Therefore the angle *ODH* will be less than *MDH*; whence the ratio of *ODH* to *HDG* will be less than that of *MDH* to *HDG*. So there will be a ratio still less than the

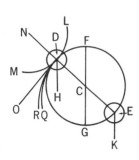

18. Several errors in the designation of angles occur in both Latin and Italian texts. In this translation such errors are emended without note. A few modifications of style have also been made by the translator or the editor in this work.

minimum, which we shall show further to be infinitely less, in this manner. Describe the circle *DR* with its center at *E* and with radius *ED;* the circumference *DR* will touch the circumference *DG* at the point *D* and the line *DO* at the point *D*. Hence the angle *RDG* will be less than the angle *ODG*, and likewise the angle *RDH* less than the angle *ODH* . . . And thus, if infinitely many circumferences are drawn between *DO* and *DG*, we shall find the ratio diminishing *ad infinitum*, and it follows thus that the ratio of the weight placed at *D* to that at *E* is not so small that one infinitely less cannot be found. And since the angle *MDG* can be divided *in infinitum*, so also one may divide *in infinitum* the excess of weight which *D* has over *E*.

Nor should it be omitted that they have assumed in their proof as a thing known that the angle *KEG* is greater than the angle *HDG*, which indeed is true if *DH* and *EK* are parallel. But since, as they likewise assume, the lines *DH* and *EK* meet at the center of the world, they are not ever parallel, and not only will the angle *KEG* not be greater than the angle *HDG*, but it will be smaller.

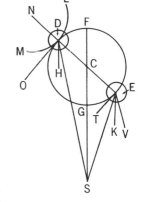

For example, draw the line *FG* to the center of the world *S*, and join *DS* and *ES*. The angle *SEG* is to be demonstrated less than the angle *SDG*. From the point *E* draw the line *ET* tangent to the circle *DGEF*, and from the same point draw *EV* parallel to *DS*. Then since *EV* and *DS* are parallel, *ET* and *DO* are parallel; the angle *VET* will be equal to the angle *SDO*, and the angle *TEG* to *ODM*, being contained between tangents and equal circumferences. Therefore the whole angle *VEG* will be equal to *SDM*. Take away from the angle *SDM* the curvilinear angle *MDG*, and from the angle *VEG* the angle *VES;* and the angle *VES* formed by straight lines is greater than the angle *MDG* formed by curved lines, so the remaining angle *SEG* is less than *SDG*. Hence by their own suppositions not only will the weight placed at *D* fail to be heavier than that at *E*, but on the contrary the weight at *E* will be heavier than that at *D*.

Nevertheless, they adduce reasons by which they attempt to show that the balance *DE* necessarily returns to *AB*, parallel to the horizon. First they show that a given weight is heavier at *A* than at any other place, and this position they call the "level position," [19] the line *AB* being parallel to the horizon. Then the closer the weight is to *A*, the heavier it will be in

19. Or "place of equality"; cf. *Quesiti*, Bk. VIII, Qu. 18.

comparison with any other position; that is, it will be heavier at *A* than at *D*, and at *D* than at *L*, and similarly heavier at *A* than at *N*, and at *N* than at *M*, only one weight on one of the arms, moved up or down, being considered. For they say that, if the support of the balance is on *CF*, the weight placed at *A* is farther from the support than at *D*, and at *D* it is farther than at *L;* for when the lines *DO* and *LP* are drawn perpendicular to *CF*, the line *AC* is longer than *DO*, and *DO* than *LP*, and the same for the points *N* and *M*. They then say that the weight is heavier where it will move more swiftly, and it moves most swiftly of all from *A;* whence it is heaviest at A.[20] Likewise, the closer it is to *A*, the more swiftly it will move; therefore it will be heavier at *D* than at *L*.

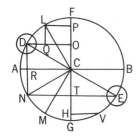

Next, they deduce another cause from the quality of straighter or more bent motion; that is, when the descent of a weight is more nearly straight along equal arcs, the weight appears to be heavier, because when it is free it moves naturally in a straight line. But at *A* it descends most straightly; therefore at *A* it will be heaviest, and they show this by taking the arc *AN* equal to the arc *LD*, drawing from the points *N* and *L* the lines *NR* and *LQ* parallel to the line *FG*, which they call the "line of direction";[21] and these cut the lines *AB* and *DO* at *R* and *Q*. From the point *N* is drawn the line *NT* perpendicular to *FG;* and they truly show *LQ* to be equal to *PO*, and *NR* to *CT*, and the line *NR* to be longer than *LQ*. Now since the descent of the weight from *A* to *N* along the circumference *AN* runs over a greater part of the line *FG* (which they call "partaking more of the straight") than [for example] the descent from *L* to *D* along the circumference *LD*, inasmuch as the descent *AN* runs over the line *CT*, while the descent *LD* covers *PO*, and *CT* is greater than *PO*, the descent *AN* will be "straighter" than the descent *LD*. Therefore the weight placed at *A* will be heavier than at any other place; and in the same way they show that the closer the weight is to *A*, the heavier it is; that is, the arcs *LD* and *DA* being equal, and the line *DR* being drawn from *D* perpendicular to *AB*, *DR* will be equal to *CO;* and thence they show that the line *DR* is greater than *LQ*, and they say that the descent *DA* partakes more of the straight than does *LD*, because the line *CO* is

20. A marginal note here cites Cardano, *De subtilitate*, I; Jordanus, Prop. 4 [corresponding to E. Moody and M. Clagett, eds., *The Medieval Science of Weights* (Madison, 1953): R1.04] and Tartaglia, [*Quesiti*, Bk. VIII], Prop. 5 [Qu. 32].
21. That is, the line of direct descent; cf. *Quesiti*, Bk. VIII, Qu. 19.

longer than *OP*. Hence the weight will be heavier at *D* than at *L*, and likewise at *N* than at *M*. Hence the assumption by which they show the balance *DE* to return to *AB* they declare as a thing known and manifest; that is, that the weight is positionally heavier in proportion as its descent from a given place is less bent, and they say the reason for the return [of the balance to the level position] is this: the descent of the weight placed at *D* is straighter than its descent at *E*, since the weight at *E* partakes less of the line of direction in descending than does the weight *D*. For if the arc *EV* is equal to *DA*, and the lines *VH* and *ET* are drawn perpendicular to *FG*, *DR* will be greater than *TH*. Consequently, by their assumption, the weight at *D* is positionally heavier than at *E*. Hence the weight at *D*, being heavier, will move downward, and the weight at *E* upward, until the balance *DE* returns to *AB*.[22]

Still another reason for this return is that, when the support of the balance *DE* is from above, the line *CG* is the *meta*,[23] and, since the angle *GCD* is greater than the angle *GCE*, and the angle greater than the *meta* renders the weight heavier, then when the support of the balance is from above, the weight at *D* will be heavier than at *E*, and thus *D* will return to *A*, and *E* to *B*.

And by the above arguments they attempt to show that the balance *DE* returns to *AB*, which arguments seem to me to be easily answered as follows.

First, when they say that the weight placed at *A* is heavier than in any other position, they deduce this from its varying distances from the line *FG*, the swiftest and straightest movement being from point *A*. To begin with, they do not truly demonstrate that the weight moves more swiftly from *A* than from any other place, nor does it follow that, since *CA* is greater than *DO*, and *DO* than *LP*, the weight placed at *A* is heavier than that at *D*, and at *D* than at *L*. Now the intellect

---

22. Marginal note: Jordanus, Props. 2 and 4; Tartaglia, Prop. 5 (see note 20, above).

23. Note by Pigafetta: "*Meta* is a Latin word employed by the ancients in games and contests held in walled circuses and theaters. The point from which the runners started moving was called *Carcere* [prison] and the finish, *Meta*, whence *meta* came to mean the finish and the goal; in another sense, it means the lowest place. Now, here the word may be understood in both senses; that is, the line *CG* is the *meta*, being the end and goal at which the weight placed in the balance must arrive, and also the lowest point on the circumference, to which the weight naturally tends. Where the author mentions the 'angle greater than the *meta*,' he means '[greater than] the angle made by the arm of the balance with the *meta CG*.' "

is not satisfied unless this can be demonstrated from some other cause, for this appears to be merely a sign rather than a cause. The same is true of their other argument, adduced from movements being straighter or more bent. Besides, all the things adduced from swifter and slower movement to persuade us that the body at *A* is heavier than that at *D* do not show that the weight at *A*, by its being at *A*, is heavier than the weight at *D*, by its being at *D*, but only by their departing from the points *D* and *A*. So, before going further, I shall first show that the closer a weight gets to the line *FG*, the less it weighs, both as to its position and as to its departure therefrom; and at the same time I shall show it to be false that the weight is heaviest at *A* of all places.

Draw *FG* to the center of the world, *S*, and from *S* draw also a line tangent to the circle *AFBG*. This line cannot be drawn from *S* to touch the circle at *A*, inasmuch as, if the line *AS* were drawn, the triangle *ACS* would have two right angles, that is, *SAC* and *ACS*, which is impossible. Still less can it touch the circle at *A* in the quadrant *AF*, for it would cut the circle. Therefore it will touch below, and let this line be *SO;* then add the lines *SD* and *SL*, which cut the circumference *AOG* at points *K* and *H;* and join also *CK* and *CH*. And thus the closer the weight is to *F*, the higher it stands above the center, as the weight at *D* presses more on and stands higher above the turning point *C* as center; that is, the weight at *D* weighs down more on the line *CD* than it would at *A* on the line *CA*, and still more at *L* on the line *CL*. For, the three angles of each triangle being equal to two right angles, and the angle *DCK* of the isosceles triangle *DCK* being less than the angle *LCH* of the isosceles triangle *LCH*, the angles *CDK* and *CKD* taken together are greater than *CLH* plus *CHL;* and the half of this, that is, *CDS*, will be greater than *CLS*. And *CLS* being lesser, the line *CL* approaches more closely to the natural movement of the weight placed at *L* when entirely free, that is to say, to the line *LS*, than *CD* to the movement *DS*. For the weight placed at *L* would move freely toward the center of the world along *LS*, and the weight at *D* along *DS*. But since the weight at *L* weighs wholly on *LS*, and that at *D* on *DS*, the weight at *L* will weigh more on the line *CL* than that at *D* on *DC*. Therefore the line *CL* will more sustain the weight than the line *CD;* and in the same way, the closer the weight is to *F*, it will be shown for this reason to be more sustained by the line *CL*, since the angle *CLS* is always less, which is obvious. For if the lines *CL* and

LS should come together, which would happen at *FCS*, then the line *CF* would sustain the whole weight that is at *F* and would render it motionless, nor would it have any tendency to descend [*gravezza*] along the whole circumference of the circle. Therefore the same weight, by diversity of position, will be heavier or lighter, and this not because by reason of its place it sometimes truly acquires greater heaviness and sometimes loses it, being always of the same heaviness wherever it is, but because it presses [*grava*] more or less on the circumference, as at *D* it presses more on the circumference *DA* than at *L* on the circumference *LD*. That is, if the weight shall be sustained [jointly] by the circumferences and the straight lines, the circumference *AD* will more sustain the weight placed at *D* than the circumference *DL* sustains the weight placed at *L*, for *CD* helps less than *CL*. Besides this, if the weight at *L* were completely free, it would move down along *LS* were it not prevented by the line *CL*, which forces the weight at *L* to move beyond the line *LS* along the circumference *LD* and in a certain sense pushes it, and, in pushing it, comes partly to sustain it; for, if it did not sustain it and give it resistance, the weight would move down along the line *LS*, rather than along the circumference *LD*. Similarly *CD* offers resistance to the weight placed at *D*, forcing it to move along the circumference *DA*. In the same way, the weight being at *A*, the line *CA* will constrain it to move outside the line *AS* along the circumference *AO*, for the angle *CAS* is acute, *ACS* being a right angle. Therefore the lines *CA* and *CD* to some degree, though not equally, offer resistance to the weight, and whenever the angle at the circumference of the circle made by the line coming from the center of the world *S* and that from the center *C* shall be acute, we shall prove the same thing to occur. Now since the mixed angle[24] *CLD* is equal to the angle *CDA*, being contained by radii and the same circumference, and the angle *CLS* is less than the angle *CDS*, the remainder *SLD* will be greater than the remainder *SDA*. Hence the circumference *DA*, which is the path of descent of the weight at *D*, is closer to the natural movement of the free weight at *D* (that is, the line *DS*) than the circumference *LD* is to the line *LS*. Therefore the line *CD* will offer less resistance to the weight placed at *D* than

24. That is, the angle of which one side is a straight line and the other a curved line. This "horn angle" or "angle of contact" gave rise to many discussions in this period.

the line *CL* to the weight placed at *L*. So the line *CD* will sustain less than *CL*, and the weight will be more free at *D* than at *L*, being moved more naturally along *DA* than along *LD*. Whence it will be heavier at *D* than at *L*. Similarly we shall demonstrate that *CA* sustains less than *CD*, and that the weight at *A* is more free and heavier than at *D*. Next, in the lower part, for the same reasons, the closer the weight is to *G*, the more it will be retained, as at *H* by the line *CH* than at *K* by the line *CK;* for, the angle *CHS* being greater than the angle *CKS*, the lines *CH* and *HS* approach closer to the [line of] direction than *CK* and *KS*, and hence the weight will be more retained by *CH* than by *CK;* for, if *CH* and *HS* meet in a line, as happens when the weight is at *G*, then the line *CG* would sustain the whole weight at *G*, so that it would remain motionless. Therefore the smaller the angle contained between the line *CH* and the line of the weight in free fall (that is, between *CH* and *HS*), the less the line *CH* will retain the weight; and where it is less retained, it will be freer and heavier. Besides which, if the weight were free at *K*, it would move along the line *KS;* but it is impeded by the line *CK*, which forces the weight to move from the line *KS* along the circumference *KH*. This restricts it in a certain way and thus comes to sustain it; for, if it were not sustained, it would move along the straight line *KS* and not along the circumference *KH*. Similarly *CH* retains the weight, constraining it to move along the circumference *HG*. And since the angle *CHS* is greater than the angle *CKS*, if we take away the equal angles *CHG* and *CKH*, the remainder *SHG* will be greater than the remainder *SKH*. Hence the circumference *KH* (that is, the descent of the weight placed at *K*) will be closer to the natural movement of the free weight placed at *K* (that is, to the line *KS*) than the circumference *HG* is to the line *HS*. Hence the line *CK* retains less than *CH*, the weight moving more naturally by *KH* than by *HG*. With similar reasons it will also be shown that the smaller the angle *SKH*, the less the line *CK* will sustain. The weight therefore being at *O*, since the angle *SOC* not only is less than the angle *CKS* but is the least of all angles that come from the points *C* and *S* and have their apex on the circumference *OKG*, the angle *SOK* will be less than the angle *SKH* and less than the others so formed. Hence the descent of the weight placed at *O* will be closer to the natural movement of a free weight at *O*, than if placed at any other position on the circumference *OKG*, and the line *CO* will sustain the weight less than if it were

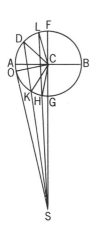

at any other place on the circumference *OG*. Likewise, since the angle of contact *SOK* is less than the [mixed] angle *SDA*, or *SAO*, or any other such angle, the descent of the weight placed at *O* will be closer to the natural movement of this weight than if placed at any other site along the circumference *ODF*. Besides, the line *CO* cannot push the motion of the weight placed at *O* when it moves down so that it will move outside the line *OS*, as the line *OS* does not cut the circle, but touches it, and the angle *SOC* is a right angle and not acute; hence the weight placed at *O* will never weigh against the line *CO*, nor will it bear upon the center, as would happen at any other point above *O*. Hence the weight placed at *O* will for this reason be free, and more completely so at this point than at any other in the circumference *FOG;* and thus it will be heavier and will bear down more here than elsewhere. And the closer it is to *O*, the heavier it will be than anywhere farther away. And the line *CO* will be parallel to the horizon, though not to the horizon of the point *C* (as they believe), but rather to that of the weight placed at *O;* for the horizontal must be taken from the center of gravity of the body. All of which was to be shown.

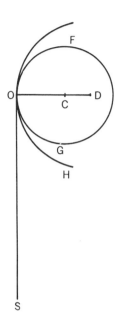

But if the balance arm were greater than *CO*, say, by the amount *CD*, the weight placed at *O* would likewise be heavier. Describe the circle *OH*, with center *D* and radius *DO*. The circle *OH* will touch the circle *FOG* at the point *O* and will also touch the line *OS* at that same point, this being the straight and natural descent of the weight placed at *O*. And since the angle *SOH* is less than the angle *SOG*, the descent of the weight placed at *O* along the circumference *OH* will be closer to its natural movement *OS* than would that along the circumference *OG*. Hence the weight at *O* will be freer, and consequently heavier, than at *C* (the center of the balance being at *D*). Similarly it will be shown that the longer the arm *DO*, the heavier will be the weight placed at *O*.

But if the same circle *AFBG* with its center *R* shall be closer to the center of the world *S*, and if a line *ST* is drawn from the point *S* tangent to the circle, the point *T* (where the weight is heaviest) will be farther from the point *A* than is the point *O*. Draw the lines *OM* and *TN* from the points *O* and *T*, plumb to *CS*, and add *RT*, the center *R* being in the line *CS*, and the line *ARB* being parallel to *ACB*. Then, the triangles *COS* and *RTS* being right triangles, *SC* will be to *CO* as *CO* is to *CM*. Similarly, *SR* is to *RT* as *RT* is to *RN*. Now, *RT* being equal to *CO*, and *SC* greater than *RS*, the ratio of *SC* to *CO* will be

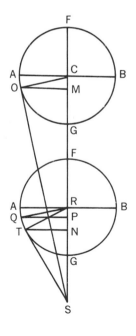

greater than that of *SR* to *RT;* whence, likewise, *CO* has a greater ratio to *CM* than *RT* to *RN*. Thus *CM* will be less than *RN*. Then cut *RN* at *P* so that *RP* shall equal *CM*, and from the point *P* draw the line *PQ* parallel to the lines *MO* and *NT*, [such that] it will cut the circumference *AT* at *Q;* and finally join *R* and *Q*. Now since *CO* and *CM* are equal respectively to *RQ* and *RP*, and the angle *CMO* is equal to the angle *RPQ*, the angle *MCO* is equal to the angle *PRQ*. But the angle *MCA* is equal to the angle *PRA*, both being right angles; hence the remainder *OCA* is equal to the remainder *QRA*, and the circumference *OA* is likewise equal to the circumference *QA*. Thus the point *T*, being farther from the point *A* than is *Q*, will also be farther from the point *A* than is the point *O*. Likewise it may be shown that, the closer the circle is to the center of the world, the farther *T* will be from *A*. Hence, as before, it may be shown that the weight on the circumference *TAF* will stand upon the center *R*, while on the circumference *TG* it will be held by the line, and it will be found heaviest at the point *T*.

And if the point *G* were the center of the world, then the closer the weight was to *G*, the heavier it would be; and hence wherever else the weight is placed than at *G*, it will always get support from the center *C;* for example, at *K*. Draw the line *GK*, along which the natural motion of the weight would be made; this will make an acute angle with the arm of the balance *KC*, because the base angles (at *K* and *G*) of the isosceles triangle *CKG* are always acute.

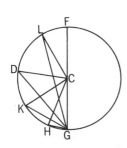

Now if the weight at *K* is compared with that at *D*, the weight at *K* will be heavier than that at *D;* for the line *DG* being drawn, and the three angles of any triangle being equal to two right angles, and the angle *DCG* of the triangle *CDG* being greater than the angle *KCG* of the isosceles triangle *CKG*, the base angles *DGC* and *GDC* taken together will be less than the angles *KGC* and *GKC* taken together; and half the sum, that is, the angle *CDG*, will be less than the angle *CKG*. Now since the weight at *K* would move in natural freedom along *KG*, and the weight at *D* along *DG*—these being the lines by which they are brought to the center of the world—the line *CD*, that is, the balance arm, will approach more nearly to the natural movement of a free weight at *D*, that is, to the line *GD*, than *CK* to the movement made along *KG*. Therefore the line *CD* will offer more support than *CK*, and therefore the weight at *K* will be heavier, by what has been said, than at *D*. Besides which, if the weight placed at

K were entirely free, it would move down along *KG* if it were not impeded by the line *CK* which forces the weight to move beyond the line *KG* along the circumference *KH;* the line *KG* will sustain the weight in part, and will make resistance to it, forcing it to move along the circumference *KH*. And since the angle *CDG* is less than the angle *CKG*, and the angle *CDK* is equal to the angle *CKH*, the remaining angle *GDK* will be greater than the remaining angle *GKH*. Therefore the circumference *KH* will be closer to the natural free movement of the weight placed at *K*, that is, to the line *KG*, than the circumference *DK* to the line *DG*. Whereby the line *CD* makes more resistance to the weight placed at *D* than the line *CK* to the weight placed at *K*. Therefore the weight placed at *K* will be heavier than at *D*. Similarly it would be shown that the closer the weight was to *F* (as at *L*) the less it would weigh, but the closer it is to *G* (as at *H*) the heavier it is.

And if the center of the world were at *S*, between the points *C* and *G*, first it will be shown in the same way that the weight, wherever it is (as at *H*), gets support from the center *C*. For the lines *HG* and *HS* being drawn, the angle at the base *GHC* of the isosceles triangle *CHG* is always acute; whereby also *SHC*, being less than this, is also acute. But drawing from the point *S* the line *SK* plumb to *CS*, I say that the weight is heavier at *K* than at any other place in the circumference *FKG*, and the closer it shall be to *F*, or to *G*, the less it will weigh. Take the points *D* and *L* toward *F*, and join *LC*, *LS*, *DC*, and *DS*, and extend the lines *LS*, *DS*, *KS*, and *HS* to the circumference of the circle at *E*, *M*, *N*, and *O*, and join *CE*, *CM*, *CN*, and *CO*. Now since *LE* and *DM* come together at *S*, the straight angle *LSE* will be equal to the straight angle *DSM*. And as is *LS* to *DS*, so will *SM* be to *SE;* but *LS* is greater than *DS*, and *SM* than *SE*. Therefore *LS* and *SE* taken together will be greater than *DS* and *SM*, and for the same reason *KN* will be shown to be less than *DM*. Moreover, since the straight angle *OSH* is equal to *KSN*, by the same reason *HO* will be greater than *KN*. And in the same way *KN* will be shown to be less than any other line passing through *S*. And since of the two isosceles triangles *CLE* and *DCM* the sides *LC* and *CE* are equal to the sides *DC* and *CM*, and the base *LE* is greater than *DM*, the angle *LCE* will be greater than the angle *DCM*. Whence the base angles *CLE* and *CEL* taken together will be less than the angles *CDM* and *CMD;* half the sum, that is, the angle *CLS*, will be less than the angle *CDS*. Therefore the weight at *L* will weigh more on the line

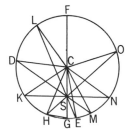

LC than that at *D* will on *DC*, and will be more supported by the center *C* at *L* than at *D*. Similarly it will be shown that the weight at *D* will be more supported by the center *C* than at *K*. Therefore the weight at *K* will be heavier than at *D*, and at *D* than at *L;* and for the same reason, since *KN* is less than *HO*, the angle *CKS* will be greater than the angle *CHS*. Whereby the weight at *H* will be more supported by the center *C* than at *K*, and in this manner it will be shown that, wherever the weight is along the circumference *FDG*, it will be less supported by the center at *K* than if placed at any other point, and the closer it is to *F* or to *G*, the more it will be supported. Then since the angle *CKS* is greater than *CDS*, and *CDK* is equal to *CKH*, the remainder *SKH* will be less than the remainder *SDK;* whereby the circumference *KH* will be closer to the straight natural movement of the free weight at *K*, that is, to the line *KS*, than the circumference *DK* to the movement *DS*. Hence the line *CD* offers more resistance to the weight at *D* than *CK* does to the weight at *K*, and for this reason the [mixed] angle *SHG* will be shown to be greater than *SKH*, and consequently the line *CH* offers more resistance to the weight at *H* than *CK* to the weight at *K*. Similarly it would be shown that the line *CL* sustains the weight more than *CD*, and for the same reasons it will be proved that the weight at *K* will weigh less on the line *CK* than at any other place along the circumference *FDG;* and the closer it is to *F* or to *G*, the less it will weigh. Therefore it will be heavier at *K* than at any other place, and it will be less heavy the closer it is to *F* or to *G*.

Finally if the center *C* is the center of the world, it is manifest that the weight placed anywhere [on the circumference] will remain fixed. Thus [if the weight is] placed at *D*, the line *CD* will sustain the whole weight, being vertical to that weight at *D*. Therefore the weight will remain at rest.

Now in the things demonstrated thus far, we have made no mention of the weight of the arms of the balance. If we next consider the weight of such an arm, we can find the center of gravity of the magnitude made by the weight and the arm; and circumferences can be described according to the distance from the center of the balance to this center of gravity, as if this contained the weight (which indeed it does). And the things we have found without considering the weight of the arm of the balance can be found in just the same way by considering this weight also.

From the things said, if we consider the balance to be removed from the center of the world as these other men have
done (and as it is in fact), then it is clearly false for them to say
that the weight is heavier at *A* than at any other place. And it is
also false that the farther the weight is from the line *FG*, the
heavier it is; for the point *O* is closer to *FG* than the point *A*,
the line drawn plumb from *O* to *FG* being less than *CA*. It is
likewise false that the weight moves more swiftly from the
point *A* than from any other place, for it will move more swiftly
from the point *O* than from *A*, since at *O* it is more free than at
any other place, and its descent from *O* will be closer to its
straight natural movement than any other descent.

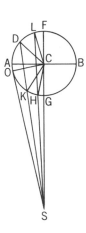

Besides this, when they argue by means of the straighter or
more curved descent that the weight is heavier at *A* than at *D*,
and at *D* than at *L*, they are certainly wrong; for if any weight
were placed at any point on the circumference, as at *D*, its
true descent would be made along the straight line *DR* parallel
to *FG*, according to its natural movement, as was first said.
For if a weight is placed anywhere, and we regard its natural
movement to that proper place to which it moves straight by
nature, taking into account the shape of the whole universe,
then the space through which it moves naturally will always be
along the line drawn from the circumference to the center.
Therefore the natural straight descents of any free weight cannot be made by parallel lines, since all the lines meet in the
center of the world. These men assume next that the motion
of a weight from *D* to *A* along a straight line toward the center
of the world is the same quantity as that from *O* to *C*, as if the
points *A* and *C* were equally distant from the center of the
world, which is likewise false. . . . Thus the assumption from
which they demonstrate that the balance *DE* returns to *AB*
turns out to be false, and all their demonstrations fall. Of
course they might say that, because of our great distance from
the center of the world, these differences are imperceptible,
and by reason of that fact may be assumed void, as all those
who have treated these matters have assumed, especially
since their being imperceptible does not alter the fact that the
descent of the weight from *L* to *D* partakes more (to use their
own phrase) of the straight than the descent *DA*. Likewise the
arc *DA* will partake less of the straight than the circumference
EV, whence [they may say] the supposition will be true, and
the other demonstrations will retain their strength. We may
even concede that the weight will be heavier at *A* than [it

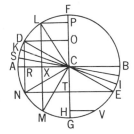

will] anywhere else, and that the straight descent of the weight must be along a straight line parallel to FG, and that any points taken in straight lines parallel to the horizon are equally distant from the center; but it will not follow from this that their demonstration is true when they say that the weight is heavier at A than elsewhere, say, at L. For if it were true that the straighter a weight descended in this sense, the heavier it would be, then it would also follow that where the same weight would descend along equal arcs partaking equally in the straight, it would have equal weights; but this may be shown to be false in the following manner.

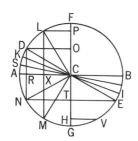

Let there be the equal arcs AL and AM, and join L and M, cutting AB at X; let LM be parallel to FG and perpendicular to AB, and XM will be equal to XL. If therefore the weight shall move from L to A along the circumference LA, its straight movement will be measured by the line LX. But if it moves from A to M along the circumference AM, its straight movement will be measured by the line XM. Hence the descent from L to A will be equal to that from A to M, by reason of the equality of arcs as well as the equal straight lines perpendicular to AB. Therefore the weight at L will weigh the same as at A, which is false; for it is far heavier at A than at L.

And although AM and LA partake equally of the straight according to these men, perhaps they would say that because the first part of the descent from L, say LD, partakes less of the straight than the first part of the descent from A, that is, AN, the weight will be heavier at A than at L. For the arc AN being equal to LD, as was assumed, it partakes (according to them) of the straight CT, but LD partakes of the straight PO; hence the weight will be heavier at A than at L. But if this were true, it would follow that the same weight at the same place, merely considered in a different way with respect to that place, would be heavier and lighter, which is impossible. That is, if we considered the descent of the weight at L with respect to its descent from L to A, it would be heavier than if we considered its descent only to D. Nor can they deny from their own statements that the descent of the weight from L to A partakes of the straight LX, or PC, and that similarly the descent AM partakes of the straight XM, taking these also in this sense, as indeed one must take them; for if they want to show, by comparing the descent of the weight at D with that at E, that the balance DE returns to AB, they must show that the straight descent OC corresponding to the arc DA is greater than the straight descent TH corresponding to the arc EV. For

if they should take but a part of the whole descent from *D* to
A, such as *DK*, and show that the descent *DK* partakes more of
the straight than the equal portion of the descent from the
point *E*, it would follow that the weight at *D*, according to
them, would be heavier than the weight at *E*, and would move
only down to *K*, so that the balance would move to *KI*. Likewise, if they wished to show, by taking a part of the descent
from *K* to *A*, such as *KS*, that the balance *KI* would return to
AB, and [they then] showed that *KS* partook more in the
straight than the equal descent from the point *I*, it would follow
in the same way that the weight would be heavier at *K* than at
I, and would move only to *S*. And, once more, if they showed
that a portion of the descent from *S* to *A* (and so on) were
straighter than the equal descent of the opposed weight, it
would always follow that the balance *SI* would get closer to
AB, but they could never show that it arrived there. Hence if
they want to show that the balance *DE* returns to *AB*, they
must assume that the descent of the weight from *D* to *A* partakes of the straight by the quantity of the line drawn from *D*
to *AB* at right angles; and thus, if we compare the equal descents *DA* and *AN*, which partake of the straight by *OC* and
CT, it will turn out that the same weight weighs equally at *D*
and at *A*, but if we take only the portion *DA*, it will be heavier
at *A* than at *D*. Thus from a mere diversity in manner of consideration, and not from the nature of the thing, it would come
about that the same weight was heavier or lighter. Moreover,
their assumption does not affirm that the positional weight will
be greater when at the same place the commencement of the
descent is less oblique. Hence the postulate [they] adopted
above, that is, that the weight is positionally heavier according
as the descent from the same place is less oblique, is not to be
conceded at all, for the reasons we have given; and not only
that, but it is not difficult to show the exact opposite; that is,
that the less oblique the descent of the same weight along equal
arcs, the less it weighs.

Let there be as before the equal arcs *AL* and *AM*, and the
point *L* close to *F*, and join *L* and *M* perpendicular to *AB*, and
LX will also be equal to *XM*. Then take the point *P* close to
M, between *M* and *G*, and let the arc *PO* be equal to the arc
AM; the point *O* will then be close to *A*. Now draw the lines
CL, *CO*, *CM*, *CP*, and *OP*, and from the point *P* draw *PN*
perpendicular to *OC*. Since the arc *AM* is equal to the arc *OP*,
the angle *ACM* will be equal to the angle *OCP*; but the right
angle *CXM* is equal to the right angle *CNP*; therefore the re-

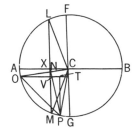

maining angle *XMC* of the triangle *MXC* will be equal to the
remainder *NPC* of the triangle *PCN*. But the side *CM* is
equal to the side *CP;* therefore the triangle *MCX* is equal to
the triangle *PCN* and the side *MX* is equal to the side *NP*, so
the line *PN* will be equal to *LX*. Draw from the point *O* the
line *OT* parallel to *AC*, which shall cut *NP* at *V*, and also from
the point *P* draw a line perpendicular to *OT*. This certainly
cannot fall between *O* and *V*, the angle *ONV* being a right
angle, making *OVN* an acute angle; wherefore *OVP* must be
obtuse. Therefore a line drawn from the point *P* between *O*
and *V* will not be perpendicular to *OT;* otherwise one of two
angles of the triangle would be a right angle and the other obtuse, which is impossible. Hence the perpendicular will fall
on the line *OT* in the segment *VT;* let it be *PT*, whence *PT*
will be the direction of descent for the arc *OP*. Now since the
angle *ONV* is a right angle, the line *OV* will be longer than *ON*,
whence *OT* will likewise be greater than *ON*. Thus the line
OP being drawn under the right angles *ONP* and *OTP*, the
square of *OP* will be equal to the sum of the squares *ON* and
NP, and likewise equal to the sum of the squares of *OT* and
TP. Whence the sum of the squares of *ON* and *NP* will be
equal to the sum of the squares of *OT* and *TP*. But the square
of *OT* is greater than the square of *ON*, the line *OT* being
longer than *ON*. Therefore the square of *NP* will be greater
than the square of *TP* and thus the line *TP* will be less than the
line *PN* and the line *LX*. Therefore the descent along the arc
LA will be less oblique than along the arc *OP*. Hence according to what these men say, the weight at *L* will be heavier than
at *O*, which is obviously false from what we have said above,
inasmuch as the weight placed at *O* is heavier than at *L*. Therefore it is not possible to deduce from the degree of straightness
or bending of the motion (taken in their sense) that the weight
is positionally heavier according as, at a given place, the fall
is less bent. And from this arises most of their error and delusion in this matter. And though at times the truth may accidentally follow from false assumptions, nevertheless it is the
nature of things that from the false the false generally follows,
just as from true things the truth always follows. So it is no
wonder that, when they assume false things as true and use
these as a basis, they deduce and conclude things that are quite
false. These men are, moreover, deceived when they undertake to investigate the balance in a purely mathematical way,
its theory being actually mechanical; nor can they reason successfully without the true movement of the balance and with-

out its weights, these being completely physical things, neg-
lecting which they simply cannot arrive at the true cause of
events that take place with regard to the balance.

Besides this, even if we concede their assumption, they are
far from the true theory of the balance when they argue from
it that the balance *DE* must return to *AB;* for they always take
one weight separately at *D,* or *E,* as if now one and now the
other were placed in the balance, but never both of them together.[25]

Indeed we must do quite the opposite; nor may we consider
directly one weight without the other when we reason about
them as placed in the balance. For when they see that the
descent of the weight placed at *D* is less bent than that of the
weight placed at *E,* the weight at *D* by their assumption must
be heavier than the weight placed at *E;* and by being heavier
(they believe) it necessarily moves downward and the balance
DE returns to *AB.* This argument is of no use whatever. In
the first place they always argue as if the weights at *D* and *E*
must descend, considering the descent of one only without its
being joined with the other. Ultimately, by a comparison of
the descents of the weights [separately], they nevertheless
conclude that the weight placed at *D* moves down, and the
weight at *E* moves up, when the two weights are joined together in the balance. But from the same principles that they
use, and from their demonstrations, one might equally well
deduce the opposite of that which they have labored to defend.
For if they compared the descent of the weight placed at *D*
with the rise of the weight placed at *E,* along the lines *EK* and
DH perpendicular to *AB,* the angle *DCH* being equal to the
angle *ECK* and the right angle *DHC* equal to the right angle
EKC and the side *DC* equal to the side *CE,* the triangle *CDH*
will be equal to the triangle *CEK* and the side *DH* equal to
the side *EK;* and, the angle *DCA* being equal to the angle *ECB,*
the arc *DA* will also be equal to the arc *BE.* Therefore the
weight placed at *D* descends along the arc *DA,* while the
weight placed at *E* rises along the arc *EB* equal to *DA,* and
the descent of the weight placed at *D* will (according to their
practice) share in the straightness of *DH,* and the rise of the
weight *E* will share the straightness of *EK,* equal to *DH.*
Therefore the descent of the weight placed at *D* will be equal
to the rise of the weight placed at *E,* and whatever the inclina-

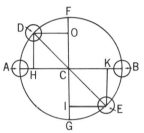

25. The criticism is important. Tartaglia sensed this, as shown by *Quesiti,*
Bk. VIII, Qu. 23, but his argument in the place cited by Guido Ubaldo suffers from the confusion charged.

tion of the one is to downward movement, such will also be the resistance of the other to upward movement. That is, the resistance to force of the weight placed at E in its ascent opposes itself to the natural power of the weight placed at D, because of their equality, so that by however much the weight placed at D goes with its natural power more swiftly downward, by so much the weight placed at E is more slowly forced upward. So that neither of the two will weigh more than the other; there being no action that proceeds from equality, the weight placed at D will not move the weight placed at E upward, because, if it did, it would be necessary that the weight placed at D should have stronger force in descending than should the weight placed at E in rising. But these things are equal; therefore the weights will remain at rest and the weighing down of the weight placed at D will be equal to the weighing down of the weight placed at E.

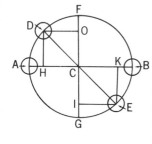

Moreover, they assume that the farther the weight is distant from the line of direction FG, the heavier it is. Now draw from the points D and E the lines DO and EI perpendicular to FG; and, as before, the triangle CDO will be demonstrated to be equal to the triangle CEI, and the line DO to be equal to the line EI. Therefore the weight placed at D is the same distance from the line FG as the weight placed at E; hence from their reasons and their assumptions the weights placed at D and E are equally heavy. And what is there to prevent our demonstrating that the balance DE will necessarily move to FG for a like reason? In the first place it may be deduced from their own demonstrations that the rise of the weight from E toward B is straighter than the rise of the weight D toward F; that is, the rise of the weight placed at D has less straightness than the rise of the weight placed at E, the arcs being equal. Thus it is assumed that the weight is positionally lighter to the extent that in a given place its rise is less straight. This assumption seems as evident as theirs, because the rise of the weight placed at E is straighter than the rise of the weight placed at D, so that by their assumption the weight placed at D will be lighter than the weight placed at E. Hence the weight placed at D will move up with respect to the weight placed at E in such a way that the balance will come to FG, and thus it would be demonstrated that the balance DE will move to FG, which demonstration is completely frivolous and suffers from the same fault. For although it may be conceded as true that the weight placed at E in rising will be heavier than the weight at D similarly rising, it does not follow from this that the weight placed at E in descending is heavier than the weight placed at *D* in rising. Therefore neither of these two demonstrations, which say that the balance *DE* returns to *AB* or that it moves to *FG*, is true.

In addition to this, if we shall examine their assumption, and the force of their argument, we shall certainly see that these have a different meaning. For since the space through which the weight moves naturally must be from the center of gravity of this weight toward the center of the world, along a straight line drawn from the center of gravity to the center of the world, it will be said that a descent of the weight made in this way will be more or less oblique according to the space designated, and that it will move more or less along the said line, always going to seek its natural place by the closest route. Thus the descent is said to be more oblique, the more it departs from that space, and straighter the more it approaches it. Now in this sense the assumption need not give rise to difficulty on the part of anyone, because this is so clear in its truth and its agreement with reason that it does not appear to need to be made evident in any way.

Therefore if the free weight located at *D* must move to its proper place, and if *S* is taken as the center of the world, it will doubtless move along the line *DS;* similarly the free weight placed at *E* will move along the line *ES*. Now if (as is indeed the case) the descent of the weight is to be called more or less oblique according to its departure from or approach to the routes designated by the lines *DS* and *ES,* it is clear that, with regard to their natural movements toward their proper places, the descent of *E* along *EG* is less oblique than that of *D* along *DA,* it having been demonstrated above that the angle *SEG* is less than the angle *SDA*. Whence the weight at *E* will weigh more than at *D,* which is completely contrary to that which they have made such an effort to prove.

Now they may rise up against us, arguing as follows: If the weight placed at *E* is heavier than the weight placed at *D,* the balance *DE* will never remain in that position, as we have undertaken to maintain, but it will move to *FG*. To which we reply that it makes a great deal of difference whether we consider the weights separately, one at a time, or as joined together; for the theory of the weight placed at *E* when it is not connected with another weight placed at *D* is one thing, and it is quite another when the weights are joined in such a way that one cannot move without the other. For the straight and natural descent of the weight placed at *E,* when it is without connection to another weight, is made along the line *ES;* but when

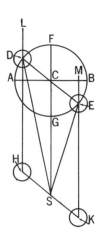

it is joined with the weight *D*, its natural descent will no longer be along the line *ES*, but along a line parallel to *CS*. For the combined magnitude of the weights *E* and *D* and the balance *DE* has its center of gravity at *C*, and, if this were not supported at any place, it would move naturally downward along the straight line drawn from the center of gravity *C* to the center of the world *S* until *C* reached *S*. Therefore the balance *DE* together with its weights will move downward in such a way that the point *C* moves along the line *CS* until *C* arrives at *S* and the balance *DE* at *HK*, the balance *HK* having the same position that it had before; that is, *HK* is parallel to *DE*. Therefore join *D* to *H* and *E* to *K*. It is evident that when the balance *DE* moves to *HK*, the points *D* and *E* will move along the lines *DH* and *EK*, equal and parallel to each other and to *CS*. Hence if we regard the weights placed at *D* and *E* with respect to their conjunction, their natural movement is not along the lines *DS* and *ES* but along *LDH* and *MEK*, parallel to *ES*. But the natural inclination of a free weight at *E* will be along *ES*, and that of a similar free weight at *D* will be along *DS*. And therefore it is not contradictory that the same weight, now at *E* and now at *D*, is heavier at *E* than at *D*. But if the weights at *E* and *D* are joined together and we consider them with respect to their conjunction, the natural inclination of the weight placed at *E* will be along the line *MEK*, because the weighing down of the other weight at *D* has the effect that the weight placed at *E* must weigh down not along the line *ES*, but along *EK*. The same is true of the weight at *E;* that is, the weight at *D* does not weigh down along the straight line *DS*, but along *DH*, both of them being prevented from going to their proper places. Therefore, since the natural straight descent of weights placed at *D* and *E* is along *LDH* and *MEK*, their natural straight rise will likewise be along the same lines *HDL* and *KEM*, and the natural rise of the weight placed at *E* will be more or less bent, according as the route shall be more or less close to the line *MK*. And in exactly this way one must take both the rise and the descent of the weight at *D* according to the line *LH*. Therefore if the weight at *E* moves downward along the line *EG*, the weight at *D* would move upward along *DF*. And since the angle *CEK* is equal to the angle *CDL*, and the angle *CEG* is equal to the angle *CDF*, the remaining angle *GEK* will be equal to the remaining angle *LDF*. Now assuming that the weight is positionally heavier to the degree that its descent from a given place is less oblique, one will also admit without doubt that the weight will be positionally heavier according as its rise at a given point will be less oblique, since this is no less evident or agreeable to reason. Therefore the descent of the weight at *E* will be equal to the rise of the weight at *D*, because the descent of the weight at *E* has as much of the oblique as does the rise of the weight at *D;* and whatever may be the inclination of the one to downward movement, this likewise will be the resistance of the other to upward movement. Hence the weight at *E* will not move the weight at *D* upward, nor will the weight at *D* move downward in such a way as to raise the weight at *E*. For, the angle *CEB* being equal to *CDA* and the angle *CEM* equal to the angle *CDH*, the remainders *MEB* and *HDA* will be equal. Thus the descent of the weight at *D* will be equal to the rise of the weight at *E*, and the weight at *D* will not raise the weight at *E*. From which it follows that the weights at *D* and *E*, considered in conjunction, are equally heavy.

Now the second reason with which they attempt to show that the balance *DE* returns to *AB* is that, when the support of the balance is *CF*, its goal [*meta*] is *CG*, and, since the angle *DCG* is greater than the angle *ECG*, the weight placed at *D* will be heavier than that placed at *E;* therefore the balance *DE* will return to *AB*. In my opinion this does not follow, and this fiction about the support and the goal should just be left out and passed over in silence; for to say anything about it only confuses the issue, the whole thing being arbitrary, since no necessary reason why the weight placed at *D* at the larger angle will be heavier, or why the greater angle is the cause of greater weight, is given anywhere. The angle *GCD* being equal to the angle *FCE*, then if the angle *GCD* is the cause of heaviness, why is not the angle *FCE* similarly such a cause? For this effect they attempt to adduce the following reasoning: Since *CG* is the goal and *CF* the support, if (they say) *CG* were the support and *CF* the goal, then the angle *FCE* would be the cause of heaviness, and not *DCG* which is equal to it. This reasoning is sheer imagination and quite arbitrary. For what can it matter whether the support is *CF* or *CG*, when the balance *DE* is always sustained at the same point *C?* But let us make their delusion still more obvious.

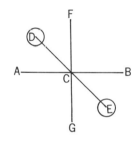

Let there be the balance *AB* with center *C* and support *FG* which remains motionless and sustains the balance *AB* at the point *C*. Now let the balance move to *DE;* and since the support is both above and below the balance, what angle will be the cause of heaviness, the balance *DE* being sustained always at the same point? Perhaps they will say that if the support is

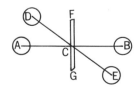

sustained by a force at *F*, then *CG* will be the goal, and the angle *DCG* will be the cause of heaviness, but if sustained at *G*, then *FCE* will be the cause of heaviness and *CF* will be the goal. For this it does not appear possible to adduce any but imaginary reasons, because the goal (as they call it) does not appear to have any kind of power that acts sometimes on the side of the larger angle and sometimes on the side of the smaller. But suppose the support to be sustained by two powers at *F* and *G*, as it might be through necessity if the power placed at *F* were so weak that by itself it could sustain only half the weight, and, the power placed at *G* being equal to that at *F*, both together could sustain the balance with the weights. Now what angle will be the cause of heaviness? Not *FCE*, because the support is at *CF* and is sustained at *F;* nor *DCG*, the support being at *CG* and similarly sustained at *G*. Therefore the angles will not be the cause of heaviness. Thus the balance *DE* will not be moved from that position for any such cause. Yet they think this opinion is confirmed in two ways. First, they say that Aristotle in his *Questions of Mechanics* posed these two questions only, and his proofs were based on the greater and lesser angle and on the position of the support of the balance. Next, they declare that experience also bears them out; that is, that the balance *DE* with its support at *CF* returns to *AB* parallel to the horizon, but when the support is at *CG* the balance moves to *FG*. But neither Aristotle nor experience favors this opinion of theirs, and indeed quite the contrary is true. They are deceived with regard to experience, since it is clear from experience that this happens when the center of the balance is above or below the balance, rather than when the support is above or below.

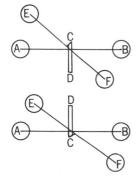

If the balance *AB* has its center *C* above and its support *CD* below the balance, then when the balance is moved to *EF* it will return to *AB* parallel to the horizon. Likewise if the balance has its center *C* under the balance while its support *CD* is above, and the balance is moved to *EF*, it is manifest that the balance will move down on the side of *F*, the support being above the balance. And in any other position of the support, it will always come out the same. Therefore it is not the support but the center of the balance that is the cause of these effects.

However, it is to be noticed here that it is difficult in fact to make an actual balance (such as we can imagine mentally) that is supported only at one point and has its arms so exactly equidistant from the center, not only as to length but as to breadth and thickness, that all its parts on both sides weigh precisely the same, because matter does not lend itself easily to such exact measurement. Hence if we consider the center to be in the balance, we need not have recourse to the senses, for artificial devices cannot be brought to such a degree of perfection. But in the other things experience may directly teach the appearances. For, although the center of the balance is always a single point, nevertheless, when it is on top of the balance, it does not matter much if the balance is not sustained precisely at that point; because so long as the center remains above, the balance will always behave the same.[26] For a like reason, that which occurs when the center is in the balance never happens when it is below the balance, because there will be a difference if it is not sustained always exactly in that center, and it is a very easy thing for that center to change its position when the balance is moved.

Now, it is certainly true that Aristotle did pose two questions only; that is, why, when the support is above, and the balance is not parallel to the horizon in equilibrium, it returns; whereas, if the support is below, it does not return, and moves farther in the direction of the lower side. But his proofs are not based on the larger or smaller angle and the position of the support, as they pretend, because in this they do not understand the philosopher's meaning when he examines the reason for various effects. And Aristotle is far from attributing these different effects to the angles; rather he says that the cause is the excess, and that [when the support is] above, more of the distance from the perpendicular along one arm of the balance is now on one side and now on the other.[27]

Now with the support above at CF, the perpendicular will be FCG, which, according to Aristotle, always points toward the center of the world and which unequally divides the [actual] balance when it is moved to DE, the larger part being on the side of D, which tends downward. Therefore on the side of D the balance will move downward until it returns to AB. But if the support is below at CG, GCF will be the perpendicular, which will likewise divide the balance DE unequally, but the larger part will be on the side of E, so that

26. The author means that the precise position of the support in this case is of no great consequence, as it is when the balance is supported from below.

27. Cf. *Aristotle; Minor Works (Problems of Mechanics)*, trans. W. S. Hett (Harvard, 1936), pp. 347–51 (cited hereafter as *Aristotle*). The second problem deals with the balancing of a beam (not a weightless arm) from above and below. Accordingly the ensuing diagram is somewhat misleading for the present purpose; see next diagram and corresponding text.

the balance will move downward on the side of *E*. Once this is fully understood, one will see that, when the support is above the balance, one must also grant that the center of the balance is on top of the balance, and if below, then the center must be below, as will be manifest farther on. Otherwise Aristotle's demonstration will not prove anything, because, if the center were in the balance, as at *C*, then the balance might be moved in any way and the perpendicular *FG* would divide the balance only at the point *C* and in equal parts. Wherefore the opinion of Aristotle not only does not help them but goes strongly against them. This is clear not only from the second and third propositions of the present book, but also because, if the center is above the balance, the higher weight acquires a greater positional heaviness, considering the return of the balance to the position parallel to the horizon. The contrary happens when the center is below the balance. These things are demonstrated in the following manner, what has been said above being assumed: that is, that the weight will be heavier in that place from which its descent is straighter, and is likewise heavier at the place from which its rise would be straighter.

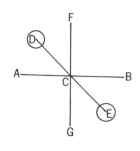

Let the balance *AB* be parallel to the horizon, with its center *C* above the balance, and let fall the perpendicular *CD*. Let the centers of gravity of two equal weights be placed at *A* and *B*. Now move the balance to *EF*. I say that the weight placed at *E* has greater heaviness than the weight placed at *F*, and therefore the balance *EF* will return to *AB*. First extend the line *CD* to the center of the world, and let this be *S*. Then draw the lines *AC*, *CB*, *EC*, *CF*, and *HS*, and from the points *E* and *F* draw the lines *EK* and *FL* parallel to *HS*. Now since the natural descent of the whole system (that is, of the balance *EF* in this position together with its weights) is that of its center *H* along the straight line *HS*, then likewise the descent of the weights placed at *E* and *F* in this position is along the straight lines *EK* and *FL* parallel to *HS*, as we have demonstrated above. Therefore the descent or rise of the weights placed at *E* and *F* may be said to be more or less oblique according to the approach or departure [from the vertical] dictated by the lines *EK* and *FL*. And since the two sides *AD* and *DC* are equal to *BD* and *DC*, and the angles at *D* are right angles, the side *AC* will be equal to the side *CB*. Now the point *C* being fixed, when the points *A* and *B* move, they will describe the circumference of a circle whose radius is *AC*. Hence let the arcs *AE* and *EF* be described with the center *C*,

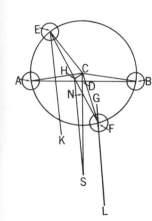

and the points *A, B, E,* and *F* will be on the circumference of the circle. But since *EF* is equal to *AB*, the arc *EAF* will be equal to the arc *AFB*. The common part *AF* being subtracted, the arc *EA* will be equal to the arc *FB*. Now since the mixed angle *CEA* is equal to the mixed angle *CFB*, and *HFB* is greater than *CFB*, and the angle *HEA* is less than *CEA*, the angle *HFB* will be greater than the angle *HEA*. When the equal angles *HFG* and *HEK* are subtracted, the angle *GFB* will be greater than the angle *KEA;* therefore the descent of the weight at *E* will be less oblique than the rise of the weight at *F*. And although the descending weight at *E* and the rising weight at *F* move through equal arcs, since the descent of the weight from *E* is straighter than the ascent of the weight from *F*, the natural power of the weight at *E* will overcome the resistance to force of the weight *F*. Hence the weight at *E* will have greater heaviness than the weight at *F*, and the weight at *E* will move downward and the weight at *F* upward until the balance *EF* returns to *AB;* which was to be proved.

The reason given by Aristotle for this effect may here be plainly seen. For take the point *N* where the lines *CS* and *EF* intersect. Since *HE* is equal to *HF*, *NE* will be greater than *NF* and the line *CS* (which he calls the perpendicular) will divide the balance *EF* unequally. So that the part *NE* of the balance is greater than *NF*. And since this must be carried downward, the balance *EF* will move down on the side of *E* and return to *AB*.

In addition to the things that have been said up to this point, it may be stated that the balance in the position *EF* will move most swiftly to *AB* when the line *EF*, if extended straight, would pass through the center of the world. Let this line be *EFS*. Since *CD* and *CH* are equal, and the circle *DHM* is described with the center *C* and radius *CD*, the points *D* and *H* will be on the circumference of the circle. And since *CH* is perpendicular to *EF*, *EHS* will be tangent to the circle *DHM* at the point *H*. Then the weight at *H* (as we have proved above) will be heavier than in any other position on the circle *DHM*. Therefore the system consisting of the weights *E* and *F* together with the balance *EF*, whose center of gravity is at *H*, will weigh more in this than in any other position of *H* on its circle. From this position, therefore, it will move more swiftly than from any other. And if the point *H* is closer to *D*, it will weigh less and will move less swiftly from that place, for its descent is always more bent and less straight. Therefore the balance *EF* will move most swiftly from this place, and the

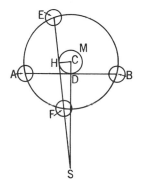

closer it approaches *AB* the less swiftly it will move. Also the more distant the point *H* is from the point *C*, the more swiftly the balance will move, which is manifest not only from what Aristotle says at the beginning of his *Questions of Mechanics* and from what has been said above, but also from the things that we are going to say in the sixth proposition below. Hence the balance *EF* will move the more swiftly, the farther it is from its center.

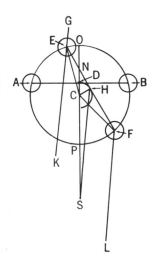

Let there be the balance *AB* with its center below and let there be equal weights at *A* and *B;* and move the balance to *EF*. I say that the weight at *F* has more heaviness than that at *E*, and therefore the balance *EF* will move downward on the side *F*. Extend the line *DC* on one side to the center of the world *S* and on the other side to *O*, and draw the line *HS*. From the points *E* and *F* draw the lines *GEK* and *FL* parallel to each other, and draw *CE* and *CF*. Then from the center *C* with a radius *CE* draw the circle *AEOBF*. The points *A, B, E*, and *F* are on the circumference of the circle, and it can be shown that the descent of the balance *EF* together with its weights would be made along the line *HS*, the descent of the weights at *E* and *F* being along the lines *GK* and *FL* parallel to *HS*. And since the angle *CFP* is equal to angle *CEO*, the angle *HFP* will be greater than the angle *HEO*. But the angle *HFL* is equal to *HEG*. Hence if we subtract the equal angles *HFP* and *HEO*, the angle *LFP* will be less than the angle *GEO*, and the descent of the weight at *F* will be straighter than the rise of the weight at *E*. Therefore the natural power of the weight at *F* will overcome the resistance to force of the weight at *E*, and thus the weight at *F* will have greater heaviness than that at *E*. Hence the weight at *F* will move down and the weight at *E* will move up.

Aristotle's reasoning is equally clear here. For let the point *N* be the intersection of the lines *CO* and *EF; NF* will be greater than *NE*, and since the perpendicular *CO*, according to him, divides the balance unequally with the larger part toward *F* (that is, *NF*) the balance *EF* will move downward on the side *F* since the greater is carried downward.

Similarly from what has been said we deduce that, the farther the balance *EF* (having its center beneath) is from the position *AB*, the swifter it will move, because, the farther the center of gravity *H* is from the point *D*, the faster will be the motion of the system composed of the weights *E* and *F* and the balance *EF*, until the angle *CHS* becomes a right angle.

And it will also move more swiftly the farther the balance is from the center C.

Besides, we may use their logic and their false assumptions to produce the effects and motions of the balance already explained, so that from this one may see the power of truth and how it forces itself to shine forth even from false things.

Assuming the same things, that is, the circle $AEBF$ and the balance AB whose center C is above the balance, move the balance to EF; I say that the weight at E has greater heaviness than the weight at F, and that the balance EF will return to AB. Draw from the points E and F the lines EL and FM perpendicular to AB, which shall be parallel, and let the point N be the intersection of AB and EF. Then, since the angle FNM is equal to the angle ENL and the angles FMN and ELN are both right angles, and the remaining angles NFM and NEL are also equal, the triangle NLE will be similar to the triangle NMF. Then as NE is to EL, so NF is to FM, and as EN is to NF, EL is to FM. But HE being equal to HF, EN will be greater than NF, and EL greater than FM. Now while the weight placed at E descends along the arc EA, the weight at F rises along the arc FB equal to EA, and the descent of the weight at E partakes (as they say) of the straightness EL, while the rise of the weight at F partakes of the straightness FM, so that the rise of the weight at F partakes less of the straight than the descent of the weight at E. Therefore the weight at E will have greater heaviness than will the weight at F.

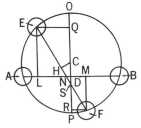

Extend the line CD to OP, which shall cut the line EF at the point S. Now since (they say) the farther the weight is from the line of direction OP, the heavier it is, then by this means also they would prove the weight at E to have greater heaviness than the weight at F. Draw from E and F the lines EQ and FR perpendicular to OP. By a like argument it would be shown that the triangle QES is similar to the triangle RFS and that the line EQ is greater than RF, and thus the weight at E will be farther from the line OP than will the weight at F, whence the weight at E will have greater heaviness than the weight at F. From this it appears evident that the balance will return from EF to AB.

But if the center of the balance is below the balance, then it will be shown by the same argument that the lower weight should have greater heaviness than the raised weight. Draw from the points E and F the lines EL and FM perpendicular

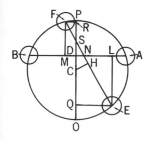

to *AB*. As before, it is proved that *EL* is greater than *FM*, and therefore the descent of the weight at *F* will partake less of straightness than the rise of the weight at *E;* hence the resistance to force of the weight at *E* will overcome the natural inclination of the weight at *F*, and therefore the weight at *E* will be heavier than the weight at *F*.

Extend also *CD* to *O* and *P* and draw from the points *E* and *F* the lines *EQ* and *FR* perpendicular to that. It will be proved in the same way that the line *EQ* is greater than *FR*, and, since the weight placed at *E* will be farther from the line of direction *OP* than the weight at *F*, the weight at *E* will have greater heaviness than the weight at *F*. From this it follows that the balance *EF* moves downward on the side of *E*.

Thus Aristotle poses only two questions and leaves out the third; that is, the case in which the center of the balance is in the balance itself. But he left this out as a thing well known, as he usually did omit obvious things. Who can doubt that, if the weight is sustained at its center of gravity, it will remain at rest? But perhaps someone will take issue with the things that we have put forth in accordance with his opinion and will say that we have not brought in his entire thought. For in the second part of the second question he asks, "Why, when the support is below, the balance being carried downward and released, it does not rise again, but remains?" Here he affirms not that the balance moves downward, but that it remains, which he seems to have deduced in the last conclusion. But not only does this not bear against us; when properly understood, it greatly assists us.

For let there be the balance *AB,* parallel to the horizon, with its center *E* under the balance. And since Aristotle considers an actual balance, it is necessary to place the support or something else under the center *E;* let this be *EF*, and this will be the support that sustains the center *E*. Let *ECD* be the perpendicular. In order that the balance *AB* may move from this position, says Aristotle, let there be a weight at *B* which will move the balance downward on the side of *B*, say, to *G*, where the obstacle prevents its moving farther downward. Now Aristotle does not say that the balance moves down on the side of *B* as far as it can, and is left there, as we say; but he would have it that a weight is placed at *B*, which by its nature will always move downward until the balance rests against its support or something else. When *B* is at *G*, the balance will be at *GH*, in which position it will remain if the weight is taken away;

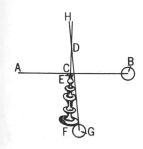

for the side of the balance from the perpendicular toward *G* (that is, *DG*) is longer than *DH;* but the balance does not move farther downward because it will be on the support, or whatever else sustains the center of the balance. And if this did not support it, in his opinion the balance would move downward on the side of *G*, and *DG*, being greater, would necessarily be carried farther downward.

But someone might add to this that, if a very small weight were placed at *B*, it would indeed move the balance downward but not all the way to *G*, and in this position, according to Aristotle, it should remain if the weight were taken away. This is evident by experience, since the balance tilts more or less when at one end of the balance only there is placed a larger or smaller weight, and this is true enough so long as the center is placed above the balance, but not when it is below or in the balance, as we shall show by way of example.

Let there be the balance *AB*, parallel to the horizon, whose center *C* is above the balance; let the *perpendicular CD* be plumb to the horizon, and let this line be extended through *D* to *H*. Now, since we consider the weight of the balance, the point *D* will be the center of gravity of the balance. But if a small weight is placed with its center of gravity at the point *B*, the center of gravity *D* of the whole system composed of the balance *AB* and the weight placed at *B* will no longer be at *D*, but it will be in the line *DB*. Say it is at *E*, so that *DE* is to *EB* as the weight placed at *B* is to the weight of the balance *AB*. Now join *C* and *E;* and since the point *C* is fixed, when the balance moves, the point *E* will describe the circumference of the circle *EFG* with radius *CE* and center *C*. But since *CD* is plumb to the horizon, the line *CE* will not be so. Hence the weight composed of *AB* and the weight at *B* will not remain in this position but it will move downward along the circumference *EFG* according to its center of weight *E*, until *CE* becomes plumb to the horizon, that is, until *CE* gets to *CDF*. The balance *AB* will then be moved to *KL*, in which position the balance together with the weight will remain, nor will it move farther downward. If a heavier weight were placed at *B*, the center of gravity of the whole system will be closer to *B*, say at *M*, and then the balance will move downward until the line joining *C* and *M* comes to the line *CDH*. Hence when a greater or lesser weight is put at *B*, the balance will be tilted more or less. From this it follows that the weight *B* will always describe an angle less than a quadrant, since the angle *FCE* is

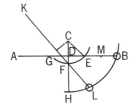

always acute, nor will the point *B* ever go all the way to the
line *CH*, because the center of gravity of the weight and the
balance together will always be between *B* and *D*. The heavier
the weight placed at *B*, the larger will be the arc described,
beginning at *E* and approaching closer to the line *CH*.

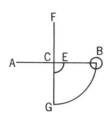

But if the balance *AB* has its center *C* in the balance, *C* will
also be the center of gravity of the balance; let the line *FCG* be
drawn perpendicular to *AB* and to the horizon. Then put any
weight you please at *B*, and let the center of gravity now be at
E, so that *CE* is to *EB* as the weight placed at *B* is to the weight
of the balance. And since *CE* is not perpendicular to the
horizon, the balance *AB* and the weight at *B* will not remain
in this position but will move downward on the side of *B* until
CE becomes perpendicular to the horizon; that is, until the
balance *AB* comes to *FG*. Whence it is clear that the weight
placed at *B* always describes a full quadrant.

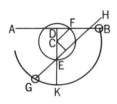

But if the center *C* is under the balance *AB*, and *DCE* is the
perpendicular, the placing of a weight at *B* will similarly make
the center of gravity of the system composed of the balance
AB and the weight at *B* be in the line *DB*, say, at *F*, in such
a way that, as *DF* is to *FB*, so is the weight placed at *B* to the
weight of the balance. Draw *CF*, and, since *CD* is perpendicular
to the horizon, the line *CF* will not be; hence the system composed of the balance *AB* and the weight at *B* will never remain
fixed in this position but will move downward if nothing impedes it, until *CF* comes to *DCE*, in which position the balance together with the weight will come to rest. Now the point
B will be at *G* and the point *A* at *H*, so the balance *GH* will
no longer have its center below, but above. And this will always happen, however small the weight placed at *B*. Therefore before *B* goes to *G*, it will necessarily happen that the
balance will strike against the support placed below, or some
other thing that sustains the center *C*, and will stop there.
From this it follows that the weight *B* always moves beyond
the line *DK* and always describes an arc greater than a quadrant, the angle *FCE* being always obtuse and the angle *DCF*
always acute. And no matter how light the weight placed at
B, it will always describe a larger arc [than a quadrant].
For the lighter the weight at *G*, the more the weight at *G* will
rise and the more closely the balance *GH* will approach a
horizontal position. All these things are obvious from what has
been said before.

These things proved, it is clear that the center of the balance is the cause of the various acts of the balance, and it is also seen that all the propositions of Archimedes in his book *On Plane Equilibrium* are true in every position, whether the balance is horizontal or not, provided only that the center of the balance is located within it, and this is the way he considers it. And even if the balance has unequal arms, the same will always happen, and it will be proved in exactly the same way that the center of the balance being situated in different manners will produce different effects.

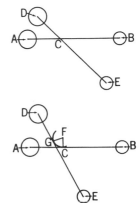

Let there be the balance *AB* parallel to the horizon, and let there be unequal weights at *A* and *B*, so that the center of gravity is at *C*, and let the balance be hung from this point *C*. Now move the balance to *DE*, and it is evident that the balance will rest not only at *DE* but at any other point.

But now let the center of the balance *AB* be above *C* at *F*, and let *FC* be perpendicular to *AB* and to the horizon. If the balance shall be moved to *DE*, the line *CF* will move to *FG*, and since this is not perpendicular to the horizon, the balance *DE* will move down on the side of *D* until *FG* returns to *FC*, when the balance *DE* will be at *AB*, in which position it will rest.

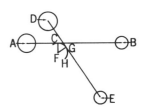

But if the center *F* of the balance shall be below the balance, and the balance is moved to *DE*, it is evident in the first place that the balance will rest at *AB* and that at *DE* it would move downward on the side of *E*, since the line *FG* is not perpendicular to the horizon.

From these things just finished, if the balance were curved or if the arms of the balance formed an angle, and the center were variously placed — although strictly speaking this would not be a balance — we might nevertheless demonstrate various effects in it also. Thus let the balance be *ACB*, which turns about the center *C*, and draw the line *AB* so that the curve or angle *ACB* is above the line *AB*, and place the centers of gravity of the weights at *A* and *B*, which will rest in this position. Next move the balance from this position as to *ECF*. I say that the balance *ECF* will return to *ACB*.

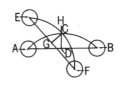

Find the center of gravity of the whole system, *D*, and join *C* to *D*. Now since the weights *A* and *B* are at rest, the line *CD* will be perpendicular to the horizon. Therefore when the balance is at *ECF*, the line *CD* will be at *CG*, and, since this is not perpendicular to the horizon, the balance *ECF* will return to *ACB*. The same will happen if the center *C* is placed above the balance as at *H*.

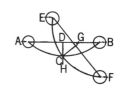

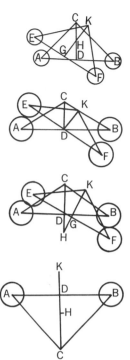

Now if the curve or angle ACB shall be beneath the line AB, we may show in the same way that the balance ECF, whether its center is at C or H, must move downward on the side of F.

And if the angle ACB is above the line AB at the center of the balance H, and if the line CH sustains the balance, when the balance is moved to EKF, it will return to ACB.

But if the center of the balance is D, one may move the balance in any way, and wherever it is placed, it will rest.

Then if the point H shall be beneath the line AB, the balance EKF will move downward on the side of F.

And for similar reasons if the angle ACB shall be beneath the line AB, and the center of the balance is at H, and the balance is sustained by the line CH, then, if the balance is moved from this position, it will move downward on the side of the lower weight. And if the center of the balance is at D, it will rest wherever it is left. And if the center is at K and the balance is moved from that position, it will return in any event to the same place. All which things are manifest from what we said at the beginning. Similarly if the center of the balance is placed in one of the arms of the balance, either within or without, or any other way, we shall find the same things.

Comment by Pigafetta

At this point it is well to notice what might have been said toward the end of page 26v, where the author wrote: "In addition to this we may consider the things that may be deduced in the same way." Now our author is the first to have considered the balance in detail and to have understood its nature and its true quality. For he is the first of all to have shown clearly the way of dealing with it and teaching about it, by propounding three centers to be considered in its theory: one is the center of the world, another the center of the balance, and finally the center of gravity of the balance: for in this was a hidden secret of nature. Without these three centers, it is clear that one could not come to a perfect knowledge or demonstrate the various properties of the balance, which were hidden in the variety of arrangements of the center of the balance — that is, whether the center of the balance is above its center of gravity, or below, or in the very center of the balance. These the author shows in the three preceding demonstrations, that is, in the second, third, and fourth propositions. In the second, he shows when the balance returns to a horizontal position, in the third when it not only does not return but moves still farther, and in the fourth that, when a balance is sustained at its center of gravity, it remains at rest wherever it is left. This last effect in particular has not been dealt with before, or seen, or even suggested by anybody besides this author. Indeed, until the present time it has been held to be false and impossible by all our predecessors, who not only have given many arguments attempting to prove the contrary, but have even assumed it to be certain that experience shows the balance never to remain fixed except when parallel to the horizon. This is quite contrary to reason, first because the demonstration of the above fourth proposition is so clear, easy, and true that I do not know how it could be contradicted in any way; and second, their view is contrary to experience, inasmuch as our author has very cleverly fashioned precise balances for the purpose of showing this truth, one of which I saw in the hands of the illustrious Giovanni Vicenzo Pinelli,[28] given to him by the author himself, and, because it was sustained at its center of gravity, it could be moved to any position and would rest at any place where it was left. True it is that in performing this experiment one might not act hastily, for it is an extremely difficult thing (as the author says above) to make a balance which is sustained precisely at the center of its arms and at its precise center of gravity. For this reason it is good to remember that, when anyone tries to perform such an experiment and does not succeed, he should not be discouraged, but rather should say that he had not been careful enough, and should try repeatedly until the balance is just and equal and is sustained precisely at its center of gravity.

And though others have touched on the other two propositions (that is, when the balance almost returns to a horizontal position, and when it moves in a contrary way), yet the truth of this has never been understood except by this author, and others have not gone far enough to have made a distinct consideration of the center of the balance in three ways, as I have explained. To the extent that they have done anything about this matter, they have done it confusedly and with poor demonstrations from which no one can draw clear and firm conclusions. These predecessors of ours are to be understood as being the modern writers on this subject cited in various places by the author, among them Jordanus, who wrote on weights and was highly regarded and to this day has been much followed in his teachings. Now our author has tried in every way to travel the road of the good ancient Greeks, masters of the sciences, and in particular that of Archimedes of Syracuse (the most famous prince of mathematicians) and Pappus of Alexandria, reading them, as he tells us, in their own language and not translated, because for the most part they are so unsatisfactorily treated that it is very hard to draw any profit from them whatever.[29]

And to the end that this new opinion of his, fully demonstrated in the aforesaid fourth proposition, should be completely clear, he has not been content to demonstrate it with vivid and certain reasoning alone, but, like a good philosopher, proceeding by the path of true doctrine and well-founded science (imitating Aristotle, who at the beginning of his books, in quest of the best doctrine, has given the contrary opinions of the ancients, analyzing the reasons which they

28. Pinelli was a wealthy patron of learning who resided for many years at Padua, where he formed a superb library and collection of rarities. He was an intimate friend of Galileo, and entertained most of the eminent scholars who visited the University of Padua.

29. Ironically, Guido Ubaldo's devotion to the ancient Greek writers and his scorn for their successors induced him to reject the correct inclined plane theorem of Jordanus and Tartaglia, accepting in its stead the erroneous theorem of Pappus in his chapters on the wedge and the screw, below.

accepted), he has wished, because there is but one truth, to propound the opinions of his predecessors and examine the reasons by which they seem to prove the contrary, and to resolve these, showing their fallacy in the present argument, which commenced on page 5v and ends at this point. Which argument will serve to clear up what is usually said to be the opinion of the ancients. And since it contains things of the highest theoretical value, especially with regard to the consideration of the place at which a single weight placed on the arm of the balance is heaviest, in order to understand it one must read it and study it with great diligence.

And certainly our author has been not only the first to discover this truth, but also the first to show in what manner one must consider and theorize concerning the whole subject; and with his theory he proves again and confirms the various effects and events of the balance already demonstrated in the adjoining three propositions, showing also how until now these things have been badly considered by others, and on false principles. Moreover, as a confirmation of the truth, he adds that they did not know how to construct their proofs; for by their own mode of theorizing and their very own reasons, he proves his opinions to be most true, supporting them always on the doctrine of Aristotle and making it clear that he is in accord with him in questions of mechanics. In dealing with this matter, the author also raises some new questions, very beautiful and curious, and then solves them clearly. Finally, in order that nothing might be lacking to the complete knowledge of this subject, he has dealt with balances whose arms are unequal and those which have curved and bent arms. In a word, it may well be said that in this argument everything is included that can be determined concerning this subject. The theories are beautiful and very subtle and are to be looked at and considered with much attention by anyone who delights in and attends to these noble and necessary studies.

Wherever the Latin word *equilibrium* is read, it means "equally counterpoised," that is, weighing as much on one side as on the other in equal scales or balances. *Librar con giuste lance,* Petrarch said.

PROPOSITION V

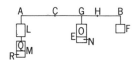

If two weights are attached to a balance and the balance is divided between them in such a way that the parts correspond inversely to the weights, they will weigh as much at the points where they are attached as if each were suspended from the point of that division.

PROPOSITION VI

Equal weights suspended from a balance have heaviness in proportion to the distances at which they are suspended.

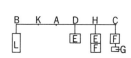

Corollary. — From this it is evident that, the farther the weight is from the center of the balance, the heavier it is, and consequently the more swiftly it will move. From this one may, furthermore, easily demonstrate the theory of the steelyard.

Comment by Pigafetta

Corollary is a Latin word employed by all Italian writers on this subject, nor did it displease Dante in the Twenty-eighth Canto of his *Purgatory.* "Also called a corollary, for example," as Varro says in his first book on the Latin language, "is anything over and above that which would normally be paid in buying something." In ancient times when the actors in tragedies, comedies, and other poems carried off their scenes well and pleased their audiences, something was given to them in addition to the fixed price — a corollary for each one — that is, a small crown to be placed on their foreheads and added to their rewards. Thus in the mathematical sciences it is customary to add certain things to propositions, as belonging to them and consequences of them, which take their rise from the things previously demonstrated, and correspond to them; and these are not propositions or problems or lemmas, but, as indicated before, are called corollaries, many of which are given with their proofs.

The steelyard may also be used in another way to make the weight of things known.

PROPOSITION VII

Problem: Given an indefinite number of weights on the balance, suspended at any places, to find a center on the balance from which, if the balance were to be hung, the given weights would be in equilibrium.

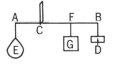

Comment by Pigafetta

Under the name of "proposition" is included "problem," also a Greek word; the problem goes beyond the proposition in that it proposes and shows how to achieve some result, whereas the proposition gives the bare theory only; and this is the difference between a proposition and a problem.

Let there be the balance AB and let there be given any number of weights C, D, E, F, and G hanging from the balance at the points A, H, K, L, and B. It is required to find the center of the balance from which, if [the balance were] suspended, the weights would be at rest.

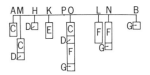

Divide AH at M so that HM is to MA as the weight C is to the weight D; then divide BL at N so that LN is to NB as the weight G is to the weight F, and divide NM at O so that MO is to N as the weights F and G are to the weights C and D; finally, divide KO at P so that KP is to PO as the weights C, D, F, and G are to the weight E. Now since the weights C, D, F, and G weigh as much at O as C and D at M and F and G at N, the weights C and D at M, F and G at N, and E at K

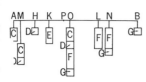

will balance if they are suspended from the point *P*. Inasmuch as the weights *C* and *D* weigh as much at *M* as at *A* and *H*, and *F* and *G* at *N* as much as at *L* and *B*, the weights *C*, *D*, *F*, and *G* hanging from the points *A*, *H*, *L*, and *B* and the weight *E* from *K*, if the balance is suspended from *P*, will weigh equally and will be at rest. Therefore *P* is the center of the balance from which the given weights will be at rest, which was to be achieved.

Corollary.— From this it is clear that if the centers of gravity of the weights *C*, *D*, *E*, *F*, and *G* were at the points *A*, *H*, *K*, *L*, and *B*, the point *P* would be the center of gravity of the whole system composed of these weights. This is obvious from the definition of the center of gravity, inasmuch as the weights will be at rest if they are sustained from the point *P*.

End of the Balance

On the Lever

LEMMA

Let there be four magnitudes *A*, *B*, *C*, and *D*, and let *A* be greater than *B* and *C* be greater than *D*. I say that the ratio of *A* to *D* is greater than that of *B* to *C*.

Since the ratio of *A* to *C* is greater than that of *B* to *C* and the ratio of *A* to *D* is greater than that of *A* to *C*, the ratio of *A* to *D* will be greater than that of *B* to *C*. Q.E.D.

PROPOSITION I

The power that sustains a weight attached to the lever has the same proportion to that weight as the distance along the lever between the fulcrum and the point of suspension has to the distance between the fulcrum and the power.[30]

From this it can easily be demonstrated that the closer the fulcrum is to the weight, the smaller the power required to sustain the weight.

Corollary.— Whence it can be quickly deduced that, *AF* being less than *FB*, a smaller power is required at *B* to sustain the weight *D;* and if they are equal, it is equal; and if *AF* is greater, the required power is greater.

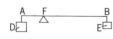

30. Hereafter, proofs are abridged or omitted in this translation unless they present some point of particular interest.

PROPOSITION II

The lever may be used in a second mode.

Proof: Let there be the lever *AB* with its fulcrum *B*, and let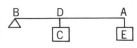
the weight *C* be attached at *D* between *A* and *B*, and let the
power at *A* sustain the weight *C*. I say, that as *BD* is to *BA*, so
is the power at *A* to the weight *C*. . . .

Corollary I.—From this also, as before, it may be shown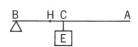
that, if the weight *E* is placed closer to the fulcrum *B*, as at *H*,
a smaller power is required at *A* to sustain the weight.

Corollary II.—It also follows that the power at *A* is always
less than the weight *E*.

Corollary III.—From this likewise it may be deduced that,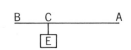
if there are two powers, one at *A* and the other at *B*, and both
sustain the weight *E*, the power at *A* will be to the power at *B*
as *BC* is to *CA*.

Corollary IV.—It is furthermore evident that the two powers
at *A* and *B* taken together are equal to the weight *E*.

PROPOSITION III

We may also use the lever in a third mode.

Let there be the lever *AB* with its fulcrum at *B*, and let the
weight *C* be hung from the point *A*, and let it be the power at *D*,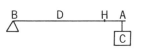
somewhere between *A* and *B*, that sustains the weight *C*. I say
that, as *AB* is to *BD*, so is the power at *D* to the weight *C*. . . .

Corollary I.—From this it is also clear, as before, that if the
weight is closer to the fulcrum *B*, as at *H*, the weight must be
sustained by a smaller force.

Corollary II.—It is likewise evident that the power at *D* is
always greater than the weight *C*.

PROPOSITION IV

If the power shall move the weight hung from the lever, the
space through which the power moves will be to the space
through which the weight is moved as the distance from the
fulcrum to the power is to the distance from the fulcrum to the
point from which the weight is hung.

Let there be the lever *AB* with its fulcrum *C*, and let the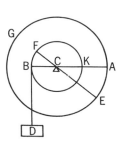
weight *D* be attached at the point *B*, and let the power at *A*
move the weight *D* by means of a lever *AB*. Then the space of
the power at *A* is to the space of the weight as *CA* is to *CB*. . . .

But let there be the lever at *AB*, whose fulcrum is *B*, and
the moving power is at *A* and the weight at *C*; I say that the

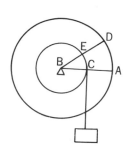

space of the moved power to the space of the weight carried is as *BA* to *BC*. . . .

Corollary. — From these things it is evident that the ratio of the space of the power which moves to the space of the weight moved is greater than that of the weight to the same power.

For the space of the power has the same ratio to the space of the weight as that of the weight to the power which sustains the same weight. But the power that sustains is less than the power that moves; therefore the weight will have a lesser ratio to the power that moves it than to the power that sustains it. Therefore the ratio of the space of the power that moves to the space of the weight will be greater than that of the weight to the power.[31]

PROPOSITION V

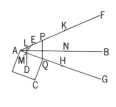

The power that sustains the weight in any way by means of the lever will have the same proportion to the weight as that of the distance from the fulcrum to the point on the lever, vertical to the center of gravity of the weight, to the distance between fulcrum and the power.

PROPOSITION VI

Let there be the straight line *AB*, and perpendicular to it the line *AD*, prolonged on the side of *D* to *C*. Join *C* and *B* and extend this to *E*. Then let there be drawn from the point *B* other lines between *AB* and *BE*, for example, *BF* and *BG*,

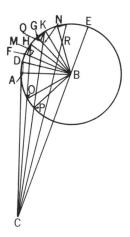

31. This statement illustrates the grounds on which pre-Galilean writers considered it impossible, and indeed illogical, even to attempt a science of dynamics. Cf. our *Galileo On Motion and On Mechanics* (Madison, 1960), pp. 141–43 and 166–67, note.

equal to *AB*. From the points *F* and *G* let there be drawn the lines *FH* and *GK* perpendicular to the above lines, and let these be equal to one another and to *AD*, as if *BA* and *AD* were moved to *BF* and *FH* and to *BG* and *GH*. Draw *CH* and *CK*, which cut the lines *BF* and *BG* at the points *M* and *N*. I say that *BN* is shorter than *BM*, and *BM* than *BA*.

If the equal triangles *BFH* and *BGK* are between *BC* and *BA* below, and if there are added the lines *HC* and *KC* which cut the lines *BF* and *BG* extended at the points *M* and *N*, then *BN* will be greater than *BM* and *BM* than *BA*.

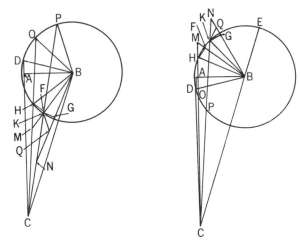

PROPOSITION VII

Let there be the line *AB* and the perpendicular *AD* extended as far as *C*. Draw *CB* and extend this to *E*. Between *AB* and *BE* draw *BF* and *BG* equal to *AB*, and from the points *F* and *G* draw the lines *FH* and *GK* also equal to *AD* and perpendicular to *BF* and *BG*, as if *BA* and *AD* were moved to *BF* and *FH*, or to *BG* and *GK*. Draw *CH* and *CK*, which cut the lines *BF* and *BG* at the points *M* and *N*. I say that *BN* is greater than *BM* and *BM* than *BA*.

And if the triangles *BFH* and *BGK* are placed between *AB* and *BC* below, and the lines *CHO* and *CKP* are drawn, which cut the lines *BF* and *BG* at the points *M* and *N*, then *BN* will be less than *BM* and *BM* than *BA*. [Cf. second statement and diagram, Prop. VI above.]

PROPOSITION VIII

If the power sustaining the weight which has its center of gravity on a horizontal lever is given, then, the more the weight

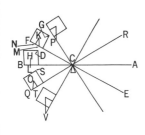

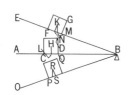

is raised from this position by means of the lever, the smaller the power required to sustain it. But if it shall be lower, the power is greater.

From this it is easily deduced that the power at *A* is to the power at *E* as *CL* is to *CM*.

In addition to this, if there is another power at *B*, so that there are two powers that sustain the weight, the power at *B*, which sustains the weight *PQ* by means of the lever *BO*, will be less than the weight *CD* on the lever *BA*. On the other hand, a greater power is required at *B* to sustain the weight *FG*, by means of the lever *BE*, than the weight *CD* on the lever *AB*.

Corollary. — From these things it is evident that, if a power raises, by means of a lever, a weight whose center of gravity is above the lever, then the more the weight is raised, the smaller becomes the power necessary to move the weight.

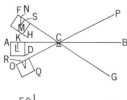

PROPOSITION IX

If the power sustaining a weight that has its center of gravity under the lever is given when that [lever] is horizontal, then the more the weight is raised from this position by means of the lever, the more power will be required to sustain it; but if it is lowered, the power becomes less.

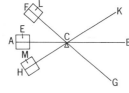

PROPOSITION X

The power sustaining a weight that has its center of gravity in the lever itself will always be the same no matter how the weight is moved by means of a lever.

PROPOSITION XI

If the ratio of the distance along the lever between fulcrum and power to the distance between fulcrum and that point on the lever vertical to the center of gravity of the weight is greater than the ratio of the weight to the power, the weight will be moved by the power.

PROPOSITION XII

Problem: To move a given weight by means of a given lever with a given power.

Let the weight *A* be 100 and the power that must move it be 10, and let the given lever be *BC*. It is required that the power of 10 shall move the weight of 100 by means of the lever *BC*. Divide *BC* at *D* in such a way that *CD* has the same ratio to *DB* that 100 has to 10, i.e., 10 to 1. . . . Take between *B* and *D* any point you wish, such as *E*, and make *E* the fulcrum. . . .

Corollary. — From this it is manifest that, if the given power is greater than the given weight, the weight can be moved whether the lever has its fulcrum between the weight and the power, or has the weight between the fulcrum and the power, or finally if the power is placed between the weight and the fulcrum. But if the given power shall be less than or equal to the given weight, it is likewise clear that the weight can be moved only if the lever is such that the fulcrum is between the weight and the power, or the weight is between the fulcrum and the power.

PROPOSITION XIII

Problem: Given an arbitrary number of weights suspended from arbitrary points of a lever whose fulcrum is also given, to find a power which will sustain these weights at a given point.

Let there be given the weights *A*, *B*, and *C* on the lever *DE* (with its fulcrum at *F*), suspended from the points *D*, *G*, and *H*, and a point *E* at which the power must be applied. . . . Divide *DG* at *K* in such a way that *DK* is to *KG* as the weight *B* is to *A;* then divide *KH* at *L* so that *KL* is to *LH* as the weight *C* is to the weights *B* and *A*. As *FE* is to *FL*, make the sum of weights *A*, *B*, and *C* be to the power which must be placed at *E*. . . .

PROPOSITION XIV

Problem: To make a given power move an arbitrary number of weights at arbitrary places on a given lever.

Let the given lever be *DE*, let the given weights be placed as above, and let *A* be 100, *B*, 50, and *C*, 30; and let the given power be 30. Find the point *L* as before; then divide *LE* at *F* in such a way that *FE* is to *FL* as 180 is to 30 (that is, as six is to one), and if *F* is the fulcrum, the power of 30 at *E* will sustain the weights *A*, *B*, and *C*. Therefore between *L* and *F* take some point such as *M*, and make this the fulcrum. . . .

PROPOSITION XV

Problem: But since in moving weights with a lever, the lever also has weight, which has not been mentioned up to this point, we shall demonstrate how to find the power which will sustain the lever in a given point, the fulcrum being likewise given.

Let there be the lever *BA* with its fulcrum at *C*, and let there be the point *D* at which the power must be applied which must

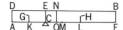

sustain the lever *AB* so that it remains at rest. From the point *C* draw the vertical line *CE*, dividing the lever *AB* into *AE* and *EF*. Let *G* be the center of gravity of *AE* and *H* the center of gravity of *EF*, and draw the vertical lines *GK* and *HL* cutting the line *AF* at the points *K* and *L*. Now since the lever *AB* is divided into two parts, that is, *AE* and *EF*, the lever *AB* is itself two weights placed in the balance *AF*, whose support is *C*. Hence the weights *AE* and *EF* are placed as if they were suspended at *K* and *L*. Divide *KL* at *M* in such a way that *KM* is to *ML* as the weight of the part *EF* is to the weight of the part *AE;* and in the proportion of *CA* to *CM* make the weight of the whole lever *AB* be to the power which, if applied at *D* (provided that *DA* is vertical), would counterpoise the lever; that is, would sustain the lever *AB* by its pressure.

Next, a weight hung from the lever is to be added, such as the weight *P* hung from *A*, and the power is to be applied at *B* in such a way that it sustains the lever *AB* together with the weight *P*.

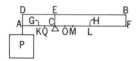

Divide *AM* at *Q* in such a way that *AQ* is to *QM* as the weight of the lever *AB* is to the weight *P*. In whatever ratio *CF* is to *CQ*, make the combined weight of *AB* and *P* be to the power placed at *B*. It is clear that the power at *B* would sustain the lever *AB* together with the weight *P*. For if *CA* were to *CM* as *AB* is to *P*, the point *C* would be their center of gravity, and hence the lever *AB* together with the weight *P* would remain at rest without the power placed at *B*. Now if the center of gravity of the weights is between *C* and *F*, as at *O*, then, as *CF* is to *CO*, so *AB* and *P* together are to the power which, when applied at *B*, will sustain the lever *AB* with the weight *P*. The same may be done when there are more weights along the lever *AB*, no matter where and in what way they are arranged. In addition to these things, one may see (as we have shown in the fourteenth proposition) the way in which we may move the given weights, placed anywhere on the lever, by means of a given power with a given lever. This we can do by considering not only the weight of the lever itself but also the other properties which have been demonstrated above independently of the weight of the lever. All may be shown by consideration of the combined weight of the lever and its weights or [the lever] without additional weights.

End of the Lever

On the Pulley

By means of the pulley things may be moved in many ways, but, since the theory is the same for all and in order to present the thing most clearly, it is to be understood in that which is about to be said that the weight is always to be moved upward at right angles to the horizontal plane.

Let there be the weight *A* which is to be raised vertically in the usual way. Held from above is the block containing two pulleys with their axles at *B* and *C*. Let another block be attached to the weight, also with two pulleys at *D* and *E;* and around all the pulleys lead the rope tied at one end, say, at *F*. Apply the power at *G,* so that, when this descends, *A* is raised, as Pappus shows in the eighth book of his *Collections,* Vitruvius in the tenth book of his *Architecture,* and others.

Now let us show how the pulley may be reduced to the lever, why a great weight is moved by a small force, in what way, in what time, why the rope must be secured at one end, what is the function of the pulley that is placed below and what that of the one above, and how one may find any given proportion between the power and the weight.

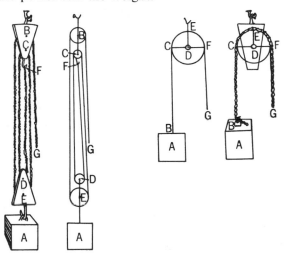

PROPOSITION I

If the rope is led around the pulley that is fastened from above, and one end of the rope is tied to the weight while the power that sustains the said weight is applied to the other end, then the power will be equal to the weight.

Corollary. — From this it is evident that the same weight can always be sustained by the same power without any assistance from this pulley.

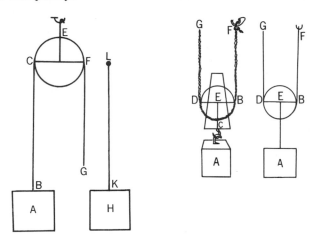

PROPOSITION II

If the rope is led around a pulley to which the weight is directly attached, and one of its ends is fastened to some place while the power that sustains the weight is applied to the other end, then the power will be one-half the weight.

. . . Since the power sustains the pulley by means of the rope, and the pulley sustains the attachment to which the weight is fixed, the whole weight will be at the center *E*. Now if we assume the power applied at *G* to act at *D* (because it is entirely the same thing), *BD* will be a lever with its support at *B* and its weight attached at *E* and the power applied at *D;* the rope *BF* being motionless, *B* serves as the fulcrum . . . . Now since the power has the same proportion to the weight that *BE* has to *BD* . . . the power at *G* will be one-half the weight *A*.

Corollary I. — From this it is evident that the weight is sustained in this way by a power of only one-half that which would be required without the aid of such a pulley.

Corollary II. — It is also evident that, if there were two powers which sustain the weight *A*, one at *G* and the other at *F*, then the two together would be equal to the weight *A*, and each of them would sustain one-half of the weight *A*.

Corollary III. — It is likewise evident why the rope must be secured at one end.

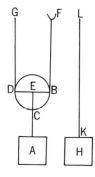

PROPOSITION III

If two pulleys are given, one attached above and the other below, to the latter of which the weight is attached, and a rope is given, led around both with one of its ends fastened at some place, then the power which sustains the weight applied to the other end will be equal to one-half the weight.

Corollary. — If there are two powers at N and L, they will be equal. For each will be equal to one-half of A.

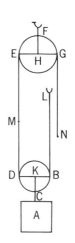

PROPOSITION IV

Let there be a lever AB with its fulcrum at A, divided into two equal parts at D; let the weight C be suspended from D, and let there be two equal powers applied at B and D which are to sustain the weight C. I say that each of these powers is one-third of the weight C.[32]

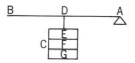
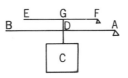

And if two levers AB and EF, bisected at G and D, have their fulcrums at A and F, while the weight C is equally supported by both levers, and if there are two equal powers at B and G, it may be shown that each of those powers is one-third of the weight C.

PROPOSITION V

If two pulleys are given, one supported from above and the other attached to the weight, with a rope led around them, one end of which is attached to the lower pulley, then a power equal to one-third of the weight applied to the other end will sustain the weight.

... The rope segment MD supports as much as HB, which in turn supports as much as FL; and since the rope MD will sustain as much as FL, it is as though equal weights were applied at D and L. Inasmuch as equal weights are sustained by equal powers, the powers at M and L will be equal, as if

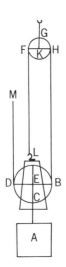

32. Though many previous writers had discussed pulleys and their combinations, the method of analysis here by which they are reduced to the lever appears to be original with Guido Ubaldo; it was adopted by Galileo and subsequent writers.

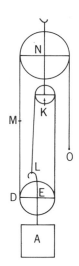

they were applied at *D* and *E*. Therefore, since the weight *A* is centrally attached to the lever *BD* and the two powers placed at *D* and *E* are equal and sustain the weight, then *B* will be the fulcrum, and each power, whether applied at *D* and *E* or *M* and *L,* will be one-third the weight *A*. . . .

Corollary. — From this it is evident that each of the rope segments *MD, FL,* and *HB* sustains one-third of the weight *A*.

. . . And in order that we need not return to say the same thing again, it should be noticed that the power at *O* is always equal to that at *M;* that is to say, if the power at *M* is one-fourth or one-fifth (or some such part) of the weight, then the power at *O* will be one-fourth or one-fifth of the weight, etc.

Comment by Pigafetta

Now some people might question these demonstrations about pulleys, for instance this fifth proposition which I select as the best example, asking whether in fact experiment is in agreement with theory as to the ratios of forces and weights. For in mathematical demonstrations, all lines are assumed to be without breadth or thickness, and all things are abstracted from actual matter, so that it is easy to persuade ourselves of the mathematical truth. But experience very often shows something different, and we find ourselves deceived, for actual matter changes things quite a bit. In this proposition it is shown reasonably that, for two pulleys and one rope, the force will be one-third of the weight. . . . Somebody might consider this very dubious, because the pulleys and their attachments, the ropes, and so on offer resistance to the force, and also have weight of their own, so that the [calculated] force may not be able to sustain the weight. We reply that these things may well offer resistance to the moving of the weight, but not to the sustaining of it; and it is necessary to note carefully that the author in these demonstrations speaks only of forces sustaining the weights so that they do not fall down; not about moving them.[33]. . . Thus we have to do only with a weight at rest, and will take into consideration only its counterweight, which is the function of the sustaining power. Hence neither in the pulleys nor anywhere else is there any resistance, and the theoretical proof will always come out very well; indeed, experience shows that the more resistance there is, the more easily the force sustains the weight. . . . But to know how much force must be added to the power in order that it may sustain the whole weight, including the lower pulley and the ropes, take the lower pulley and part of the rope as an additional weight. . . .

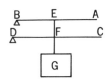

PROPOSITION VI

Let there be two levers *AB* and *CD*, bisected at *E* and *F,* with their fulcrums at *B* and *D;* and let there be the weight *G,*

---

33. Pigafetta's comment applies only through Prop. IX of this book; see particularly Props. X, XIV corollary 2, XV corollary, XVIII corollary, and XXVI and its corollary.

suspended in such a way that it weighs equally on *E* and *F*, and two equal powers at *A* and *C* which sustain the weight. I say that each of the powers is one-fourth of the weight *G*.

But if there shall be three levers *AB*, *CD* and *EF*, bisected at *G*, *H*, and *K*, with their fulcrums *B*, *D*, and *F*, and if the weight is similarly suspended from *G*, *H*, and *K*, while three equal powers *A*, *C*, and *E* sustain the weight, it will be likewise seen that each is one-sixth of the weight *L*. And in the same manner, if there were four levers and four weights, each power would be one-eighth of the weight; and so on.

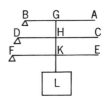

PROPOSITION VII

If three pulleys are given, one of which is suspended from above and two from below, and to these latter a weight is attached, and rope is wound around them and one end secured, then a power applied at the other end equal to one-fourth of the weight will sustain the weight.

Corollary I. — From this it is evident that each of the ropes *EF*, *GK*, *LN*, and *OP* sustains one-fourth of the weight *A*.

Corollary II. — It is also clear that the pulley whose center is *C* sustains no less than that whose center is *B*.

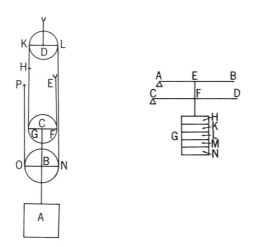

PROPOSITION VIII

Let there be two levers *AB* and *CD*, bisected at *E* and *F*, with their fulcrums at *A* and *C*; and let the weight *G* be suspended in such a way that it weighs equally on *E* and *F*; and let there be three equal powers at *B*, *D*, and *E* which sustain the weight *G*. I say that each of these powers alone is one-fifth of the weight *G*.

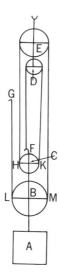

PROPOSITION IX

If there are four pulleys, one of which is attached from above and one of which is attached to the weight, and the rope is led around with one of its ends tied to the lower pulley, while the force is applied to the other end, the force that sustains the weight will be one-fifth of the weight.

Comment by Pigafetta

In this treatise on the pulley (as in the others), the author assumes that all readers of his book on mechanics understand arithmetic and geometry; he has therefore always adhered to that precise and demonstrative style which is customary among good mathematicians, using the special words of the science, some of which I have been able to popularize so that anybody can easily understand them . . . . But for certain Latin terms used for special proportions, we have no equivalent in our language, and I have been obliged to leave them untranslated for lack of ordinary words to express them. . . .

Corollary. — From this it is evident that the [lower] pulleys to which the weight is attached make it possible for the weight to be sustained by a lesser power than itself — something which is not accomplished by the upper pulleys.

PROPOSITION X

Let the rope be led around a pulley suspended from above, and let the weight be attached to one end, while the power that moves it is applied to the other. Then the said power will always move the weight as by a lever always parallel to the horizon.

Under the above assumption, the space of the power that moves the weight is equal to the space of the weight that is moved.

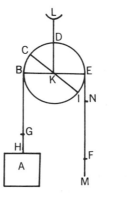

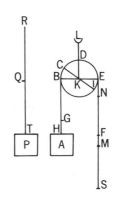

Moreover, the power moves the weight through an equal space in an equal time, whether the rope is wound around a pulley supported from above, or [lifts the weight] without any pulley at all, provided that the movements of this power are equal in speed.

PROPOSITION XI

Let the rope be led around a pulley to which the weight is attached, and let one end of the rope be sustained at some place while the power that moves the weight is applied at the other. Then the power will always move the weight as with a lever parallel to the horizon.

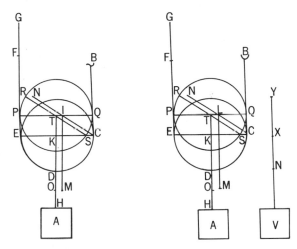

In these circumstances the space of the power that moves the weight is double the space of the weight that is moved.

The power will move the weight in an equal time through half the space . . . which would be traversed without the pulley, provided that the speeds of the power are equal.

PROPOSITION XII

If the rope is wound around many pulleys, and one end is fastened at some place, and the force that moves the weight is applied to the other end, the force will always move the weight as by a lever parallel to the horizon.

PROPOSITION XIII

If the rope is wound around a pulley supported from above and another to which the weight is attached, and one end of the

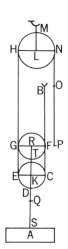

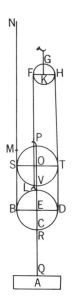

rope is tied to the lower pulley while the power that moves the weight is applied to the other end, the space passed through by the power is three times that of the space of the weight moved.

PROPOSITION XIV

If the rope is led around three pulleys in two blocks, with a single pulley attached from above and two below attached to the weight, and one end of the rope is secured at some place while the power that moves the weight is applied to the other, then the space traversed by the power will be four times that through which the weight moves.

Corollary I.—From these things we see why the ratio of the weight to the power that sustains it is the same as the ratio between the space of the moving power and the space of the weight moved.

Corollary II.—It is likewise evident from what has been said that the pulleys to which the weight is attached have the function of reducing the space passed through by the weight with respect to the power that moves it, and that the weight goes through the same space in a longer time than [it would if moved] without the pulleys; which function does not belong at at all to the pulley attached from above.

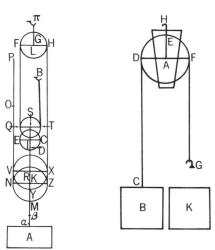

PROPOSITION XV

If the rope goes around the pulley of a block held from above by the power, and one end of the rope is fastened at some place while the weight is attached to the other, then the power will be twice the weight.

PROPOSITION XVI

Under the same assumptions, if the power that moves the weight shall be at *H,* it will move as if it were a lever parallel to the horizon.

Under these assumptions the space of the weight moved is twice that of the space of the power that moves.

Corollary.—From this it is evident that a given weight is drawn with this sytem of pulleys by the same power in an equal time through twice the space traversed without pulleys, provided that the movements of the power are equal in speed.

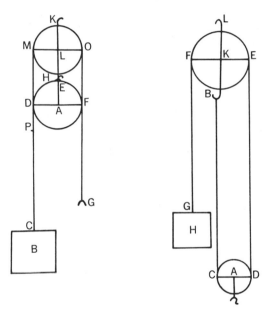

PROPOSITION XVII

If an upper pulley is suspended from above by the power and a lower pulley is securely fixed, while the rope is wound around them with one of its ends tied to the pulley above and the other attached to the weight, then the power will be three times the weight.

PROPOSITION XVIII

If there are two blocks of two pulleys each, one of which is held from above by the power while the other is secured below, and a rope is wound around them with one of its ends attached elsewhere than to the upper block and the other end holds the weight, then the power will be four times the weight.

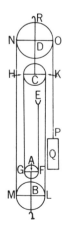

Corollary.—From this it is evident that if the rope were fastened at *G* and wound around the pulleys whose centers are *B, C,* and *D,* then the power at *R* would sustain four times the weight *Q;* for the pulley whose center is at *A* does nothing. . . .

Corollary.—In these things it is evident that the pulleys of the upper block constitute the reason for which the weight is moved by a greater power than itself, through a greater space than that of the power or through equal space in less time; but this is in no way caused by the lower pulleys.

PROPOSITION XIX

If there are two pulleys, one of which is held from above and the other is held by the power [*O*] that sustains [weight *M*], while the rope around them has one of its ends secured and the other attached to the weight, then the power will be twice the weight.

Corollary.—From this it is evident that the pulley below in this case causes the weight to be moved by a greater power than itself and through a greater space than that of the power (or through equal space in less time), which cause does not belong to the pulley attached from above.

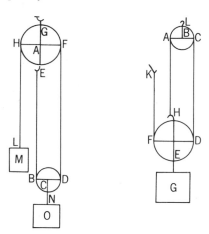

PROPOSITION XX

If there are two pulleys, the upper one of which is sustained by the power and that below is attached to the weight, while the rope is secured at one end and the other end is attached to the pulley below, the weight will be one and one-half times the power.

PROPOSITION XXI

If there are three pulleys, one of which is sustained from above by the power while the other two are below and attached to the weight, and one end of the rope is fastened while the other is attached to the upper pulley, the weight will be one and one-third times the power.

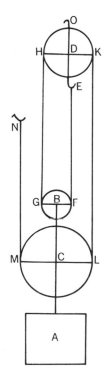

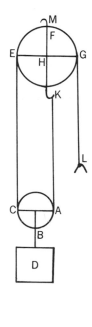

PROPOSITION XXII

If there are two pulleys, of which one is sustained from above by the power and the other is attached to the weight, while one end of the rope is fastened and the other is attached to the upper pulley, then the power will be one and one-half times the weight.

PROPOSITION XXIII

If there are two pulleys, one of which is sustained by the power from above while the other is attached to the weight below, and the rope has both ends fastened elsewhere than to the pulleys, then the power will be equal to the weight.

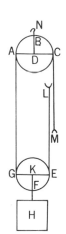

[Propositions 24 and 25 deal similarly with variant systems.]

PROPOSITION XXVI

Problem: To find [systems of pulleys such that] the ratio of the weight to the power that sustains the weight shall be as five to three.

[The author also gives analogous demonstrations for various other problems.]

Corollary. — From these things it is evident that the space of the power that moves has always a greater ratio to the space of the weight moved than that of the weight to the same power. This is evident from what was said in the corollary to the fourth proposition concerning the lever.[34]

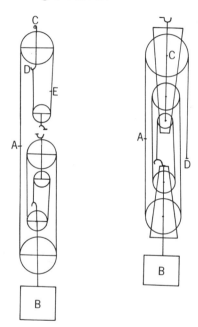

PROPOSITION XXVII

Problem: To move a given weight with a given power by means of pulleys.[35]

The power is greater than the weight, or equal to it, or less than the weight. If the power is greater, the given weight may

---

34. See notes 31 and 33, above. The words "moved" and "moving" are to be understood in all foregoing propositions merely as descriptive adjectives. Here they take on a dynamic character, but Guido Ubaldo was unable to bridge the gap from his statics to a valid dynamics.

35. This statement, for pulley systems, corresponds to the fundamental problem of mechanics as solved for levers by Archimedes: to move any given weight with any given force.

be moved without any instrument, or a rope around a pulley supported from above will move it . . . .

But if equal, it will move the weight by a rope around a pulley attached thereto, because the power that would sustain the weight is one-half the weight. . . . If it is less, suppose the weight is 60 and the power is 13. Find the power at *A* that will sustain the weight, which [power] is one-fifth the weight; so the power at *A* that sustains the weight is 12 . . . so a power of 13 at *A* will move the weight.

It is also to be noted that, in the moving of weights, sometimes it is more convenient to move the power downward rather than upward, so the rope may be carried over the pulley whose center is *C* . . . .

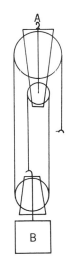

PROPOSITION XXVIII

Problem: To provide that the power moving the weight and the weight itself shall move through given spaces which are commensurable.

. . . We can achieve the result with a single rope by what was said in the twenty-second and twenty-fifth propositions. And if we wish to use more ropes, we may do it in an infinite number of ways. . . .

Corollary I.—From these things it is manifest that in an infinite number of ways, by means of pulleys, we can achieve any given proportions between the weight, the power, and the spaces through which they are moved.

Corollary II.—It is also evident that the more easily the weight is [to be] moved, the greater will be the time [required]; and the greater the difficulty with which the weight is moved, the shorter the time; and conversely.[36]

End of the Pulley

On the Wheel and Axle

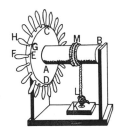

The construction and nature of this instrument is explained by Pappus in the eighth book of his *Mathematical Collections*. *AB* is called the axle, *CD* the drum on the same center (which we will call the wheel), and those rods that are inserted in the holes of the wheel and designated *EF, GH,* and so on, we shall call handles. The power or the force is always applied to the handles, as at *F,* and this turns the wheel, which in turn moves

---

36. That is, the greater the mechanical advantage, the more time is required for pulley systems to move a given weight through a given space, etc.

the axle, which draws up the weight *K* suspended by the rope *LM* around the axle. It now remains for us to show how weights may be moved by a small force with this instrument, and in what manner, and, moreover, to show the rule of the times and spaces of the moving power and of the weight moved; and finally, to reduce this instrument to the lever.

PROPOSITION I

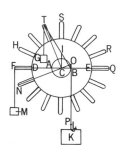

The power sustaining the weight by means of the wheel and axle is in the same ratio to the weight as the radius of the axle to the radius of the wheel including the handle.

Corollary. — It is evident that the power is always less than the weight.

. . . It should be noted that, if the weight were applied to a different handle, say *T*, and should sustain the weight *K* so that the weight applied at *T* and the weight *K* suspended from around the axle should remain motionless, then the weight at *T* must be heavier than the weight *M* applied at *F*. . . .

If in place of the weight at *T* one were to apply a living force to sustain the weight *K*, acting as if it wished to reach the center of the world, as did the weight applied at *T* by its own nature, then this power will be equal to the weight at *T* . . . but if each power able separately to sustain the weight, at *T* as well as at *F* around the circumference *THFN*, were to move as if the handle were pressed by the hand, then the same power placed at *F* or at *T* would be able to sustain the same weight *K*. . . .

Now the power moves the weight by means of the lever *FB;* that is, when the power at *F* rotates the wheel, the axle also rotates, and *FB* serves the function of a lever with its fulcrum at *C*, its moving power at *F*, and the weight applied at *B*. . . .

. . . Therefore let the power be where it will, the space of the power will be to the space of the weight moved as *CF* is to *CB;* that is, as the radius of the wheel to the radius of the axle.

Corollary I. — It is evident from the above that, as the weight is to the power sustaining the weight, so is the space of the moving power to that of the weight moved.

Corollary II. — It is also evident that the space of the moving power has always a greater ratio to the space of the weight moved than that of the weight to the power.[37]

37. The apparent contradiction of this corollary with the one preceding is to be explained as in note 34, above.

Corollary [*III*]. — It is further evident that the more easily the weight is moved, the longer time it will take [to raise the weight a given distance], and conversely.

PROPOSITION II

Problem: To move a given weight by means of the wheel and axle, with a given power.

Let there be given a weight of 60 and a power of 10. Draw the straight line *AB*, divided at *C* in such a way that *AC* is to *CB* as 60 is to 10. And if *CB* were the radius of the axle and *CA* the radius of the wheel with its handles, it is clear that the power at *A* would counterpoise the weight at *B*. Therefore take between *B* and *C* any other point such as *D*, and make *BD* the radius of the axle and *DA* the radius of the wheel and handles. . . .

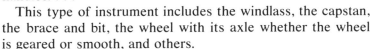

This type of instrument includes the windlass, the capstan, the brace and bit, the wheel with its axle whether the wheel is geared or smooth, and others.

But the brace and bit partakes somewhat of the screw, because when the weight moves (that is, when it makes a hole) by its very nature it always travels onward; hence the worm of the screw is described around a cone. But since it has a sharp point, it may also be reduced to the wedge.

Comment by Pigafetta

Here the author has given us five figures, representing five instruments for moving weights which may be reduced under a single property, in order that one may see each to be the same as the wheel and axle already explained.

He has put the letters *A*, *B*, and *C* with their lines, in order that one may understand that the weight has the same proportion to the power that sustains it as *AC* has to *CB*, and, as the weight shall be moved by a power, the space of the power will likewise be to the space of the weight as *AC* is to *CB*. In each case the power is to be understood as placed at the end of the handles at the distance *CA* from the center. The weight is to be understood as tied to a rope wound around the axle, at the distance *CB* from the center. And thus, for the reasons given above, the power that sustains will have the same ratio to the weight as *CB* has to *CA*. Similarly the figure with the drum is to be considered as if the force were at the outside of the drum and the weight were attached to the axle. As to the bit and brace, or auger, as it is called, since this is an instrument not designed to sustain things but to move them, the power must have a greater proportion to the weight than that of *CB* to *CA*, in accordance with the eleventh proposition in the section on the lever.

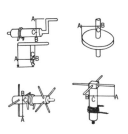

End of the Wheel and Axle

On the Wedge

Aristotle in the seventeenth of his *Questions of Mechanics* declares that the wedge performs the function of two opposed levers in the manner indicated below.

. . . Let the wedge be struck as usual on *AC; AB* is a lever whose fulcrum is at *H* and weight at *B*, and in the same way *CB* is a lever whose fulcrum is at *K* and weight similarly at *B*. But when the wedge is struck, it enters into *DEFG* in a ratio greater than that which was before; let this be the portion *MBL*. And since *MB* and *BL* are greater than *HB* and *BK, ML* will also be greater than *HK.* . . .[38]

But since there are three kinds of levers, as set forth previously, it will be perhaps more convenient to consider the wedge in the following manner.[39]

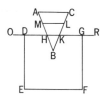
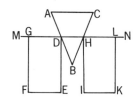

We may regard *AB* as a lever with its fulcrum at *B* and the weight at *H* . . . and similarly the lever *CB* with its support at *B* and the weight at *K*. . . .

Thus let the wedge be *ABC*, and let there be two separate weights *DEFG* and *HIKL* with the part *DBH* of the wedge between them. . . . Now while *DG* is moved by the wedge toward *M* . . . it is by the lever *AB* with fulcrum *B*. . . . Similarly *HL* is moved from *H* by the lever *CB*. . . .

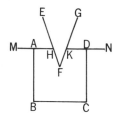

If one must split the rectangle *ABCD* and there are two equal levers *EF* and *GF* . . . and it is necessary with these levers to split *ABCD* without striking, then the moving powers at *E* and *G* are equal.

But since the whole wedge is moved in the splitting, we may consider it also in another manner; that is, when it enters into the thing to be split and is the same as moving a weight upon an inclined plane.

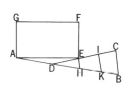

. . . In this example, if we consider the wedge as moving like a lever, it is evident that the wedge *BCD* moves the weight

38. Cf. *Aristotle*, p. 375.
39. The abridged version which follows shows sufficiently the difficulties into which Guido Ubaldo was led by his attempt to deal with a problem in dynamics and by his incorrect theorem regarding inclined planes, adopted from Pappus.

AEFG by means of a lever *CD*, so that *D* is the fulcrum and the weight is placed at *E*, rather than with the lever *BD* with its support at *H* and the weight placed at *D*. But in order that this may be more clear we shall use another example.

. . . This movement is easily reduced to the balance and to the lever, since that which is moved on an inclined plane is reduced to the balance by the ninth proposition of the eighth book of the *Mathematical Collections* of Pappus. For it is the same thing whether the wedge stands still and the weight moves on the side of the wedge, or the wedge is moved and the weight moves along its side, as upon an inclined plane.

Comment by Pigafetta

The proposition from Pappus, cited here by our author, I have withheld for a more convenient place in the section on the screw, for it is my opinion that perhaps it is more apropos there and serves more clearly than with regard to the wedge. This proposition was sent to me by the author, and, though there was nothing wrong with it, I have compared it carefully with the Greek edition of Pappus owned by Signor Pinelli, that it might be most useful and pleasant to those who have never seen anything of Pappus and have never read that marvelous writer on mechanics.[40]

Now we shall see how things that are split move as upon inclined planes.

Next we shall consider what two things are necessary in order that the thing may be more easily moved, or split.

First, as to things being more easily split, the chief essential of the wedge is the angle at the point, for, the more acute the angle, the more easily the wedge moves and splits.

We can also show this by another theory, considering the wedge as moving by two opposed levers, as mentioned previously.

Let there be the lever *AB* that has its fixed support at *B* and must move the rectangle *CDEF*, so arranged that it cannot move downward on the side *FE*. Let the point *E* be motionless, and consider it as a center, so that the point *D* will move along the circumference of the circle *DH*, whose center is *E*, while *C* moves along the circumference *CL*, and the line

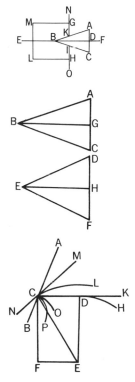

40. A Latin translation of Pappus by Commandino was published posthumously in 1588; it was certainly known to Guido Ubaldo, who is said to have done the final editing. This theorem, however, was not published in the Latin *Mechanics* of 1577. Pigafetta appears to have applied to Guido Ubaldo for it, perhaps because he was (justifiably) puzzled and uncertain about this section of the book. But having compared it with the Greek, and found the Latin to agree, he appears to have been fully satisfied. See further under Prop. II of the "Screw."

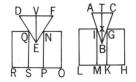

CE is its radius. . . . Now let there be another lever *MCN*, which also moves *CDEF*, and this has its fixed support at *N*. . . . I say that *CDEF* is more easily moved by the same power with the lever *AB* than with the lever *MN*. . . .

Corollary. — From this it is clear that the smaller the angle *BCF* or *BCE* or *BCD*, the more easily the weight is moved. But this may be demonstrated in the same manner.

The second thing is the reason why a thing is split more easily by striking in order to move the wedge.

Let there be the wedge *A* which has to split *B*, and let this be struck by *C*, which moves and strikes either by itself or by some power that governs and moves it. If by itself, first one must notice that the heavier it is, the greater the stroke will be. In addition to this, the greater the distance *AC*, the greater the stroke; for any heavy object when moved takes on more heaviness than when standing still, and the more so, the farther it moves.

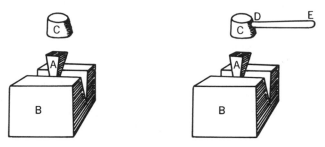

Now if *C* is moved by some power, as, for example, the handle *DE*, then first by the greater weight of *C*, and second by the greater length of *DE*, the stroke will be made greater. For if the moving power is placed at *E*, *C* will be more distant from the center and therefore will move more, as Aristotle demonstrates in his *Questions of Mechanics*.[41] And it may also be clear from what was said in the section on the balance that the farther *C* is from the center, the more it will weigh, and it will strike with greater impetus, the force at *E* being more potent.

Now here is the second thing, which is the reason that great weights can be moved and split with this instrument. Percussion is a very strong force, as is evident from the nineteenth of Aristotle's *Questions of Mechanics*;[42] for if a very heavy weight shall be placed upon a wedge, the wedge will accomplish

41. Cf. *Aristotle*, pp. 367–68.
42. *Aristotle*, p. 375. Problems of percussion also baffled Galileo.

nothing compared with its [work by] being struck. And even if one were to add a lever or a screw to the wedge, or some other instrument to drive the wedge or screw into the weight, nothing would happen of any importance compared with [that caused by] a stroke. Thus if the body *A* were a stone from which someone wanted to remove a certain part, say the corner *B,* then he might break it easily with an iron hammer and without any other instrument; but this he could do only with great difficulty by means of an instrument such as the lever or screw which does not use percussion. Thus percussion is the cause that great weights are split, and if to the great force of percussion we add some instrument suitable for moving and splitting, we shall achieve marvelous things. This instrument is the wedge, of which two properties as to its form must be considered: one, that the wedge is suited to receive and stand the blow; the other, that by the thinness of one of its edges it may easily enter into bodies. Thus the wedge is operated by a blow, and we see almost miracles in the splitting of bodies. To the same property as that of this instrument one may also easily reduce all those things that cut, divide, make holes, or do other things by means of percussion, such as spades, swords, knives, and the like. The saw may also be reduced to this, because its teeth strike and resemble wedges.

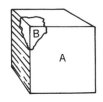

End of the Wedge

On the Screw

Pappus[43] in his eighth book deals with many matters of the screw, showing how it should be made and how great weights

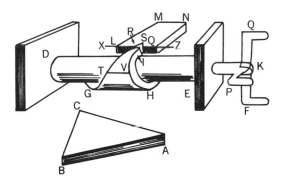

43. The *Mechanics* of Hero of Alexandria was preserved in fragmentary form by Pappus, from whose hint Guido Ubaldo reconstructed his method

may be moved with such an instrument; moreover, he gives many useful theories for its understanding. Now, since among other things he promises to show that the screw is nothing but a wedge used without percussion, which makes its movements by means of a lever, and this is lacking in his book, we shall attempt to show this and, moreover, to reduce the screw to the lever and the balance in order that ultimately we shall understand it completely.

PROPOSITION I

If the wedge is adapted in the following manner to the cylinder, it is precisely a screw which has two worms joined together at one point.

Corollary. — From this it is evident how one may describe the worms on the screw.

We shall now show how weights are moved on the worms of the screw.

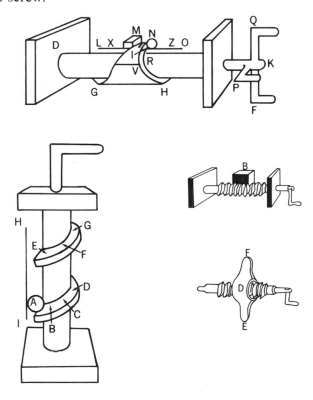

of analysis of the screw. His analysis was vitiated by the defective theory of the inclined plane, taken also from Pappus.

Now if to the screw in the next diagram below is applied
the gear *C* with twisted teeth, as Pappus shows in the same
eighth book, or even with straight teeth made in such a manner
as to fit the screw, it is evident that, with the movement of

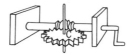

the screw, the gear *C* will also turn, and the teeth of the gear
will move on the worms on the screw; and this is called the
perpetual screw, because both the screw and the wheel will
go on turning in the same way.

PROPOSITION II

Let there be the screw *AB* with the worm *CDEFG;* I
say that this is nothing but an inclined plane wound around
a cylinder.

But as to the manner in which this is reducible to the balance, that is evident from the ninth proposition of the eighth
book of Pappus.

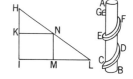

Comment by Pigafetta

The author, in all his books on mechanics, has wished not to insert
anything said by others, or which is not entirely his own. Therefore
he has omitted the proposition of Pappus which he cites here. But
this is so admirably relevant to the explanation of what he says here
that I have judged it advisable to add this.

The Problem of Pappus of Alexandria in the Eighth Book of His Mechanical Collections[44]

A given force is needed to draw a given weight along a horizontal
plane. It is required to find the force needed to draw the weight up
another plane inclined at a given angle to the horizontal plane.
Let the horizontal plane pass through *MN,* and let the plane
inclined to the horizontal at the given angle *KMN* pass through *MK.*
Let *A* be the weight and *C* the force required to move it over the
horizontal plane. Consider a sphere on the inclined plane passing
through *M* and *K.* The sphere will be tangent to the plane at *L,* [as

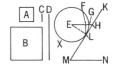

44. The translation is based on that in M. Cohen and I. E. Drabkin, *Source
Book in Greek Science* (New York, 1948), pp. 144ff. The analysis is obviously
defective, implying that an infinite force would be required to lift any weight
vertically. Note that the problem as stated by Pappus is not a static but a
dynamic one. It was Galileo who first attacked the assumption of Pappus that
some force was required to move a body horizontally. Previous writers
(Jordanus, Tartaglia, Cardano, and Stevin) dealt only with the problem of
equilibrium on inclined planes. See Drabkin and Drake, *Galileo On Motion
and On Mechanics,* pp. 172ff.

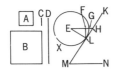

shown in the third theorem of the *Spherics* of Theodosius]. *EL* will therefore be perpendicular to the plane (for this is [also] shown in the *Spherics*, Theorem IV), and also to *KM*. Draw *EH* parallel to *MN*, and draw *LF* from *L* perpendicular to *EH*. Now since the angles *EHL* and *KMN* are given equal, angle *ELF* is also given, for the angles *ELF* and *EHL* are equal (since triangles *EHL* and *ELF* are similar). Therefore the triangle *ELF* is given in form. Hence the ratio *EL* : *EF*, that is, *EH* : *HF*, is known, as is also (*EH* − *EF*) : *EF*, that is, *HF* : *EF*.

Let weight *A* be to weight *B* and force *C* to force *D*, as *HF* is to *FE*. Now *C* is the force required to move *A*. Therefore the force required to move *B* on the same plane will be *D*. Since weight *A* : weight *B* equals *HF* : *FE*, it follows that if *E* and *H* are the centers of gravity of weights *A* and *B* respectively, the weights will be in equilibrium if balanced at the point *F*. But weight *A* has its center of gravity at *E* (for the sphere represents *A*). Therefore if weight *B* is placed so that its center is at *H*, it will so balance the sphere that the latter will not move down because of the slope of the plane, but will remain unmoved, as if it were on the horizontal plane. But weight *A* required force *C* to move it in the horizontal plane. Therefore, to be moved up the inclined plane it will require a force which is the sum of the forces *C* and *D*, where *D* is the force required to move the weight *B* in the horizontal plane. . . .

We must now consider the two reasons for which weights are easily moved by this instrument.

First, what makes the weight easy to move and what especially belongs to the nature of the screw is the worm; for, if around a given screw *AB* there should be two unequal worms *CDA* and *EFG*, I say that the same weight is moved more easily on *CDA* than on *EFG*.

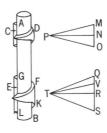

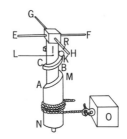

The other reason for which weights are easily moved consists in the [length of] the radius or handles by which the screw is turned.

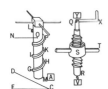

Corollary.—From these things it is evident that the more turns there are to the worm, and the longer the rods or handles, the more easily, but more slowly, the weight is moved. And finally we shall explain the strength of the power that moves it as applied to the rods.

Corollary. — From this it is clear that a given weight can be moved by a given power by means of the screw.

Thus it has been demonstrated that a weight is moved by the screw as by a wedge without percussion; for, in place of a blow, it moves by means of a lever, that is, the rod or handle.

These things demonstrated, it is evident one can move a given weight by a given power. For if we wish to achieve the effect by a lever, we can with a given lever lift a given weight with a given power. This cannot be done entirely by any of the other devices, whether by the screw, or the wheel and axle, or the pulley; for with a given pulley [system], wheel and axle, or screw, in order to move a given weight, the [required] power is always determined. Therefore, if the power that must move the weight is less than this, it will never move the weight. Yet given the wheel and axle (without handles), we can move a given weight with a given power, since we can arrange handles in such a way that the radius of the wheel together with the length of the handle has the given ratio to the radius of the axle. Now this can also be done with the screw; that is, a given weight can be moved with a given screw by means of the given power. For, if the power is known which must move the weight on the worm, we may arrange the handle in such a way that the given power on the handle has the same force as the power moving the weight on the worm. Although this cannot be done with a given pulley system, yet we can move a given weight with given pulleys by a given power in an infinite number of ways.[45] With the wedge, it seems clear that one can never move a given weight by means of a given power, because a given power cannot move a given weight along an inclined plane.[46] Nor by a given power can a given weight be moved by opposed levers, such as those of the wedge, inasmuch as the levers in the wedge cannot maintain the true and natural ratio of the lever because the fulcrums of the levers are not motionless when the entire wedge moves.

Anyone will be able then to construct machines and compound several together, such as pulleys and windlasses, or many gears, or in various other ways, and from what we have said one may easily find the relation between the weight and the power.

45. The sense of the passage is that pulleys can be arranged to do the job, though a given arrangement may be inadequate for it. The translation here is a free one.

46. This succinctly illustrates the distance that still separated the most systematic sixteenth-century treatment of mechanics from the inertial concept.

Comment by Pigafetta

Here one may note that the author has not gone into these last instruments—that is, the wedge and the screw—as he did the lever, the pulley, and the wheel and axle, for which he has very exactly shown the ratio of the force and weight. This is because these two instruments in themselves are not suitable to be considered as sustaining a weight, but rather as moving it. Now, since the powers that move may be infinite, one cannot give a firm rule for them as may be done for the power that sustains, which is unique and determined. That the wedge is not suitably considered as sustaining is clear in itself, and that the same is true of the screw is evident in the ordinary uses of the screw to move weights. . . . Thus the author has dealt with the two last instruments as suits their nature, as he says in comparing together all five of the instruments for moving weights as a conclusion to his work.

The End

Galileo Galilei

De motu dialogus

Unfinished and untitled dialogue
composed at Pisa about 1590

and
Related memoranda left by Galileo
in connection with his
Sermones de motu

As published in the National Edition
of the

Opere di Galileo Galilei

edited by Antonio Favaro from
manuscripts preserved at Florence

Translated and annotated by
I. E. Drabkin

Dialogue on Motion

Composed by Galileo Galilei

Pisa, *ca.* 1590
Left unfinished and untitled

Alessandro: Where are you going in such a hurry, my dear Domenico?

Domenico: Ah, greetings, my good friend!

Alessandro: Please stop a while. I've become so tired trying to catch up with you that I can hardly supply my over-heated heart with as much of the breath of life as it needs.

Domenico: On the other hand, though I've been walking briskly, I cannot overcome the persistent cold. In my case there isn't any verification of that old saying "motion is the cause of heat." Let us, then, proceed as slowly as you please and take a walk beyond the city limits, as we usually do. In fact, I had planned to go there even if I had no company. But what shall we talk about this morning?

Alessandro: Let's talk about whatever first comes into our heads so long as it isn't an unpleasant topic for conversation.

Domenico: Well, then, take the matter I just mentioned.

Alessandro: What was that?

Domenico: I had just mentioned that old saying . . .

Alessandro: Oh, yes, now I remember.

Domenico: Now I think that a conversation about motion would not be unpleasant. Yet to discuss motion in general terms, its essence and several properties, would take too long and would be, in fact, unnecessary. For when I want thorough treatment of this subject, I'll consult Aristotle himself, in his *Physics,* and all his commentators. And so, I'll now confine myself to only one type of motion, namely, the motion of heavy and light bodies. But since this subject, too, has been thoroughly treated by many, and most thoroughly by Girolamo Borro,[1] I shall be glad to know your opinion on certain particular questions, omitting all others, and to hear your solution of certain

---

1. Hieronymus Borrius Arretinus, *De motu gravium et levium,* Florence, 1576. [Borro was probably one of Galileo's teachers at Pisa. — *S. D.*]

problems on which my mind is not quite satisfied with the view and solution of others. These problems are as follows:

(1) Whether you believe it true that at the turning point of motion[2] there must be an [interval of] rest. [*cf. 323–28*]

(2) What reason you give for the fact that, if two bodies of the same size, one, let us say, of wood, the other of iron, that is, one heavier than the other, are let fall from a height at the same moment, the wooden one moves through the air more swiftly than the iron one, that is, the lighter more swiftly than the heavier—*if*, indeed, you admit this is a fact.[*cf. 333–37*]

(3) How it happens that natural motion is swifter at the end than at the middle or at the beginning, while forced motion is swifter at the beginning than at the middle, and swifter at the middle than at the end. [*cf. 315–23*]

(4) Why the same body falls more swiftly in air than in water; why, indeed, some bodies fall in air which do not sink in water. [*cf. 262, 268–70*]

(5) (A query of our dear friend Dionisio Font,[3] most worthy knight.) What reason you give for the fact that cannons used against fortifications, as well as manual arms, shoot lead balls farther along a straight line if they are shot at right angles to the horizon than if on a line parallel to the horizon, although in the former case the motion is more opposed to natural motion. [*cf. 337–40*]

(6) Why those same guns shoot heavier balls more swiftly and farther than lighter ones (e.g., iron as compared with wood), although the lighter are less resistant to an impelling force. [*cf. 335*]

It will, then, be my great pleasure to hear your opinion on these problems and on similar ones that depend on these. For I know that on this subject you will either say nothing or will adduce something new and very near the truth. For since you have become used to very rigorous, clear, and subtle mathematical proofs, such as those of divine Ptolemy and of most divine Archimedes, you cannot at all agree with certain crude ways of reasoning. And since the subjects I proposed are not far removed from mathematical considerations, with eager ears I await something beautiful from you.

2. I.e., at the highest point, in the case of a body thrown vertically upward. Note that the bracketed italic numbers following this and other passages refer to pages of *Opere di Galileo Galilei* (ed. A. Favaro [Ed. Naz., 1890]), Vol. I, where, in other early works, Galileo expressed the same or similar ideas. The pagination in the running heads is also that of Favaro, *Opere*.

3. According to Favaro (*Opere*, XX, 442), this was the Dionisio Font who died December 5, 1590.

Alessandro: My friend Domenico could not help saying something worthy of his lofty intelligence.

But, in order for me to be able to elucidate my opinion on these questions, I must first make certain statements. Thus, three things must be taken into consideration in motion, so far as the present inquiry is concerned. [*cf. 418.16*] These are the mover, the moving body, and the medium through which the motion takes place. The last two are the same in both natural and forced motion. But the first, i.e., the mover, is not the same in both kinds of motion, for in natural motion it is the characteristic heaviness[4] or lightness, and in forced motion it is a certain force impressed by a mover. In a medium . . .

Domenico: But wait. We must proceed step by step. We don't want to build our house on unstable foundations, so that, when it comes to putting the roof on, the whole structure collapses. Now you just said that in natural motion the body was moved by its weight or lightness, but in forced motion by an impressed force. But these two propositions, before I can grant or believe them, require clarification. In the first place, on what reasoning do you rely when you confidently assert that in forced motion the body is moved by a force impressed on it by a mover, despite the fact that Aristotle adduces a different cause for this motion, holding that the body is moved by the medium?[5] Do you believe, then, that Aristotle's opinion on the matter is false?

Alessandro: There is no necessity that I too should adopt this view, which they ascribe to Aristotle, a view that many maintain, or rather strive to maintain. For the other view also has its adherents, and, indeed, very learned adherents. And yet, if you want to hear by what reasoning I am motivated in rejecting Aristotle's view, I shall set forth reasons that completely refute that view, reasons that are not the product of imagination or hallucinations, but are derived from the senses themselves.

Domenico: To hear these arguments will afford no less pleasure than will the solutions of the problems. And if there is no time to complete these solutions before the dinner hour, we'll save the rest of the discussion for tomorrow. But perhaps it isn't a bad idea for us to go to the seashore and get some good food there from the fishermen and sailors and stay until late

---

4. I.e., specific gravity.

5. Marginal note (*Opere*, I, 369): "Aristotle *Physics* 8.82, *De Caelo* 3.28." Cf. *Physics*, 267a, and *De Caelo*, 301b.22–31.

in the day. For, as the sun climbs higher above the horizon, the air becomes warm enough to temper the harshness of the wintry chill. And since in that way there will be enough free time, I shall be able to hear from you all that I want.

Alessandro: Let us go now wherever you wish.

Domenico: But in the meantime don't hesitate to state your arguments.

Alessandro: Before I come to my arguments, it will be well first to make clear the view of those who share Aristotle's opinion.[6] [*cf. 308.3–8*] Now these people say that, when a body is thrown by a projector, e.g., when a stone is thrown by a man, the air next to the mover's hand is first set in motion by the mover's hand, and this air also sets the other parts of the air in motion; and that after the body is released from the hand and is in the air, the air already in motion keeps the body moving along with itself. This is the reason Aristotle gives [for the motion of projectiles]. Some also add that, as the body moves, the parts of air move up to fill the void which the body leaves behind it; and that the rear portion of the body is driven forward by the massing of these parts [of air].[7] But I shall try to show by the following arguments that this view is completely false.

(1) Let this, then, be my first argument. [*cf. 308.25*] If the body is moved by the medium, the body will necessarily move in the same direction as the medium; but we often see the opposite of this. Therefore, [etc.]. The minor premise is obvious. For if, while a very strong wind is blowing, a body is hurled against it, though the wind is blowing from the south, yet the body, if hurled in that direction, will move toward the south. Clearly, therefore, the body is moved not by the medium, but by another mover. And we must not say that, though we observe the wind moving in the opposite direction, still the parts of the air nearest the body move in the same direction as the body. For here is a very clear example on this point. [*cf. 308.29*] Do you see that small boat which a solitary boatman is rowing toward Pisa against the direction of the current? We observe that the boat, if given just one impulse by the boatman, is then carried for some distance against the force of the current; yet it is perfectly clear to the eye that the parts of water that touch the boat move in the opposite direction. And

---

6. Marginal note (*Opere*, I. 370): "Aristotle's opinion in *De Caelo* 3.28." See *De Caelo,* 301b.22–31.

7. This is the doctrine of *antiperistasis.* [Cardano, among others, supported this doctrine. — *S. D.*]

one should not by any chance suppose that the last portions of water, which touch only the stern of the boat, drive the boat on, whereas the other parts of water are observed to move in the opposite direction. For, apart from the fact that this view is ridiculous, experience actually indicates the opposite. Because, if someone at the stern of the boat lets a small piece of wood hang from a cord and lowers it into the water, he will see clearly that the wood is carried in the direction opposite [to that of the ship] and resists the pull of the cord. But the contrary would be the case if the water there moved in the same direction as that in which the boat is being rowed.

(2) If it is the medium that carries moving bodies along, how does it happen that when one shoots an iron ball and, with the same shot of the cannon, [a ball of] wood, or tow, or something light — the heavy object being the first out — how, I say, does it happen that the iron is flung a very great distance, while the tow, after following the iron for some distance, stops and falls to the ground? [*309. 2–10*] If, then, it is the medium that carries along both of them, why does it carry the lead or iron so far, but not the tow? Or will it possibly be easier [for the air] to move the very heavy iron than the very light tow or the wood?

(3) If the [projected] body is moved by the medium, then that body with which a greater number of impelling parts [of the medium] are in contact will move more swiftly and over a greater distance. But experience indicates the opposite of this. For, if a very thin shaft sharpened at both ends is shot from a bow, it will be carried over a longer distance than will a piece of wood that is thick, but of the same [total] weight as the shaft, and is shot from the same bow. And yet fewer parts of air strike the sharp point [at the rear] of the shaft.

Domenico: There would be an easy answer to this. For the shaft, since it is sharp, is better able to divide the medium than is the blunt piece of wood, and the air will therefore resist it less.

Alessandro: Do you by any chance suppose that the other supporters of this view [that the medium causes the motion of the projectile] will also make the same answer to my argument?

Domenico: Yes, indeed; and I, too, think it is the best solution.

Alessandro: Then how will you avoid asserting that the medium does not move in the same direction as that in which the body moves? For, if it did move in the same direction, the air would not have to be divided by the moving body. But it is clearer than daylight that the medium does have to be divided, since sharp objects move [through it] more easily than blunt objects. Hence it follows that the medium does not move in the direction in which the body moves.

It remains, then, merely for me to prove that the medium does not move the body by coming into the parts vacated [in the rear] and driving the body forward. For, if the parts of air move to fill up the void left by the body, why does not the body, in like manner, move backward to fill up that same void?

And suppose, for example, that the moving body is a part of a cylinder, with one of its bases in front, and the direction of motion is toward the north. Clearly some parts of air will rush into the void from the east, and some from the west, but none from the south. For the space into which the forward parts of the cylinder enter must be vacated by air (for otherwise there would be a case of interpenetration of bodies). And these portions of air must enter into the space which is vacated by the cylinder. But the space which the cylinder empties of air is always equal to the space which the cylinder leaves behind itself. Hence, to fill this space, the portions [of air] which formerly were in front of the cylinder are sufficient. But these parts,[8] entering into the void from the circumferential edge of that [rear] base, all come from the east or the west. And if this is the case, how will they drive the cylinder on toward the north?

To these arguments add another: if we suppose that the moving body is a cone, with its base in front and its vertex in the rear, then no parts of air will be able to move to fill the void.

Finally—and this is the strongest argument—consider a marble or iron sphere perfectly round and smooth, which can rotate on an axis, the ends of which rest on two supports. [*cf. 309. 18–24*] Then suppose a mover comes who twists both ends of the axis with his fingertips. Surely in that case the sphere will rotate for a long interval of time. Yet the air was not set in motion by the mover, nor can the medium ever move into the spaces left behind by the moving body since the sphere never changes its position. What, then, are we to say about this case of forced motion? By whom will the sphere be moved

---

8. Presumably Galileo would include also other parts directly or indirectly displaced by these parts.

when it is outside the hands of the mover? What are we to say, except that it is moved by an impressed force?

Domenico: I cannot but give approval to your arguments. Yet, on this last argument, which you seem to consider very important, I have reason to entertain some doubt. For those who maintain the opposite view could, perhaps, reply to an argument of this kind by saying that that motion, since it is circular, is not forced. For, since natural motion is the contrary of forced motion, but there is no motion contrary to circular motion, circular motion will under no circumstances be forced motion. And since it is not forced motion, the conclusion which you draw from the motion of the sphere will have no weight.

Alessandro: It will not be difficult to refute this answer. For when they say that forced motion is produced by the medium, they are thinking not only of that motion which is diametrically opposed to natural motion, but of any motion whatever which is not natural, i.e., of forced motion as well as mixed motion.

For it would be childish and ridiculous to say, for example, that a stone thrown [upward] at right angles to the horizon, because its motion will be diametrically opposed to natural motion, is moved by the medium when it leaves the hand, but that, if the same stone is thrown at an oblique angle, it would then be moved by a different agent. Yet the second motion will be a mixture of natural and forced motion.

Therefore whatever motions are different from natural motions are all included under the motion of projectiles. But the motion of a sphere such as we described is not natural but mixed motion. Therefore [etc.].[9] That the motion is mixed is proved as follows: Mixed motion is that which is compounded of natural and forced motion; but such is the motion of the sphere. Therefore [etc.]. And the rotation of the sphere is compounded of natural and forced motion, for some of its parts recede from the center of the universe (toward which the sphere would move in natural motion), while other parts approach that center. Nor should one say that the parts of the sphere which move down draw up the parts that ascend, alleging as a reason that, since the sphere is in equilibrium, the ascending parts do not resist the descending parts. For such an

---

9. Marginal addition (*Opere,* I, 373): "This is also obvious in the case of the bow whose very thin cord cannot move enough air to carry the arrow forward." The relevance of this remark at this precise point is not clear.

argument comes to naught. For if all parts of the sphere are
equally heavy, there will be no reason why the parts on the
right should draw up the parts on the left rather than the reverse. Then consider a sphere whose parts are not equally
heavy. In its motion you will see the heavier parts lifted by the
lighter, and yet there will be resistance in that motion. Tell me,
is there not always in such motion a resistance at the axis,
which, burdened by the weight of the sphere upon the supports,
resists the motion? Finally, I would have you note this; that
when it is said that circular motion is not forced motion, the
reference is to circular motion which takes place about the
center of the universe, e.g., the motions of the heaven. [*cf. 305.
8*] If, then, there were a marble sphere at the center of the
universe, so that the center of the universe and the center of
the sphere were the same, and if a beginning of motion were
imparted to the sphere by an external mover, the sphere would
perhaps move not with forced but with natural motion. For
there would be no resistance at the axis, and the parts of the
sphere would neither approach the center of the universe nor
recede from it. Now I said "perhaps." For, if the motion were
not forced, it would endure forever; but such eternity of motion
seems quite out of keeping with the nature of the earth itself,
to which rest seems more congenial than motion.[10]

And so, for all the reasons that have been adduced, it appears quite clear that the medium not only does not help motion
but, rather, on the contrary, opposes it. We must therefore
conclude that, when a body moves with other than natural motion, it is moved by a force impressed on it by a mover; but
what that force is is hidden from our knowledge.[11]

Domenico: The view [of the Peripatetics] may at last be
taken to be satisfactorily refuted. And since I have been convinced by your arguments, it may be conceded that the body
is moved not by the medium, but by an impressed force. But
now, before you say anything about the medium, I should
like you to explain to me how you conclude that in natural
motion the body moves by reason of heaviness and lightness.

Alessandro: What has to be said about the agent of natural
motion and about the medium can conveniently be set forth
together, so far as the present discussion is concerned. Let
us, then, first assume that it has so been arranged by nature

10. I.e., in the Ptolemaic system. [It is evident from this remark and from
the preceding phrase "center of the universe" that Galileo was not a Coper-
nican when he wrote this. —*S. D.*]
11. Galileo had added but deleted: "So also hidden from our knowledge is
the nature of the force [*virtus*] that causes strings to resound."

that heavier bodies remain at rest under lighter. And it is quite clear to the senses that this is the case.

Domenico: We do, of course, grasp by our senses that what you assume is perfectly true. But I would like to understand the reason why nature has maintained this arrangement rather than the opposite.

Alessandro: To give the reason for this arrangement cannot contribute anything useful for our purpose, since it is clear that those are the facts. [*cf. 252–53, 342–46*] And it would, no doubt, be very difficult to give the precise cause. Indeed, I should not be able to give any other except that things had to be disposed in some arrangement, and it has pleased nature to dispose them in this one.[12] Unless, perchance, we should wish to say that heavier bodies are nearer the center than lighter bodies because somehow those bodies seem to be heavier which contain more matter in a smaller space.[13] For example, suppose there is a bag filled with wool, the wool not having been forcibly stuffed into it. If, now, much more wool is forcibly compressed into the bag, the latter will be heavier than before because more matter is compressed in the same space. Since, then, the spaces which are nearer the center of the universe are always more constricted than those which are more remote from the center, it was reasonable that these narrower spaces should be filled with that form of matter whose greater weight, in contrast with that of any other [form of matter], would fill the narrower spaces.

Domenico: Though that argument is not to be considered a conclusive reason for this disposition of the elements, still it has in it some appearance of truth, to which the mind readily gives assent. And so, not only because what you assume is self-evident, but also because the remarks you have just made on the same subject give, in a sense, a reason for the arrangement, I shall confidently concede that heavier bodies are naturally situated under lighter bodies. Therefore, please go on with the rest of the discussion.

Alessandro: We must now observe that bodies are not called heavy or light except by way of comparison.[14]

12. Marginal note (*Opere*, I, 374): "And this is Aristotle's reason, *Physics* 8.32." Cf. *Physics* 255b.13–17.

13. Marginal note (*Opere*, I, 374): "This is clear in the case of vapors that ascend and descend."

14. Marginal note (*Opere*, I, 375): "Aristotle holds the contrary view in *De Caelo* 4. Hence it will be appropriate to refute Aristotle's view at this point. It is refuted at *b*." The reference is to separate manuscript sheets, inserted here. The next five speeches of the "Dialogue" must therefore be considered a later addition.

Domenico: Just wait, please. We must proceed step by step. You say that there is no such thing as the absolutely heavy or light, but merely the relatively heavier or lighter. This first point I will not grant, especially since Aristotle's view throughout all of *De Caelo,* Book 4, is to the contrary. There he shows, against the view of the ancient philosophers, that earth is absolutely heavy and fire absolutely light. Therefore, unless you first dispose of the arguments set forth by Aristotle, I shall never embrace your opinion.

Alessandro: Our discussion will be too long if Aristotle will have to be refuted in connection with all the arguments opposed to Aristotle's doctrine that I shall set forth.

Domenico: Our discussion will be too short if by any chance you wish to lay foundations by merely asserting your doctrines without argument. For in that case I don't care to hear any more from you. Therefore, unless you make clear by what reasoning you are led to reject Aristotle's view, you may stop speaking; for you would be talking to deaf ears (as the saying goes).

Alessandro: Since I have undertaken this task, I shall, in order to please you, indicate the reasons that have compelled me to reject Aristotle's view. [*cf. 289-94*] I shall do so briefly, though I could discuss the matter at great length.

Now, in the first place, Aristotle asserted that earth is the heaviest of all substances; for he wrote (*De Caelo* 4.29) that all things have lightness except earth, and that mixed substances have more heaviness the more they contain of earth. If, then, earth is the heaviest of all substances, it is clear that none of the mixed substances will be heavier than earth itself, since they are composed also of water, air, and fire, which have less heaviness than earth. But this is false. For who cannot see that all the metals are heavier than earth itself, e.g., mercury, on which earth floats? And a substance which is at rest above something else is, according to Aristotle himself, lighter than that upon which it rests. And who will doubt that a vessel filled with lead is heavier than one filled with earth? How, then, is earth the heaviest of all substances?

But Aristotle adduces another indication of the [absolute] heaviness of earth when he says: "If air or water is removed, earth will never rise to the place [emptied] of air or water. This is obvious in the case of the physicians' cups, which draw up water and flesh, but certainly not earth. Therefore earth is the heaviest substance." [*cf. 292.32-293.4*] But what more childish argument can be imagined? For if earth is not a fluid body, how will one part of it be drawn up over the other part? Therefore, this [failure of earth to rise in the cupping glass] comes about not from the absolute heaviness of earth but from its solidity. For neither will ice be drawn up nor expand in the cup,[15] and ice is not earth. On the other hand, quicksilver, though it is far heavier than earth, will be drawn up, because it is fluid.

Next consider this argument: "The center is the contrary of the extremity. Therefore it is reasonable that that which is at the center is the contrary of that which is at the extremities. And this will not be the case unless earth is assumed to be absolutely heavy, and fire absolutely light." [*cf. 292. 23–30*] Now such an argument not only does not lead to a necessary conclusion, but, in my opinion, has little force. For if [Aristotle] takes the sphere of the moon[16] as the extremity, and the center of the universe as the center, then surely the sphere of the moon will no more be contrary to the center of the universe than will the sphere of air or the sphere of water.[17]

And if spatial contrariety is to be understood in this way, earth itself will be in contrary places, since it is both at the center and also in the sphere of air and of water. Aristotle reasons in similar fashion about fire, saying that, "If the air [beneath it] is removed, fire will not move downward; on the other hand, air will move downward, if the water [beneath it] is removed."[18] [*292.30–32*] Such a statement requires proof; but Aristotle did not furnish a proof. Unless, possibly, you believe that there is a proof in what he sets forth at section 39 of the same book [*De Caelo* 4] when he says that fire will not move downward because it has no heaviness.[19] But it would be circular reasoning if, in an attempt to prove that fire has no heaviness because it does not descend, he were to assert that

15. Reading in *cucurbitula* (as in *Bull. bibl. e storia sci. math. e fis.*, 16 [1883], 28.6) for *incudo* (*Opere*, I, 367.25) the meaning of which is not clear.

16. I.e., the sphere on which the orbit of the moon is traced. Below this sphere are the spheres of the several terrestrial elements in the Aristotelian system. To each of these elements there corresponds a sphere having as center the center of the universe, in the order of earth, water, air, and fire, proceeding from the center upward to the region of the moon. The region of air, for example, would be the region above the surface of the sphere of water and below the surface of the sphere of air; and the phrase *concavum aeris*, "the concave surface of [the sphere of] air," the upper extremity of the region of air, might stand for "region of air." I have so translated it in this paragraph.

17. Marginal note (*Opere*, I, 376): "And I say this especially since Aristotle himself says (*Physics* 8.75) that the spatial contraries are up and down." The statement is not in the passage cited, but see, e.g., *De Caelo*, 273a.8–9.

18. Cf. *De Caelo*, 312b.5–11.

19. Cf. *De Caelo*, 312b.15.

it did not descend because it had no heaviness. Thus this last proposition [that fire has no heaviness] lacks proof, and all the more because Aristotle himself, so far as I know, did not make a test to determine whether fire moves downward if the air [beneath it] is removed. And how did Aristotle know that there is nothing lighter than fire? [*cf. 294.2*] Cannot there be certain exhalations that will float aloft above fire? [*293.12–20*] And, finally, how will anyone ever be able to conceive of fire, a substance linked with quantity, as not having heaviness? Such an idea seems to me to be wholly unreasonable.

And when we say that earth is the heaviest of all substances because it lies below all others, we must, whether we wish to or not, declare that earth is heaviest *in comparison with others,* for the reason that it lies below all the others; and that it lies below all others for the reason that it is heavier than those below which it lies. This much is clear: if it is heaviest because it lies below all others, then, if everything else is removed, it will no longer be able to be called heaviest, since it does not lie below anything. And therefore it is said to be heaviest only in comparison with the less heavy. And we must make the corresponding statement about fire.[20] [*cf. 294.9–13*] Now one should not say: "If fire had any heaviness, it would move downward." For does not air have heaviness? And yet air does not move downward below water. So also fire has heaviness, but it does not move downward below air; for it has less heaviness than air.

Therefore, to put the matter succinctly, I say that, in the nature of things, there is some one substance that is the heaviest of all, and some one substance that is the lightest of all, i.e., possessing the least heaviness. But I deny that those substances are necessarily earth and fire. And similarly I deny that we can speak of an absolutely heaviest or an absolutely lightest substance, without reference to the less heavy or the less light. We can say only that such and such a substance is the heaviest of all [known] heavy substances, but not of all substances that could possibly exist. These are the brief remarks I would make in support of my viewpoint.

Domenico: Your arguments satisfy me completely, and all the more because this problem, whether[21] earth is absolutely or relatively heavy, is of small importance; and, in my opinion,

20. I.e., that it is lightest only in comparison with the less light.
21. Transposing *utrum* and *an* (*Opere,* I, 377.36).

in whatever way it is resolved, it does not contribute much to our present discussion. Let us therefore proceed to the rest of our problems.[22]

Alessandro: I assert, then that heavy and light substances cannot be spoken of except in comparison. And this comparison takes two forms: either we compare with each other two bodies that are in the same medium, or we compare a body with the medium in which it moves. [*cf. 251, 341*] Now in the first comparison those substances are said to be equally heavy which, when they are of equal size [volume] and in the same medium, have the same weight. From this it is clear that, if we take two pieces, one, let us say, of wood and the other of iron, and the pieces are of equal weight, those substances will still not be considered equally heavy; for the piece of wood will be much greater in volume than the piece of iron. Thus, one substance will be said to be heavier than a second if, when equal volumes of the two substances are taken and they are weighed in the same medium, the first weighs more than the second. And one substance is called lighter than a second if, when equal volumes of the two substances are taken and they are weighed in the same medium, the first turns out to weigh less than the second.

In similar fashion, media also will be said to be heavier or lighter in comparison with one another. And solid bodies likewise will be said to be heavier or lighter in comparison with the medium in which they move. And all the elements except earth serve as media through which motion may take place. Earth, since it is the most solid [of the elements], cannot be divided by any other body: but the other elements, namely, water, air, and fire, since they are fluids, permit motion to take place in their midst.

Having clarified these matters, we shall readily understand that heavy bodies move by reason of their heaviness, and light bodies by reason of their lightness. [*cf. 253–54*] Thus, those bodies that are heavier than the medium through which they move move downward; for it has so been arranged by nature that heavier bodies remain at rest under lighter. But if a body heavier than water remained at rest above the water, then the lighter would be under the heavier. Thus heavy bodies move downward insofar as they are heavier than the medium through which they move; hence their heaviness in comparison with the medium is the cause of this downward motion. And similar

22. Here the original manuscript resumes. See note 14, above.

reasoning will lead to corresponding conclusions about bodies lighter than the medium.

Domenico: What you have just said is not completely satisfactory, and I have reason to entertain some doubt. [*cf. 254.8–14*] For if we take, let us say, a very tiny pebble and throw it into the sea, the pebble will no doubt move downward through the aqueous medium. But I do not at all understand for what reason the pebble is to be considered heavier than the water of the sea, particularly since [all] the water of the sea is certainly heavier than almost any number of pebbles.

Alessandro: Have you so quickly forgotten what I said just now? Did I not say that one substance is heavier than a second only when a given volume of the first weighs more than an equal volume of the second? If, then, we take a portion of water whose volume is equal to that of the pebble, we shall find that the pebble is heavier than the water. Hence, it is not strange that the pebble sinks in the aqueous medium.

Domenico: This is all very true. But I do not yet understand why, in the case of a pebble in water, we must consider only a volume of water equal to the volume of the pebble, and not the whole quantity of water.

Alessandro: Now at last I can no longer avoid demonstrating some theorems to you. From the comprehension of these theorems you will clearly understand not only the answer to your question, but also the ratio of the speed or slowness of the motion[23] of bodies both heavy and light, as well as the ratio of the heaviness or lightness of one and the same body weighed in different media. All these theorems had to be demonstrated when I tried to find the true reasoning by which, in a mixture of two metals, we could indicate the exact amount of each separate metal.[24] For these theorems (though they are not different from those demonstrated by Archimedes) I shall adduce proofs that are less mathematical and more physical. And I shall use assumptions that are clearer and more obvious to the senses than those which Archimedes embraced.

Domenico: Are you saying, then, that you also have worked out the precise reasoning by which Archimedes is said to have detected the goldsmith's theft in the case of the gold crown? But wasn't this discussed by many writers, and by Vitruvius in particular?[25]

23. I.e., natural motion in a given medium.
24. The Archimedean problem of the crown. This was taken up in Galileo's *Bilancetta* (1586).
25. Cf. Vitruvius, *On Architecture,* IX, Preface, trans. M. H. Morgan (Cambridge, 1914), p. 254. The hydrostatic balance was described in the medieval *Carmen de ponderibus,* 11. 124–208.

Alessandro: I could show how thoroughly unreliable is the method commonly described, in which people speak of a vessel filled with water, etc. On the other hand, the method I have discovered is completely accurate and is, I believe, the same as Archimedes' method both because it is so elegant and because it depends on Archimedes' own demonstrations.

Domenico: Indeed, if that beautiful discovery of yours depends on the demonstrations that you are soon to set forth, it will be a great pleasure to hear about it too. But now, if you need line drawings for the purpose of your demonstrations, here is a smooth and broad level area of the finest sand, and you will be able to make suitable drawings in it with that rod.

Alessandro: Before I come to my demonstration, I must make this point: though there are, as I have said, three especially common media through which motion takes place, yet, since [the region of] fire is too far distant from us, and since we do not have ready to hand objects that rise in air, I shall set forth my demonstrations with water as the medium. And nobody will deny that what is demonstrated in the case of water is true of other media also.

First, then, I say that, if a solid body has the same weight as a portion of water equal to the said body in volume, and if the solid body is let down in water, it is completely submerged. [*cf. 254.19-22*] And when it is all under water, it no longer moves either upward or downward. Now this is the same as saying that solid bodies of equal heaviness as water, if let down in water, are completely submerged, and yet do not continue to move either downward or upward. [*254.31-255.24*]

Thus, let *CED* be the first position of the water, before the body is let down into it. Suppose that body *AB*, which is of the same heaviness as water, is let down into water and, if possible, is not completely submerged, but some part of it protrudes above water, namely *A*. Now suppose that the surface *CD* of the water is raised up to *FG* and that both the water and the body are at rest in this position. Clearly, then, the volume of water which is raised, i.e., the volume bounded by surfaces *FG* and *CD*, is equal to the volume of the part of the magnitude which is submerged, namely, *B*. For it is quite clear that the said volume of water cannot be smaller [than the volume of the submerged part of the solid], for otherwise there would be interpenetration of bodies. Nor can it be greater, for otherwise some empty space would be left. Now the water bounded by surfaces *FG* and *CD* strives by its heaviness to return downward to its original state, but cannot attain this end unless solid *AB* is first raised by the water and forced out of it. And

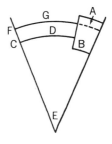

the solid with its own whole weight resists being raised. But both the solid and the water are assumed to be at rest in this position. It therefore follows that the weight of water *FGCD*, with which it strives to raise the solid, is equal to the weight with which the solid resists and presses downward. (For if the weight of the water *FGCD* were greater than the weight of the solid *AB*, then solid *AB* would be raised by the water and forced out of it. If, on the other hand, the weight of solid *AB* were greater, the water would be raised. But the whole system is assumed to be at rest in this position.) If, then, the weight of only a part of water *FGCD* is equal to the weight of the whole solid *AB*, then the volume of a part of water *FGCD* will be equal to the volume of the whole solid *AB*. But this is impossible. For the volume of water *FGCD* is equal to the volume of that portion of the solid which is under water, namely, portion *B*. And *B* is less than *BA*. Therefore it is not true that some portion of the solid body will protrude outside the water. Hence the whole solid will be submerged.

And yet, when the whole body is under water, it will not sink farther, but will remain in whatever position it is placed. [*256.8-13*] Indeed, there is no reason why it must descend; for, since it is assumed to be of the same heaviness as water, to say that it sinks in water would be the same as saying that water, when placed in water, sinks underneath this water; and then that the water which rises above the first mentioned water again moves downward, and that the water thus continues to move alternately downward and upward forever. This is impossible.

Having thus concluded this proof, we must next show that solid bodies lighter than water, if let down into water, are not completely submerged, but that some part of them protrudes from the water. [*cf. 256.14-34*] And we must therefore recall that one solid body is said to be lighter than a second body if, when there are equal volumes of both, the first is lighter in weight than the second. Suppose that the first position of the water is *EF*, before the solid body is let down into it. And suppose that *AB*, the body let down into the water, is lighter than water. I say, then, that the solid is not entirely submerged. For, suppose that, if possible, it were completely submerged and that the water were raised up to surface *CD*. The volume of water *CF* will therefore be equal to the volume of body *AB*. Suppose, if it is possible, that both the body and the water remain in this position. Now the weight with which water *CF* presses downward to raise the solid body is equal to the weight with which body *AB,* pressing downward, resists being raised (for they are assumed to be at rest in this position); and, furthermore, the volume of water *CF* is equal to the volume of body *AB.* For these reasons it follows that the solid body is of the same heaviness as the water. But this is impossible: for it was assumed to be lighter. Therefore, the body *AB* will not be completely submerged, but some part of it will protrude from the water.

Now that it has been proved that solid bodies lighter than water are not completely submerged, it is appropriate for us to show what portion of such bodies is below water. I say, then, that solid bodies lighter than water, if let down into water, sink as far as the point where a volume of water equal to the volume of the submerged part of the body has the same weight[26] as the whole body.

Suppose that the first position of the water is *AB,* and that the solid body *CD,* lighter than water, is let down into the water. It is clear, from what has already been said, that the body will not be completely submerged. Suppose, then, that part *D* is submerged and that the water is raised so that its surface is *ET.* I say, therefore, that a volume of water equal to the volume of the submerged portion of the solid has the same weight[27] as the whole solid. For since water *EB* presses down with as much weight as that with which body *CD* resists (for they are assumed to be at rest in this position), the weight of water *EB* is consequently equal to the weight of the whole body *CD.* But the volume of water *EB* is equal to the volume of the submerged part of the solid, namely, *D.* Therefore, a volume of water equal to the volume of the submerged part of the solid has the same weight as the whole solid. Which was to be proved.

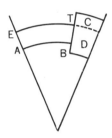

And now, before we come to the demonstration about solids heavier than water, we must show with how much force a solid body lighter than water is lifted upward if it is completely submerged by force under water. [*cf. 257–60*]

I say, then, that solid bodies lighter than water, if forced under water, are lifted upward with a force equal to the excess of the weight of a volume of water equal to the volume of the submerged solid over the weight of the solid itself.

Suppose, then, that the water, in its first position before the solid is let down into it, has as surface *AB.* And suppose that solid *CD* is forcibly submerged in it, the water being thereby

26. I.e., in a medium in which both have weight.
27. In a medium in which they both have weight.

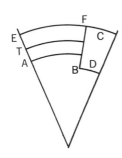

raised so that its surface is *EF*. Since the water *EB*, which is raised, has a volume equal to the volume of the entire submerged solid, and the solid is assumed to be lighter than water, the weight of water *EB* will be greater than the weight of solid *CD*. Suppose, then, that *TB*, a portion of the water, has a weight equal to that of solid *CD*. We must prove, therefore, that the solid body is carried upward with a force equal to the weight of water *TF* (for this is the amount by which the weight of water *EB* exceeds the weight of water *TB*, i.e., exceeds the weight of solid *CD*). Now, since the weight of water *TB* is equal to the weight of solid *CD*, water *TB* will press down[28] to raise the solid with as much force as the solid will resist being raised. That is, the weight of a part of the water that presses, namely, *TB*, is equal to the resistance of the solid body. But the weight of all the water that presses, namely, *EB*, exceeds the weight of water *TB* by the weight of water *TF*. Therefore the weight of all the water *EB* will exceed the resistance of solid *CD* by the weight of water *TF*. Therefore, the weight of all the water that exerts pressure will impel the solid upward with a force equal to the weight of a part of the water, namely, *TF*. Which was to be proved.

Now, from what has been proved, it is quite clear that solid bodies heavier than water [continue to] move downward if they are let down into water. For if they do not [continue to] move downward, either some part of them will protrude or they will remain at rest under water and will move neither up nor down. But no part of them will protrude; for in that case they would, as has been proved, be lighter than water. Nor will they remain at rest under water, for then they would be of the same heaviness as water. They can, therefore, only continue to move downward.

Let us now show with what force [*vis*] they move downward. I say, then, that solid bodies heavier than water, if let down into water, move downward with a force equal to the amount by which a volume of water equal to the volume of the solid body is lighter than that body.

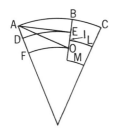

Thus, suppose that, in its first position, the surface of the water is *DE*. And suppose that a solid body *BL*, heavier than water, is let down into the water, so that the surface of the water is raised to *AB*, water *AE* having a volume equal to the volume of the solid. Since the solid is assumed to be heavier

---

28. Reading *deorsum* (*Opere*, I, 383.16), which Favaro emended to *sursum*.

than water, the weight of water *AE* will be lighter than the weight of the solid body. Suppose, then, that *AO* is the volume of water that has a weight equal to the weight of *BL*. Since water *AE* is lighter than *AO* by the weight of *DO*, we must prove that solid *BL* moves downward with a force equal to the weight of water *DO*.

Consider a second solid, lighter than water and joined to the first solid. Let the volume of this second solid be equal to the volume of water *AO*, and its weight equal to the weight of water *AE*. Suppose that this second body is *LM*. Since the volume of *BL* is equal to the volume of *AE*, and the volume of *LM* is equal to the volume of *AO*, it follows that the volume of the combined bodies, *BL* and *LM*, is equal to the sum of the volumes of water *EA* and *AO*. But the weight of water *AE* is equal to the weight of body *LM*, and the weight of water *AO* is equal to the weight of body *BL*. Therefore the total weight of both bodies *BL* and *LM* is equal to the combined weight of water *OA* and *AE*. But the volume, too, of the combined bodies *ML* and *LB* has been shown to be equal to the [sum of the] volumes of water *OA* and *AE*. Therefore, by our first proposition, bodies so combined will move neither upward nor downward. And consequently the force with which body *BL* presses downward is equal to the force with which body *ML* presses upward. But, according to our previous proposition, body *ML* presses upward with a force equal to the weight of water *DO*. Therefore, body *BL* will press downward with a force equal to the weight of water *DO*.

And from this the conclusion is clear that a solid heavier than water is lighter in water than in air by the weight, in air, of a volume of water equal to the volume of the solid. For solid *BL* moves downward in air with a weight which is assumed to be equal to the weight of water *AO;* and it moves downward in water with a weight equal to the weight of *DO*. But the weight of *AO* exceeds the weight of *DO* by the weight of *AE*. And that is the weight of a volume of water equal to the volume of solid *BL*.

Now all the demonstrations made with respect to water as medium must also be understood as applying to air as medium. And from this the general conclusion is obvious; that [in natural motion] bodies heavier than the medium through which they must move move downward, while bodies lighter than the medium through which they must move move upward.

Domenico: From your very clear and definitive proofs I have attained so perfect and exact an understanding of these

motions that I need never again entertain any doubt about
them. Previously, however, I did have doubts because I had
never understood these motions except in a sort of confused
manner.

Alessandro: It also follows that [in natural motion] heavy
bodies move downward more readily to the extent that the
medium through which they must move is lighter; and that
light bodies move upward more readily to the extent that the
medium through which they move is heavier.

Domenico: Hence we may consider false the assertion of
practically all philosophers when they say, in their attempt to
prove that air is more heavy than light, that air must be considered more heavy than light for the reason that it more
readily carries heavy bodies downward than light bodies upward.[29] [*cf. 285.31–286.4*]

Alessandro: O ridiculous hallucinations! O inept ideas of
men, ideas which not only do not approach truth, but are opposed to truth itself! Heavens! How, I ask, are we to believe
the fantastic notions of these men by which they profess to explain the most recondite secrets of nature, if in matters that
lie completely open to their senses they rashly assert the very
opposite of the truth? And who, I ask, dreamed up the notion
that those media are more heavy than light if they carry heavy
bodies downward more readily than they carry light bodies
upward, when the facts are just the reverse? [*cf. 287.10–
288.3*]

Indeed, if that conclusion were sound, surely air would be
heavier than water itself. For, whatever heavy body it is that
moves downward, it will always move more readily and more
swiftly in air than in water. In fact, there are many bodies
which in air move downward very swiftly and readily, but in
water not only do not move downward, but float on the surface
and, if forcibly submerged, tend to move upward. This may
be made clear by an example. A gourd, for instance, readily
moves downward in air, but it cannot be made to move downward in water except with the greatest difficulty and force.
Hence [according to the view under discussion], since air
carries bodies downward more readily than does water, it will
have to be considered heavier than water. What absurdity!

29. Marginal note (*Opere,* I, 385): "Aristotle (*De Caelo* 4.39 [*De Caelo,*
312b.2–4])asserts that water and air are heavy in their own region. And he
says the same thing in section 30 of the same book [*De Caelo,* 311b.8–10]
where he cites the example of the inflated bladder. In *De Caelo* 3.28 [*De
Caelo,* 301b.23–25] he declares that air, insofar as it is light, aids upward
motion, and insofar as it is heavy, aids downward motion."

But now please listen. It has been proved that heavy bodies that move downward in water move downward with a force equal to the excess of their weight over the weight of an equal volume of water. [*287.16–288.1*] Suppose there is a heavy body *A* whose weight is, let us say, 8, and that the weight of a quantity of water *B*, whose volume is equal to the volume of *A*, is 4. In that case the solid will move downward with a speed and facility represented by 4. But if the same body *A* were to move [in natural motion] through a lighter medium, a medium such that a volume *C* of it, equal to the volume of *A*, has a weight of 2, then solid *A* would, of course, move downward in this second medium with a speed and facility represented by 6. It is clear, therefore, that the same body *A* moves downward more readily through lighter than through heavier media. And hence it follows that a medium is to be considered lighter in proportion as heavy bodies move downward more readily in it. But general opinion, for the most part, asserts the opposite of this view. To whom, then, is it not perfectly clear that, if the air were still lighter, heavy bodies would more readily move downward [through it]? And if that is true, it follows that air is very light just because heavy bodies easily move downward in it.

A	B	C
8	4	2

Moreover, we must reason in opposite fashion about light bodies which move upward. And our conclusion will be that that medium is to be considered heavy through which light bodies move upward more readily, and that medium is to be considered light through which light bodies move upward with difficulty. Therefore, both because light bodies move upward with greater difficulty in air, and because heavy bodies move downward with more facility in air, it follows that air is rather to be considered light than heavy. And yet I should wish my remarks to be understood as directed only to the refutation of the view of those who have said that air is more heavy than light.[30]

Domenico: What, then, is your view about the heaviness and lightness of the elements?[31]

Alessandro: If we are to speak about absolute heaviness or lightness, I say that all bodies, whether they are mixed or unmixed,[32] have heaviness. But if we are to discuss relative

30. Galileo's view is that everything that exists (including fire) has weight. But he has digressed to indicate why he takes air to be among the lighter substances.

31. Marginal note (*Opere*, I, 386): "Aristotle asserts that air and water have weight in their own natural places, and he adduces the example of the inflated bladder." See *De Caelo*, 311b.8–10.

32. I.e., non-elements or elements.

heaviness or lightness, I say again that all bodies have heaviness, but that some have more and others less, and that it is
this lesser heaviness which we call lightness. And thus we say
that fire is lighter than air, not because it lacks heaviness, but
because it has less heaviness than air; and, similarly, we say
that air is lighter than water.

Domenico: But, tell me, if you ascribe heaviness to fire,
how will you explain the fact that fire does not move downward,
since heavy bodies are those that move downward? And everyone asserts that it is impossible for fire to move downward,
even if the air under it were removed.

Alessandro: Ah, ah! New hallucinations, new fantasies!
Only those things move downward which are heavier than the
medium through which they must move. But fire is not heavier
than air and for that reason cannot move downward [in air].
Yet if the air were removed from under it, so that a void were
left under the fire, who can doubt that fire would move down
into the place of air? For, since there is nothing in a void, that
which *is* something is heavier than nothing. Therefore, since
fire is something, it would surely move below nothing; for it
has been arranged by nature that heavier bodies remain at rest
under lighter bodies. But do not suppose that the fire will move
downward merely for the purpose of filling the void. For in
descending it would leave a void under the sphere of the moon,[33]
but it would not be able to rise to fill this void, since heavier
bodies do not rise above lighter.

Domenico: But if you assert that all the elements, air as
well as fire and water, are heavy, how are we able to sustain
the weight of air? [*cf. 288.4–17*] And, similarly, when we
swim under water, how are we able to sustain the immense
mass of water by which we do not at all feel weighed down?

Alessandro: The solution of this uncertainty is extremely
easy, and depends on what we have proved above. But the
solutions given by others are not entirely satisfactory.

Thus some[34] say that fish and men under water are like mice
in a wall. The mice [they say] do not feel the weight of the
bricks, because the bricks rest on other bricks, not upon the
mice. How mistaken this solution is is perfectly clear. For, I
ask, what analogy can there be between hard, solid bodies like
stones, and fluid, liquid bodies like water? Thus, if we remove

33. The region of fire lies beyond the sphere of air and reaches as far as
the sphere that contains the moon's orbit.
34. Marginal note (*Opere,* I, 387): "Simplicius on *De Caelo* 4.30." Simplicius, *Commentary on Aristotle, De Caelo,* ed. Heiberg, p. 710.

the mice from the wall, an opening in which there was a mouse will still remain. Hence it is clear that the stones do not rest upon the mice, but upon other stones. But if we move a fish from the water, do you suppose that the place in which the fish was will remain [unoccupied]? Will not water flow into it immediately? And this is the clearest indication that the water rests upon the fish.

Now others have said that the elements are neither heavy nor light in their own proper place, and that therefore swimmers under water are not weighed down by the water. But those who say this do not yet remove the uncertainty. For, in the first place, there will have to be proof of what they assume, namely, that elements are neither heavy nor light in their own proper place. But if this is so, why do they later say that air is more heavy than it is light? But if they wish to support the proposition [that the elements are neither light nor heavy in their proper place], by what reasoning will they maintain the view that water does not have great weight? And, then, even if this is conceded,[35] the problem is nevertheless not yet solved. For if in their proper place the elements are neither heavy nor light, I ask of them what the proper place of water is. They answer, I suppose, [that the proper place of water is] under air. But if we climb up a high tower and there is a bathtub at the top of it, the same thing will happen to us if we are immersed in the water there as if we are immersed in the sea. For we will not be weighed down by the water, even if the water has air underneath it and is consequently outside its proper place. That is, the whole error of those people comes from the fact that they failed to take account of the heaviness of the medium through which the heavy bodies must move, and considered only the heaviness or lightness of the bodies themselves. But listen to my own solution, that I may give a satisfactory answer to your question.

We are said to be weighed down when some weight rests upon us which, by its heaviness, tends to move downward, while we must exert force to resist its further downward motion. [288.18–289.6] It is that resistance which we call being weighed down. But it has been shown that bodies which are heavier than water descend when let down into water; and, though they are heavy in water, they are less heavy than in air by the weight of a volume of water equal to the volume of the body. And it has also been shown that bodies lighter than water, if they are forcibly submerged under water, are lifted

---

35. I.e., that water has no weight in its proper place.

upward by a force equal to the amount by which the weight, in air, of a volume of water equal to the volume of the body exceeds the weight of the body. Finally, it has been shown that bodies of the same heaviness as water, when immersed in water, move neither upward nor downward, but remain at rest where they are placed, provided they are completely under water.

Now from all this it is clear that, if some weight heavier than water [e.g., a stone] rests upon us while we are under water, we shall be weighed down, to be sure, but less than if we were in air, since the stone is less heavy in water than in air. On the other hand, if, while we are in the water, a body lighter than water is attached to us, not only will we not be weighed down, but we will actually be lifted by it. Now this is clear when we float with the help of a gourd, although otherwise, e.g., when we are in the air, we are weighed down by the gourd. And the reason for this is that, when the gourd is thrust under water, it tends to move upward and has a lifting effect, whereas in air it moves downward and exerts downward pressure.

But if, while we are in the water, a body of the same heaviness as water rests upon us, we will be neither weighed down nor lifted by it, since such a body would move neither downward nor upward. But no body can be found that is more nearly equal to water in heaviness or lightness than water itself. It is therefore not strange that water does not sink or exert downward pressure in water, and that it does not rise or exert a lifting force. And we said that the putting forth of force to resist a body that tends to move downward is what we mean by being weighed down. A completely parallel argument is applicable to air.

Domenico: What a beautiful discovery! How true and accurate a solution! By your discussion you have so completely removed all the clouds of uncertainty that you have left no room for any further doubt about these matters. But what about the problems I proposed?[36]

Alessandro: From the observations which we have made concerning media, moving bodies, and motions, the solution of at least one of your problems is already completely clear; and the solutions of the others will soon become clear both from what has been set forth and from what is still to be discussed.

36. At the beginning of the "Dialogue."

Thus the answer to the question why the same body moves downward more rapidly in air than in water is obvious. [*cf. 261.35–262.12, 272.20–29*] For the difference [*proportio*] between the weight of the body and the weight of the air is greater than the difference between the weight of the same body and the weight of water; for water is heavier than air.[37] Hence it follows that the same body will fall with greater force [*vis*] in air than in water.

Domenico: I had already attained a perfect understanding of your solution of this problem from the propositions you had set forth and proved. But the problem of the turning point of motion . . .[38]

Alessandro: Before I come to the explanation of my view of this problem, we must look into certain matters. First, then, we assert (as we have abundantly corroborated above) that a body which moves in non-natural motion is moved by a force [*virtus*] impressed on it by a mover. Now let it be assumed that the same body, when projected by equal forces along straight lines that make equal angles with the horizon, always moves over equal distances.[39] Secondly, let it be assumed that a body moved by a finite impressed force cannot be moved in forced motion over an unlimited distance. Thirdly, let it be assumed that a body is not moved in forced motion unless the impelling force is greater than the resistance exerted by the body's own weight; but that, if the resisting weight is greater than the impelling force, then the body no longer moves in forced motion, but changes over to natural motion. From this it clearly follows that, when the body is at rest, its own weight is equal to the impelling force. For if the weight were greater, the body would move downward; but if the impelling force were greater, the body would then move with forced motion. From this it is clear that the body will remain at rest as long as there is equality between the resisting weight and the impelling force. Let our fourth assumption be that the same heavy body is sus-

---

37. [The use of *proportio* here suggests that Galileo had not yet seen the pseudo-Archimedean medieval treatise published with Tartaglia's edition of Jordanus in 1565, where the term "specific gravity" was used. Galileo had adopted that term by 1612. —*S. D.*]

38. I.e., whether there is an interval of rest at the turning point in reflected motion, e.g., between the upward and downward motion of a projectile thrown vertically upward.

39. The assumption is made that the initial part of the trajectory of a projectile, even if thrown other than vertically upward, is a straight line. See "Essay," *De motu,* Ch. 23 (in our *Galileo On Motion and On Mechanics* [Madison, 1960], pp. 110–14).

tained [*substineri*]⁴⁰ by equal forces [*virtutes*] over equal
intervals of time. [*327.1–3*]

On the basis of these assumptions it can now be proved that
the force impressed by a mover is continuously weakened in
forced motion, and that in any given motion no two points can
be assigned in which the impelling force is the same.

Let *AB* be a line on which a forced motion from *A* to *B* is
produced by a finite force. [*cf. 327–28*] Since this motion,
by our assumption, cannot be without end, let it terminate at
point *B*, and let the body move no farther. I say, therefore, that
the impelling force continuously becomes weaker in such motion and that no two points can be assigned on line *AB* at
which points the impelling force is of the same strength.

For, if it is possible, let there be two points *C* and *D* with the
force at *D* no weaker than that at *C*. The force will therefore
be either the same or greater. Suppose, first, that it is the same.
Now, since the moving body remains the same, and the impelling force is the same at *D* as at *C*, and the line over which the
motion takes place makes the same angle with the horizontal
(since it remains one and the same line),⁴¹ it follows that the
body will move over equal distances as measured from *C* or
D. But it moves from *C* as far as *B*; therefore it will move from
D to a point beyond *B*. But this is absurd; for it was assumed
that it does not move beyond *B*.

And now even greater absurdity would follow if we were to
say that the force is greater at *D* than at *C*. If, then, in motion
AB no two points can be assigned at which the impelling force
is equal, it is also clear that, in the course of the time during
which such motion takes place, no two moments can be assigned
in which the impelling force is the same. On the basis of these
assumptions and demonstrations it follows necessarily that
there is no [interval of] rest at the turning point.⁴² For if there
is a state of rest that lasts for some interval of time, there will
also be, between the weight of the body and the impelling
force, a state of equality that lasts over some interval of time.
But it has been shown that the impelling force is always and

----

40. The use of this assumption in the proof that follows (*Opere,* I, 391.12–
18) shows that Galileo is thinking in particular of bodies projected vertically
upward. The force which prevents their falling is said to sustain them.

41. The initial part of the trajectory of a projectile is considered to be a
straight line, regardless of the angle of projection. But Galileo's demonstrations, as in the "Essay," *De motu* (see *Galileo On Motion,* pp. 13–114),
except for Ch. 23, deal specifically with the case of vertical upward projection.

42. E.g., at the point where a projectile thrown vertically upward turns
downward.

continuously diminishing. It is therefore impossible that the force should remain, for an interval of time, in a state of equality with the weight of the body; and, for that reason, it is impossible for the body to be at rest for an interval of time.

This will become clearer from the following demonstration, based on the same diagram as above. [*327.12–328.4*] For if the body, when it is at *B*, is at rest for an interval of time, suppose that the end moments of such an interval of time are *C* and *D*. If, then, the body is at rest for time *CD*, the impelling force is equal to the weight of the body throughout the whole time *CD*. But the weight of the body is always the same. Hence the force at moment *C* is equal to the force at moment *D*. Now the heavy body remains the same. Therefore it will be sustained over equal intervals of time by equal forces. But the force at moment *C* sustains the body over the interval of time *CD*. Therefore the force at moment *D* will sustain the same body over an interval of time equal to interval *CD*. Hence the body will remain at rest for double the interval *CD*. But this is a contradiction; for the body was assumed to be at rest [only] over interval *CD*. And by a repetition of the same method of argument, it will also be proved that the body remains forever at rest at point *B*. Yet our assumption is that the body changes over to natural motion.[43]

Domenico: I cannot but agree that your demonstrations lead to a necessary conclusion, since they depend on the clearest and most certain principles, which cannot possibly be denied. And yet there is a certain something that bothers me. For, if you assert that the impelling force is at some time equal to the resisting weight, how can you avoid the conclusion that the body is, at some time, at rest?

Alessandro: It will be easy to remove this doubt. For it is one thing to say that the weight of the moving body comes, at some time, into equality with the impelling force; it is quite another thing to say that that weight remains in such equality over an interval of time. This will become clear by an example. For in the motion of a body, since (as has been shown) the impelling force is always becoming smaller, while the weight always remains the same, it follows necessarily that, before they arrive at the relation of equality, innumerable other ratios occur; but it is impossible for the force and weight to

43. Marginal addition (*Opere*, I, 391): "And if it is impossible to pass from contrary to contrary without a mediate interval, it will follow from this that, after upward motion, there is no interval of rest, for rest is the contrary of motion."

remain over an interval of time in any of those ratios. For it
has been shown that the impelling force never remains at the
same level for any interval of time but is continuously diminished. And so it is true that the force and weight pass, for example, through the ratio of two to one, of three to two, of four
to three, and innumerable others, but it is completely false to
say that they remain for any interval of time in any one of
these ratios. Suppose, for example, that a body is moving over
a surface. It will touch all the lines of the surface and will
cross over all of them. But it is false to say that it remains at
rest for any interval of time on any of the lines; for it is never
at rest, but always in motion. Now for the same reason it is
true [in the case of forced motion] that the force and the
weight at some time arrive at a relation of equality, just as
they arrive at innumerable other ratios; yet it does not follow
that the body therefore remains at rest in that relation [of
equality between impressed force and weight] and continues
so for any length of time, any more than that it persists at any
of the other ratios.

Domenico: Now my doubts are completely removed, and
I am compelled to agree with your demonstrations. But what
about the arguments of Aristotle?

Alessandro: Aristotle's chief argument was the following,
as he stated it: "If there is actually no [interval of] rest at
the turning point,[44] it follows that two contrary motions are
continuous and are therefore only one, since they have only
one terminus. But this would be absurd." [*cf. 323.22–26*]

Now our answer to this argument is that such motions are
contiguous, not continuous, and so no error arises; that between the terminus of the upward motion and the [initial]
terminus of the downward motion nothing intervenes, for such
is the nature of contiguous things, according to Aristotle himself. But some have at times objected, on the basis of Aristotle's own views, by saying that air is also a cause of the
body's being at rest [for an interval of time] at the turning
point; for, since the body has, at that point, been considerably
weakened, it encounters resistance from the air which hinders
its motion.

To this I say that it is not my duty to answer. For even if [an
interval of] rest did occur at the turning point for this reason
alone [i.e., the resistance of the air], still one would not have
to say that therefore [an interval of] rest *necessarily* occurs at

44. E.g., in the motion of a body projected vertically upward.

that point. For such a cause would be accidental. But if Aristotle had thought that, from the nature of contraries, contrary motions had the property of being able to be joined together directly, because, from their nature, [an interval of] rest between them was not required at the turning point, he would surely have said that [an interval of] rest was not an absolute necessity at the turning point. And then if such [an interval of] rest did occur by reason of the intercession and resistance of the air, Aristotle would have considered such resistance among accidental causes and would have completely disregarded it in his discussion of the nature of contrary motions.

And yet, so that no one may possibly suppose that [an interval of] rest really occurs at the turning point for this reason [i.e., air resistance], I have decided to remove this cause completely from the discussion by saying that no [interval of] rest need be asserted even for accidental reasons.

For, in the first place, the inconsistency of my adversaries would argue for a view opposed to theirs. For according to their needs they declare that the motion is helped by the medium, or else that this motion is hindered by the medium, obviously because the medium is in the way. But if the medium does help the motion, how can it destroy the motion at the turning point?

Secondly, the resistance of the medium is proportional to the [speed of] motion.[45] Hence the swifter and stronger is the motion, the more will the medium resist it, since the medium must be more swiftly divided. Hence, reversing the proportion, the [speeds of] motion will always be in the same ratio as the resistance of the medium, so that the slower the motion, the smaller will be the resistance [of the medium]. Thus, since the motion at the turning point is slowest, the resistance of the medium will also be smallest; and just as the swiftest motion overcomes the greatest resistance, so also the slowest motion will overcome the smallest resistance, since the [speed of] motion is always in the same ratio to the resistance.

Thirdly, consider, please, how weak is our adversaries' method of reasoning. In their very attempt to deny motion,[46] they must necessarily assume it. For the air must resist the

45. Galileo does not here discuss this notion, which later plays a part in his development of the idea of terminal velocity. In the *De motu* the resistance of the medium is considered solely in terms of its specific weight, and does not vary with the speed of the body moving through it.

46. I.e., in asserting an interval of rest at the turning point.

body either while it [the body] is at rest or while it is in motion. But surely it does not resist while the body is at rest. For such resistance on the part of the air means that the air is subjected to a certain effect. Yet surely the air is not subjected to anything by a body at rest, but only by a body that moves. Hence the air will not resist motion except while the motion is underway. For, if the air exerted resistance and did so before the motion, then surely the motion would never take place. We must therefore say that the air resists while the body moves. It therefore necessarily follows that those who say that the medium resists the moving body at the turning point of the motion assume a body in a state of motion. Thus in trying to deny motion they admit it. What could be more inept?

I could also frame arguments based on Aristotle's own views. Indeed, he asserts that motion cannot take place without resistance on the part of the medium. Hence he denies the existence of motion in a void, since in a void there would be no resistance of a medium; and so he declares that if there were a void, a motion in it would surely be instantaneous. But since my own view is different, I shall content myself with the arguments I have already adduced.

Domenico: Your view on motion at the turning point has been fully and abundantly proved, not only by your own demonstrations, but by your refutation of the arguments for the contrary view. But now that you have incidentally mentioned the void, do not hesitate to say something about it.

Alessandro: I could say many things on the subject of the void; but I shall omit them, so that the discussion of that subject may not distract us from our main purpose. I shall merely discuss that which depends on what has already been said. And so, I hold that, in a void, motion would not take place in an instant. This becomes clear from what has been proved. For it has been shown that the speed of bodies moving [naturally] is dependent on the excess of the heaviness of the bodies over that of the medium through which they move.

Thus, suppose A is the volume of the moving body and B a volume of the medium through which the motion is to take place, equal to the volume of A. [*cf. 279.15–280.30*] And suppose that the weight of A is 8 and of B 3. Then the speed of descent will have a measure of 5. But if the weight of B were 2, the speed of A would be 6; and if the weight of B were 1, the speed of A would be 7. Now if the weight of B were zero, the speed of A would be 8, and not infinite. For the weight of A exceeds the weight of [B], which is zero, by its own entire weight, which is finite. But the excess weight determines the speed of the motion: and this excess weight is finite. Therefore the speed of the motion will also be limited, and not infinite. That is, the speed would be infinite and [the motion] instantaneous whenever the body's weight was infinite; in that case the motion would be instantaneous not only in the void but in a plenum, provided, however, that the plenum was penetrable and not of infinite resistance. And so, we shall be perfectly right in saying that infinite weight moves instantaneously wherever it moves, but that finite weight moves with finite speed wherever *it* moves.

Hence just as the argument against us runs as follows: "If motion took place in the void, it would take place instantaneously. But motion does not take place instantaneously. Therefore motion cannot take place in the void"; so we shall reason in converse fashion as follows: "If motion were to take place in the void, it would certainly not be instantaneous. But motion does take place, provided it takes place in time. Therefore motion will take place in the void."

From this we can also gather that the medium does not help the motion, but that it undeniably hinders it; for in the absence of a medium the motion would be swifter.

Secondly, even from the argument of those same adversaries, in which they try to deny the existence of a void, one may elicit this conclusion, namely, that motion in a void takes place in time. For they themselves argue as follows: "If you take two perfectly smooth stones whose surfaces fit together so closely, when they are applied to each other, that no other substance is left between them, then, if you try to separate them in such a way that the surfaces are always kept parallel, you will fail. For nature strongly abhors the void which would at some time be left between the surfaces." And from this they conclude that a void cannot exist.

But if this is true, as certainly appears to be the case, I then argue as follows: "The stones cannot be separated; therefore instantaneous motion does not take place in the void. For, [if we say that] the stones cannot be separated, on the ground that otherwise empty space would be left [between them], [we may also say] that the stones will be able to be separated when empty space is no longer left between them. For the surrounding air will instantaneously fly into the void and thus there will not be empty space at any time. But the fact that the stones cannot for a time be separated is an indication that for some interval of time a void is left between them. But the fact that this void persists over an interval of time abundantly proves that motion in it does not take place instantaneously but in time."[47] And yet I should like these remarks on the void to be considered merely as incidental, especially since they do not contribute to our primary purpose.

Domenico: But it gave me the greatest pleasure to hear them, since, in my judgment, they achieve truth. Therefore, please do not hesitate to dwell longer on this discussion. For I am eager to hear your opinion on a certain question I have. Thus, if your view is sound, as your demonstrations seem to show, the contrary view will necessarily be false. And it follows then that Aristotle fell into some error when he sought to prove the contrary view, as he, of course, did in Book 4 of the *Physics*. But since he employed a kind of geometric demonstration, I am surprised that there is a fallacy in that demonstration, and for that reason I am somewhat insistent in asking you to uncover that fallacy.

Alessandro: If we wish to examine Aristotle's demonstration carefully, our discussion will be too long and will go beyond our primary purpose. But, since you wish, I shall show the fallacies in his argument; and that they may appear more clearly, I shall set forth his demonstration.

He asserts, to begin with, that slowness and speed of motion depend on a twofold cause, i.e., either on the body itself or on the medium. [He holds that they depend in some way] on the body itself, for a heavier body will move more swiftly than a less heavy body through the same medium. And he holds that speed and slowness of motion occur in two ways by reason of

47. At this point the manuscript contains the following passage which Galileo deleted (*Opere*, I, 395): "But listen to some fantasies concocted by a certain person against this view. For this person answers that the air would, in that case, move instantaneously, but only over one dimension, namely, the breadth of the stones. [In reality the two-dimensional stone surface is what is meant, as the context shows. — *I. E. D.*] But he denies that the air could move instantaneously over two dimensions, i.e., the breadth and depth. Now, although he does not merit a reply — first, because he assumes what not only has never been proved but, so far as I know, has never been thought of by anyone else, and, second, because no one can conceive of air as moving in length alone, since it is a body that occupies all dimensions — still I shall make one point by way of reply. In the view of this person, air could move in a void over only one dimension, e.g., over the surface of one stone, and this instantaneously. But when air has occupied one dimension, there remains but one dimension, namely, from the surface of one stone to the surface of the other. And since this is only one dimension, air will be able to move through it instantaneously. Consequently, it will move through both lengths in two instants. But two instants of time, so far as divisibility or indivisibility is concerned, are the same as one instant. Therefore air will move instantaneously through the whole void. But we have already shown that this is false."

the medium: (1) when we have one and the same medium in
view, and it is either at rest or else moving in the same direction as the moving body or in the opposite direction (for the
motion of the same body will be swifter if the medium is moving in the same direction as the body than if the medium is at
rest; and swifter if the medium is at rest than if it is moving in
the direction opposite that of the body); (2) when there are
different media, the body will move more swiftly through the
rarer medium than through the denser one, e.g., more swiftly
through air than through water.

Having made these preliminary observations, since he saw
that the same weight moved more swiftly through rarer than
through denser media, he assumed that the ratio of the speeds
of motion [in two different media] was equal to the ratio of the
rareness of one medium to the rareness of the other. [*277.
10–278.2*]

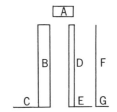

Now, on the basis of these assertions, turning to the proof,
he argues as follows: Suppose that body *A* traverses medium
B in time *C*, and a medium rarer than *B*, namely, *D*, in time *E*.
Clearly, time *C* is to time *E* as the density of *B* is to the density
of *D*. Now suppose that *F* is a void, and that body *A* traverses
F, if that is possible, in time *G*. Therefore the void will have to
a plenum the same ratio [in density] as *G* has to *E*. Now let
us consider another medium, rarer than *D*, such that the ratio
of the density of *D* to that of the new medium is equal to the
ratio of time *E* to time *G*. In that case, from what has been
posited, body *A* will move through the new medium in time *G*,
since the ratio of [the density of] medium *D* to that of the new
medium is equal to the ratio of time *E* to time *G*. But body *A*
also moves in time *G* through void *F*. Hence *A* will in the same
time traverse two equal distances, one of which is a plenum
and the other a void. And this is impossible.

That is Aristotle's demonstration. And it would have led
to a sound and necessary conclusion, so far as the form of the
demonstration is concerned,[48] if Aristotle had proved the things
that he assumed, or if, at least, they were true, even though not
proved. But Aristotle was deceived (in my judgment, at any
rate) in that he assumed, as if they were known axioms, propositions that not only are not supported by evidence of the
senses, but have never been proved and are in fact not provable,
since they are utterly false. That this may become perfectly
clear, we shall examine his various assumptions one by one.

---

48. This phrase is a marginal addition (*Opere*, I, 397).

To begin with, he assumes that the cause of the speed or slowness of motion is the rareness or density of the medium through which the motion takes place. But this assumption is false, not only because of what was proved above (where it was shown that the *heaviness* of the medium, not its density or rareness, is the cause of the slowness or speed of [natural] motion), but also because of what we shall now add.

For, please note, if the rareness of the medium is the cause of the speed of motion, then surely the motion in the rarer medium will be swifter. [*cf. 260.21–261.17*] Now, according to Aristotle himself, air is rarer than water; and yet some bodies move more swiftly in water than in air. For example, if we consider an inflated bladder, it will move more slowly, in natural motion, in air than in water; for, if it were forcibly held and tied down in deep water, then, upon being released from the bonds, the bladder would swiftly rush upward. And if we took a still lighter body, it would move more slowly[49] in air and more swiftly[50] in water. And we might in this way be able to arrive at something which scarcely moved at all in air, but moved very swiftly in water. Indeed, I shall add the following. If we take a body so light that it ascends in air, such a body will, of course, be lighter than the bladder. Now if that body is forcibly kept under water and then released, who will doubt that it will rise more swiftly in water than in air, although the air is rarer than the water? How, then, will it ever be true that natural motion is necessarily swifter in a rarer medium than in a denser one?[51]

In this connection, it does not escape me that there are a large number of modern philosophers whose knowledge of what they profess to know is based on faith and the authority of others rather than on proof. Now, if they heard my arguments, they would immediately try to answer me, and they would be satisfied to get two words in, even if what they said had nothing to do with the subject. For a little later they would say: "This view has been fully and completely refuted before." And they would utter similar inanities by which alone they would convince themselves and their hearers, who are even more ignorant than they, of the truth of their opinions. That is, if they heard my arguments and answered that my reason-

49. I.e., downward.
50. I.e., upward.
51. Marginal addition (*Opere,* I, 398): "Natural upward motion is swifter in denser media, and natural downward motion is swifter in rarer media. For upward motion results from lightness, and downward from heaviness."

ing was not conclusive, because I speak now of upward motion and now of downward, which is contrary to Aristotle's purpose, or if they uttered similar, ineffectual remarks, I am sure that they would really believe they had thoroughly refuted my opinion. But let us say no more about them. I am satisfied that I forestalled this answer of theirs when I added the second example, in which only one direction of motion is involved. We must conclude, therefore, that it is entirely false to say that speed or slowness of [natural] motion depends [merely] on the density or rareness of the medium.

Secondly, Aristotle assumed, as if it was known, that the speed of the first motion has the same ratio to the speed of the second motion as the rareness of the first medium has to the rareness of the second medium. Aristotle did not prove this, and in fact skillfully avoided proving it. For he would have struggled in vain, since it is not demonstrable, and not only not demonstrable, but actually false. For, even if it is granted that rareness [of the medium] is a cause of speed, it will follow, to be sure, that in [a medium of] greater rareness the speed will be greater; but it still will not follow that the speeds and the rarenesses increase in the same ratio.

Moreover, to argue according to Aristotle himself, the rareness of the air would have no ratio to the rareness of water. [*cf. 269.11–24*] For wood, to give an example, falls in air, but not in water; hence the speed in air will have no ratio to the speed in water. That this may appear with perfect clarity, I shall first show that, with regard to speed, motions do not have the same ratio to each other as do the rarenesses of the respective media. And I shall prove this by a demonstration in all respects similar to that with which Aristotle sought to prove that motion in a void would occur instantaneously. And then I shall also show what ratio these motions do maintain, so that the truth may be more clearly revealed.

If, then, as Aristotle himself said, the speed [of natural motion in the first medium] has to the speed [in the second] the same ratio as the rareness of the first medium has to that of the second, suppose there are a body *O* and two media, *A* and *B*, of which *A*, let us say, is water and *B* air. [*cf. 268.10–269.24*] Let the rareness of the air be 8, and thus greater than that of water, which we represent as 2. Suppose that the body does not sink in water but floats, and that the speed [of its natural motion] in air is 4. Now consider the ratio of the rareness of air *B*, which is 8, to the rareness of water *A*, which is 2; and let the speed in air, which has been assumed to be 4,

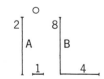

have that same ratio to another speed, which will, of course, not be zero, but 1.

And so, since body O moves with a speed of 4 in a medium of the rareness of B, and since the speed of 4 is to the speed of 1 as the rareness of B is to the rareness of A, clearly, the speed of O in a medium of the rareness of A will be 1. But it was assumed to be zero; and this is a contradiction. Therefore the ratio of the speeds will not be the same as the ratio of the rarenesses [of the respective media], as Aristotle assumed.

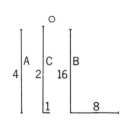

But I wish to adduce still another proof. Consider the same situation as above, but let the rareness of B be 16 and that of A 4. And suppose that body O does not sink in A but floats on its surface; but that the speed of this same body in medium B is 8. Now again let there be another speed 1, and let the rareness of B have the same ratio to the rareness of another medium, C, as speed 8 has to speed 1. Thus the rareness of C will be 2. Since speed 8 is to speed 1 as the rareness of B is to the rareness of C, and since body O moves in the rareness of medium B with a speed of 8, it follows that the same body O will move in the rareness of medium C with a speed of 1. That is, body O will move [naturally downward] in medium C. But medium C is denser than medium A; for the rareness of medium A, 4, is greater than that of medium C, which is 2. And it has been assumed that body O does not move [naturally downward] in medium A. Hence body O will move [naturally downward] through a denser medium, but not through a rarer one. But this is quite absurd and entirely unworthy of Aristotle.

It is clear, therefore, that the speeds of [natural] motions are not in the same ratio to each other as the rarenesses of the media. But in order to learn what *is* the ratio of the speeds, consider the true cause of the speed or slowness of the body, namely, as we proved above, the lightness or heaviness of the medium with respect to that of the moving body.

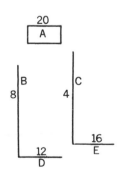

Suppose there are a body A and a medium C twice as light as medium B. [*cf. 279.15–280.30*] The time in which A moves [naturally] through B will certainly not be double the time in which A moves through C. That is, the speed in C will not be double the speed in B.[52]

For suppose that the time in which A moves through B is D, and the time in which A moves through C is E. Now the speed of A in the space of B will be measured by the excess of the heaviness of A over that of B, as was shown above. And

---

52. This sentence is a marginal addition (*Opere*, I, 400).

so, if the heaviness of *A* is 20, and that of *B* 8, the speed *D* will be 12; but, for the same reason, if the heaviness of *C* is 4, the speed of *E* will be 16. Therefore, speed *E* will not be double speed *D*. And therefore, since distances *B* and *C* are equal in length, time *D* will not be double time *E*. Clearly, therefore, the speed in medium *C* is not as great as Aristotle held it to be, but is much smaller. For in Aristotle's view speed *E* should have been 24 in comparison with the speed of *D;* actually it is 16.

It is clear, therefore, that, the lighter the medium, the swifter will be the motion through it that is due to weight. But since the ratio of speed [in the rarer medium] to speed [in the denser medium] is smaller than the ratio of the rareness [of the first medium] to the rareness [of the second], it follows that, though the rareness of a void as compared with that of a plenum is the greatest of ratios,[53] still the ratio of the speed in a void to the speed in a plenum is not the same as that ratio, as Aristotle erroneously believed it was.

And, from what has been set forth, it follows that the relation of speed to speed is the same as that of the lightness of the one medium to the lightness of the other medium, but in an arithmetic, not a geometric, sense.[54] For, if, to revert to our example, the heaviness of *C* is 2, so that its lightness is four times that of *B,* the speed of *E* will surely not be four times the speed of *D* but will be three-halves of it. For speed *E* will be 18; but it will have the same arithmetic relation [to *D*] as *C* has to *B,*[55] since the differences are equal, namely, 6.

If, now, the heaviness of *C* is 1, so that the lightness of *C* is eight times that[56] of *B,* surely speed *E* will not be eight times as great as speed *D,* but will be far less than eight times; it will be nineteen-twelfths as great, for speed *E* will be 19. And there will be the same arithmetic relation [between the two speeds] as between the two lightnesses, since the difference in each case is the same, namely, 7.

But if the heaviness of *C* is zero, so that the lightness of *C* has no ratio to the lightness of *B,* speed *E* will be 20, having the same arithmetic relation [i.e., difference] to *D* as 8 to 0. For the excess of 20 over 12 is the same as the excess of 8 over 0, namely, 8.

53. I.e., greater than any assignable ratio.
54. I.e., if *v* is velocity and *l* is lightness, $v_1 - v_2 = l_1 - l_2$, not $v_1/v_2 = l_1/l_2$.
55. It would have been more strictly accurate to say "as *B* has to *C.*"
56. We should expect *levitatis* for *gravitatis* (*Opere,* I, 401.8; cf. *Opere,* I, 401.4).

And so, contrary to what Aristotle says, it is not a contradiction for one number to have to another the same relation as a third number has to zero, provided we are talking about an arithmetic relation [i.e., difference]. For 20 has the same relation to 12 as 8 has to 0, the excess of 20 over 12 being the same as the excess of 8 over 0.

Domenico: What a subtle discovery, and how beautifully worked out! Surely they should be silent who assert they can attain philosophy without a knowledge of divine mathematics. And will anyone ever deny that the true can be distinguished from the false with mathematics alone as guide, and that with its help the keenness of genius can be activated, and that, finally, with its guidance whatever is really known among mortals can be apprehended and understood?

Alessandro: Now listen to this, if you please. Aristotle deduced a second argument from his hypotheses, namely, that, if motion in a void required time, then lighter and heavier bodies would all move with the same speed, since there would be no resistance on the part of the medium, either for lighter or for heavier bodies, and that this is a contradiction. Aristotle was similarly deceived in this argument, in that he assumed that speed or slowness of motion arose only from the resistance of the medium, whereas in fact the whole matter depends on the heaviness or lightness of the medium as well as of the moving body.[57] And so I hold that, in a void, heavier bodies fall more swiftly than do lighter, since, in comparison with the weight of the medium, the excess [of the weight] of the heavier bodies is greater than is the excess [of the weight] of the lighter bodies.

Nor is there any truth in what Aristotle said about the ratio of [the speed of] motions as compared with the [ratio of the] weights of the moving bodies, namely, that the ratio of the speeds [of natural motion of two bodies] in the same medium is equal to the ratio of their weights: e.g., if A is twice as heavy as B, its speed [of natural motion] would also be double that of B. We shall prove the falsity of this in a manner similar to that employed above.

For if the weight of A is 4, and that of B is 2, and in a medium of water A sinks with a speed of 2, then B will not sink. In that case it is obvious that the speed of A will not be double that

---

57. I.e., the speed of natural motion of a body in a given medium is proportional to the difference between the specific weights of the body and of the medium.

of *B*, since *B* does not move. But here, too, the ratio of the speeds will be arithmetic, i.e., will depend on the excess of the weights [per unit volume] of the bodies over that of the medium.[58] Thus, if *A* is 4, and *B* is 2, and the medium is 1 in weight, the speed of *A* will be 3 as compared with that of *B*, which will be 1.

Therefore, to summarize, the slowness or speed of every downward[59] motion comes, in the first instance and essentially, from the characteristic heaviness of the moving bodies. Now the fact that this heaviness is reduced by the heaviness of media is the reason why the actual motion is weaker. But if the heaviness of the medium is equal to that of the bodies, in that case, since the bodies have no heaviness in such a medium, no motion takes place. On the other hand, if the heaviness of the medium is greater [than that of the bodies], the heaviness of the bodies in comparison with the heaviness of the medium is, in that case, [relative] lightness, and there is upward motion. But if the heaviness of the medium is zero, then bodies will move in it according to their own precise weight, and in the [speed of their natural] motions they will maintain the same ratio as their characteristic[60] weights bear to each other.

And from this another serious error is revealed, Aristotle's view being precisely the opposite of what it should have been. For he held that the ratio of the speeds of [natural] motion of heavy bodies in a plenum is equal to the ratio of their weights, but that in a void this is not the case, all bodies moving in the same time.[61] [*cf. 294.17–296.4*] But, on the contrary, it is in a void that the ratio [of the speed] is actually equal to the ratio of the weights, since the excesses over the [weight of the] medium are the whole weights of the bodies themselves; but this ratio does not hold in a plenum, as has been shown above.

However, as we have said again and again, one must always understand and assume that the diverse bodies of which we speak differ only in weight, when they are equal in size. Otherwise one might argue as follows: "Suppose there is a body [*A*] whose weight is 8, and that the weight of a volume of water equal to the volume of the body is 3. Clearly, according to what

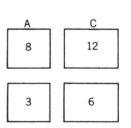

A	C
8	12
3	6

---

58. If *v* is velocity, *w* specific weight, and the subscripts refer to the first body, the second body, and the medium, we have $v_1/v_2 = (w_1 - w_m)/(w_2 - w_m)$, where $w_1 \geqq w_m$, and $w_2 > w_m$.

59. Galileo had originally written "upward or downward."

60. I.e., specific. [For the void, a debatable distinction. — *S. D.*]

61. I.e., instantaneously.

has been said, the speed of the body [in natural motion in water] will be as 5. Now, if you consider another body, such as *C,* whose volume is double that of *A,* but whose [total] weight is less than double that of *A,* let us say 12, the weight of a volume of water equal to the volume of *C* would be 6, and in that case the speed of *C* would be as 6."

But it is wrong to say that *C* will move downward more swiftly than *A.* For the ratio would not hold good in this case, since the bodies differ in size. However, if we wish the ratio to apply, we must consider half of *C,* so that the volume of that half will equal the volume of *A* and its weight will now be 6. And the weight of a volume of water equal to half [the volume of *C*] will be 3, so that the speed of half of *C* will be 3, as compared with the speed of *A,* which is 5. Therefore *A* will move downward more swiftly than the whole of *C* (for the whole moves with the same speed as its half). It will therefore be correct to say that the speed of *A* is 5 as compared with the speed of half of *C,* which will be 3; or else that speed of all of *C* is 6 as compared with the speed of the double of *A,* which will be 10.[62]

It is clear, therefore, that, when we reason about the speed or slowness of bodies, we must have in mind those bodies that differ solely in weight when their sizes are equal. And when the ratios of the speeds of such bodies are compared, it will be found that the same ratio is maintained in the case of bodies of the same material, no matter how much they differ in size.[63] For a 10-pound piece of lead falls with the same speed as a 100-pound piece of lead.

Domenico: This surely seems strange, and it is contrary to Aristotle's view.[64] It will be difficult for me to believe it unless you convince me somehow.

Alessandro: The demonstrations we adduced above should

62. The numbers, of course, are not absolute speeds. Only the *ratio* (5:3) of the speeds of bodies *A* and *C* is indicated. [In a void, on the other hand, it appears that Galileo considered total weight, rather than density, to govern the true speed of fall. Cf. *Galileo Galilei On Motion and On Mechanics* (Madison, 1960), p. 45. For one case, then, he would be forced back into the basic Aristotelian law that he was attacking. Ultimately that reflection may have turned him from the idea that there are constant speeds in free fall, and diverted his attention to the essential role of acceleration. —S. D.]

63. I.e., *v* being velocity, and *A* and *B* being bodies of different material moving downward in the same medium, v_A/v_B is constant, despite the size of *A* and *B*. (Likeness of shape is assumed.)

64. Marginal note (*Opere,* I, 403): "*De Caelo* 4.26 *fin.;* and, most clearly, *De Caelo* 3.26, 27." See *De Caelo,* 311a.21; 301a.31–32.

suffice to convince you of it.[65] Even if they do not explicitly prove this result, still the latter depends on those demonstrations. But if they are not enough for you, I shall adduce others.

For me to achieve my purpose, I require that you grant the following: namely, that if there are two bodies, one of which moves more swiftly than the other, then, if those bodies are joined, they will together move more swiftly than the one which by itself moved more slowly, and more slowly than the one which by itself moved more swiftly. [*cf. 264.34–266.8*] For example, if body *A* moves more swiftly than body *B,* I say that the body composed by uniting both *A* and *B* will move downward more slowly than *A* alone, but more swiftly than *B* alone. Of course, this is perfectly obvious. For who can doubt that the slowness of *B* will retard the speed of *A*, and, on the contrary, that the speed of *A* will accelerate the motion of *B*, and that a motion will result that is between the speed of *A* and the slowness of *B* ?

Domenico: I shall never venture to deny it.

Alessandro: On this assumption, then, let us suppose, if my adversaries consider it possible, that a large mass of a substance moves [in natural motion] more swiftly than a small mass of the same substance. Suppose that the large mass is *A* and the small mass *B.* If, then, *B* moves more slowly than *A,* it follows from what was assumed above that the combined body of *A* and *B* will move more slowly than *A* alone. Now *A* and *B* are of the same substance. Therefore the larger mass of a substance will move more slowly than the smaller mass of the same substance. And this is diametrically opposed to their own view and is contrary to their assumption. It is therefore not true that a great mass moves more swiftly than a small mass, if they are of the same substance, though Aristotle, in opposition to ancient philosophers, assumed this throughout Book 4 of *De Caelo* as something known. (Note, therefore, on what firm foundations Aristotle relied when he sought to refute the view of those who did not posit, as they very properly should not have posited, absolute heaviness and absolute lightness, but merely the comparatively lighter and heavier. And, consequently, you can see how much strength there is in the arguments with which Aristotle tried to assign absolute heaviness to earth and absolute lightness to fire, and

65. Marginal note (*Opere,* I, 403): "and those who have doubts on this point will also doubt, when they see a small piece of wood floating, whether a large piece will also float."

even tried to attribute heaviness to water and air[66] in their own natural places.)

But leaving these matters aside and returning to the matter before us, I hold that the same argument applies also in the case of the void, i.e., that bodies of the same substance, though unequal in size, move with the same speed. And this can be proved in exactly the same way as in the case of a plenum.

Domenico: You may now return to the solution of the other problems which I await with attentive ears.

Alessandro: Listen now to my solution of the problem in which the reason was sought why natural motion is swifter at the end than in the middle, and swifter in the middle than at the beginning—a solution which may be understood merely from what was set forth above. [*cf. 318.3–319.22*] Recall, then, what we have already shown; that, when a body is moved by forced motion, it so moves as long as the force impressed by the mover is greater than the resistant weight. Hence it follows that, when the heavy body ceases to move upward, the force impressed on it is equal to its weight. And from this it clearly follows that, when the heavy body begins to move downward, it does not at that time move simply by natural motion. For at the beginnning of this [downward] motion there is still in the body some of the impressed force which impelled it upward. And because this force is smaller than the weight of the body, it does not drive the body upward any farther. It does, however, continue to resist the downward motion of the body, inasmuch as it has not yet been completely destroyed. For it has been shown that this force is continuously diminishing. That is why the body at the beginning of its natural motion moves slowly. But later, since the counter-force is weakened and diminished, the body, encountering a smaller resistance, moves more swiftly.

Consider, for example, a body whose weight is 4, moving from *A* to *B* in forced motion. Clearly, the force which impels the body will, at any point of the line *AB* traced by the forced motion, be greater than 4. But at point *B* itself the force will not be greater than 4 (for if it were greater, the body would be driven by it beyond *B*); nor will that force be less than 4 (for in that case it would have been equal to 4 at some point before *B;* but we have shown that, before the body reached *B*, the force was always greater [than 4]). Therefore the force at *B* will be

---

66. Reading *aeri* for *aeris* (*Opere*, I, 404.27).

equal to the weight of the body; i.e., it will be 4. And when, therefore, the body recedes from *B,* the force which had been 4 begins to be reduced [below that figure] and the body consequently begins to encounter less resistance to its weight. And since this resistance is continually diminishing, the result is that the natural motion is continually accelerated.

Domenico: This solution thoroughly pleases me; and yet it seems to be applicable only in the case of a natural motion which is preceded by a forced motion.[67] [*cf. 319.32–320.11*] But suppose someone has a stone in his hand, which he does not throw upward but merely lets fall; what will be the cause of acceleration [*intensio*] in this motion, which is not preceded by forced motion?

Alessandro: Some doubts on this very point had also occurred to me when I was thinking out the solution of the problem. But when I looked at these doubts more carefully, I realized that they were of little weight. That is, in both instances of [natural] motion, that preceded by a state of rest and that preceded by forced[68] motion, acceleration [*intensio*] takes place, and for the same reason. For in the case of natural motion preceded by forced motion, the body also[69] begins [its downward motion] from a relation of equality,[70] which is the relation of rest.

Pay attention now, so that you may understand this more clearly. [*cf. 321.1–322.5*] Suppose there is a body *O* whose weight is 4, and that the line over which the forced motion takes place is *OE.* Clearly, then, a force can be impressed on body *O* great enough to move the body as far as *R.* This force will be greater than 4, which is the weight of the body. Also a force can be impressed which will move the body only as far as *T;* and this force will also be greater than 4, but less than that force which impelled the body as far as *R.* Also such a force can be impressed as will move the body only as far as *S;* and this force will still be greater than 4, though smaller than that which impelled the body as far as *T.* And so on, indefinitely, a force can always be impressed which will impel the body over a distance, however small; yet that force will always

67. Cf. Sagredo's discussion of this question in the *Discorsi . . . , Opere,* VIII, 202 (see Galileo, *Dialogues concerning Two New Sciences,* trans. H. Crew and A. DeSalvio [New York, 1914], p. 165). [See also "Memoranda," 10, below, and *Galileo Galilei On Motion and On Mechanics,* pp. 85–94, esp. p. 89. —*S. D.*]

68. Reading *violentus* for *naturalis* (*Opere,* I, 405.34).

69. As in the case of free fall from rest.

70. I.e., equality between upward and downward tending forces.

be greater than 4. The conclusion remains that the force which impels the body over no distance at all in forced motion is 4.

Hence it is clear that, when body *O* leaves the hand [i.e., falls from rest], it leaves it having an [impressed upward] force represented by 4. And since that force is continuously consumed by the weight of the body, an acceleration [*intensio*] of the motion is produced.

What I have said will appear even more clearly if we consider the case of a body at rest in the hand. [*cf. 320.14–35*] Since the body presses downward with its weight, it must at the same time be impelled upward by something, namely, by the hand, with a force equal to its own weight, which presses downward. Otherwise, if the body were not impeded [from falling] by another force, equal [to its weight] and impelling it upward, it would move; and the direction would be downward, if the resistance were less [than the weight], and upward, if the resistance were greater. It is clear, therefore, that when the body is let fall by that which holds it up, it takes its departure with a force impressed on it equal to its own weight. Hence it follows, [etc.].

Domenico: What you say is quite convincing. But there is still something that troubles my mind. [*cf. 328.15–329.24*] For, if the slowness at the beginning of natural motion comes about from the resistance due to the impressed force, that force will ultimately be consumed, since you assert that it is continuously diminished. And, therefore, once that force has become nil, the natural motion will not be further accelerated.[71] But this is contrary to the view of many.

Alessandro: The fact that it is contrary to the view of many does not concern me, so long as it is in harmony with reason and experience, even though at times experience seems rather to point to the opposite. For, if a stone falls from a high tower, its speed is observed always to be increasing. But this happens because the stone is very heavy in comparison with the medium through which it moves, i.e., the air. Since it starts its fall with an impressed force equal to its weight, it starts, of course, with a great deal of impressed force; and the motion from the tower's height is insufficient for the using up of all this impressed force. Thus it happens that the speed continues to increase all through the single tower's distance. But if we were to take an object that has weight, but whose weight did

71. [For Galileo's mature understanding of equilibrium speed in actual fall, see *Discorsi*, trans. Crew and DeSalvio, pp. 88–94, esp. p. 94. —*S. D.*]

not so very far exceed the weight of air, we should then surely
see with our own eyes that, a little after the beginning of its
motion, the body would move uniformly, so long as the air
remained quite calm. And we would observe the same thing
happen in the case of the stone if it were dropped from very
high places and we were so placed as always to observe the
line of motion under the same conditions.[72] For our position,
too, keeps us from observing the uniformity of the motion.

Thus, suppose that there is a uniform motion from *B* to
F, and that distances *BC, CD, DE,* and *EF* are equal. Let the
eye of the observer be at *A*, and draw the lines of sight, *AB,
AC, AD, AE,* and *AF.* Since the motion is assumed to be uniform, and distances *BC, CD, DE,* and *EF* are equal, the body
will traverse these distances in equal times. Thus the time of
transit from *B* to *C* will be the same as that from *C* to *D*. But
the motion from *C* to *D* will appear swifter to the observer,
since distance *CD* appears greater than distance *BC;* for it is
seen under a greater angle.[73] So, too, the motion from *D* to *E*
will appear swifter than that from *C* to *D*, since distance *DE*
appears greater than distance *CD* but is traversed by the body
in equal time. Similarly, the motion from *E* to *F* will appear
swifter than that from *D* to *E*. Hence also the whole motion
from *B* to *F* will appear not to be uniform, but to be always
accelerated to the very end, although it is assumed to be uniform.

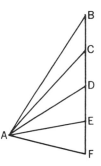

Hence, for a distinction to be made between uniform motion
and accelerated motion, the distance must be great enough
for the body to be able, in the course of it, to use up the entire
[impressed] force that resists [the downward motion], and
the observer's eye must be so placed as not to be deceived by
the disparity of angles.

Domenico: I fully understand your very elegant explanation. And so, I have but one further question to ask on this
subject — that is, whether, in your opinion, a heavy body which
is projected downward with some motive force is accelerated
in its motion in the same way as a heavy body, which, as it
begins to fall, has received from a mover no force impelling
it downward.

Alessandro: From what has been set forth above, it is clear
that a heavy body falling from a state of rest undergoes accele-

---

72. I.e., so that equal distances on that line subtended equal angles in the
eye of the observer.
73. I.e., *CD* subtends a greater angle in the eye of the observer than does
BC.

ration up to the point where the impressed force that resists [the downward motion] is completely consumed. But if that force is destroyed by an external mover, the body will not then undergo further acceleration.

If, for example, a body with a weight of 4 falls from a state of rest, it will, of course, start with a resistance of 4. Since this resistance has to be destroyed by the weight of the body, the natural motion will be slower at the beginning. [*cf. 329.30–34*] But if the aforesaid [impressed] force of 4 is destroyed by an external mover which impresses on the body a downward pressing force of 4, then surely the body's motion will no longer be accelerated, since at the very beginning it is not retarded by any force resisting [the downward motion].

But if the force of downward projection impressed on the body by the external mover is less than 4, i.e., less than the [contrary] force which was impressed on the body while it was at rest, then surely the body will be accelerated [in its downward motion]. For some portion of the contrary force will still have to be consumed, since it was not entirely destroyed by the external mover.

And if the force of downward projection impressed on the body is greater than 4, then the natural motion will be swifter at the beginning [than subsequently], for the body will be moving with greater than natural speed, one that exceeds the speed required by its characteristic weight.[74] [*329.34–330.13*] In fact, its own heaviness would in that case have the effect of lightness, since that heaviness, unadulterated and simple, would produce a slower fall than when mingled with the [downward] force. And so, the characteristic and natural slowness of the [freely] falling body would oppose the force of that which [extraneously] impels[75] it downward.

This will be clarified by the example of the common experience of every swimmer. For, clearly, if a man is in the water, his weight is sufficient for him, if he wishes, to sink to the very bottom of the water. In that case, he will go down with uniform motion under the pull of his own weight. But if he is impelled downward by an external mover, with no matter how great a force, e.g., if he is thrown down from a high place, at first his motion in the water will certainly be forcibly quickened

74. I.e., its speed at the beginning is greater than the greatest speed which a body of its specific gravity could ever attain in that medium in a free fall from rest.

75. Leaving *impellentis* (*Opere,* I, 408.22–23) unemended (as at *Opere,* I, 329.36).

and swifter than natural. But that motion will be retarded by the body's absolute[76] natural weight; for then the weight, by comparison with the combination of weight and impressed force, amounts to lightness. And the motion will be retarded until the sinking body arrives at its natural slowness. Moreover, if the water is quite deep, the diver will not suffer any greater injury at the bottom than if he went down merely from the surface of the water with his own natural motion.

76. I.e., uncompounded by the externally impressed force.

Memoranda on Motion

Composed by Galileo Galilei

Pisa, *ca.* 1590

[In our volume entitled *Galileo Galilei On Motion and On Mechanics* (Madison, 1960) there was a brief discussion (pp. 129ff.) of the series of short Latin notes that Galileo made, mainly on the subject of motion, presumably at the time he was teaching at Pisa, from 1589 to 1592. The substance of many of these notes, often couched in identical language, is also to be found in the "Dialogue" and the "Essay" *De motu,* which Galileo wrote during those years.

Since the "Memoranda," as they will be called here, throw considerable light on Galileo's reading and thinking in the early years, it was decided to include a translation of them in the present volume. Text annotations show, by references to the page and line numbers of Favaro's text of the "Dialogue" and "Essay," precisely where Galileo made use of material that is present in these "Memoranda."

The latter part of the "Memoranda" (*Opere,* I, 414–17) is particularly relevant to the last rewriting of the "Essay" *De motu* (see *Galileo Galilei On Motion and On Mechanics,* p. 129). Also the presence among the "Memoranda" of a portion of the "Dialogue" *De motu* may throw some light on the chronology of the early writings (see *Isis,* 51 [1960], 273–74).

For convenience of reference the "Memoranda" have been numbered serially. Details about the form in which this material is preserved in Galileo's manuscripts are given by A. Favaro, the first editor of this text, in *Bullettino di Bibliografia e di Storia delle Scienze Matematiche e Fisiche,* 16 (1883), 25–35. Favaro's text in *Opere,* I, 409–17, is followed; the pagination of that edition is indicated in the heading at the top of each page.

The themes of these notes are the same as those found not only in Galileo's early works on motion but in those of practically every writer on the subject of motion from the fourteenth century on—the heavy and the light, natural and violent motion, acceleration, the motion of projectiles, the plenum and the void, the theory of impetus, and analogies between alterative changes and locomotion.]

1.*a* So that you may not, by any chance, suppose that Aristotle considered the absolutely heavy to be not earth but, in general, that which is at rest beneath everything else, see *De Caelo* 4.29 [311b.5], where he says that everything except earth has lightness.[1]

b Unsound is the following statement of Aristotle, *De Caelo* 4.32 [311b.24–25]: "It is impossible for fire to have weight, for it would then be at rest underneath another body." If the elements are transmuted into one another, how can fire not have weight, when it is made out of air, which has weight?[2]

2. Aristotle says, *De Caelo* 3.27 [301b.11–13], that in forced motion, if the impelling force is the same, the speed of the smaller body is to the speed of the larger as the [weight of the] larger is to the [weight of the] smaller.

3. It is a mistake to doubt that a body projected is moved by an impressed force. For its motion is like that found in other cases of motion, e.g., like the alterative motion of iron when heated by fire. For the fire impresses heat [on the iron]. And then, when the iron is removed from the fire, the heat still remains, but not because of the force and heat of the surrounding [air], if the iron is brought into very cold air. The iron then gradually moves toward coldness as it cools off. In the same way the stone is moved by a man; and, when let go by the man, it continues to move until it comes to rest, the medium being unmoved. Similarly, if one strikes a bell with a hammer and deprives it of its silence. The bell then continues to move, though the striking agent is removed, and that sonorous quality runs through it, without the medium being moved. (And even if the medium were moved, this has no effect on the sound. Thus, no matter how strong a wind is blowing, the bell [unless struck] remains silent.) But the sound in the bell gradually diminishes, and the bell returns to silence.[3]

4. Just as heat is impressed more sharply in a dense and cold substance, such as iron, than in a rare and less cold substance even if both are heated by the same heat, so also a heavier body is at times moved farther and more swiftly by the same force.[4]

5. We must not say that it is the air that sounds in the bell. Only fools would say this; for different bells would then have

1. Cf. *Opere*, I, 376.7–8.
2. Cf. *Opere*, I, 293.5–6.
3. Cf. *Opere*, I, 310.4–22 (example of heated iron); 310.25–311.18 (example of bell). Read ⟨*si*⟩ *quis*, 409.19.
4. Cf. *Opere*, I, 335.13–336.18.

the same sound, and a wooden or leaden bell would sound just as would a bronze one.[5]

6.   There is a threefold classification of motion: first, with respect to space, i.e., one straight, another circular; second, with respect to the terminus toward which [the motion is directed], i.e., one upward, another downward; third, with respect to the effective principle, i.e., one natural, another forced.

7.   Aristotle asserts, *De Caelo* 3.27 [301b.1–4], that things in motion must be either heavy or light; for, if they were neither heavy nor light, they would, if moved by force, move a distance without limit. And in the next section he says [301b.23–27] that air supports both motions, i.e., upward and downward motions, and that projected bodies are carried along by the medium. If, then, the medium air is impelled by force and, since it is neither heavy nor light, moves a distance without limit, it will carry projected bodies a distance without limit.

8.   Philoponus, Avempace, Avicenna, St. Thomas, Scotus, and others who try to maintain that motion in a void would be accomplished in [an interval of] time do not reason well when they posit a twofold resistance in the moving body, one accidental and coming from the medium, the other intrinsic and due to the body's own weight. For these two resistances are clearly one, since the same medium, if heavier, offers more resistance and makes the body lighter.[6]

9.   Averroës, *Commentary on De Caelo* 1.32 [in, e.g., Aristotelis *Opera Omnia* (Venice, 1560), V, 27], says that a sphere is not *physically* tangent at a point.

10.   Alexander thought that he had successfully refuted the opinion of Hipparchus on acceleration toward the end of natural motion, when he adduced in opposition the case of natural motion not preceded by forced motion. But indeed even Hipparchus did not observe that forced motion preceded *every* case of natural motion, as we have made clear.[7]

11.   We must refute the false opinion of those who declare that, when a pebble is thrown into the water, the water thereafter moves by itself in a [widening] circle.[8]

5. Cf. *Opere*, I, 311.8–11.
6. Cf. *Opere*, I, 284.16–29.
7. If the view set forth in *Isis*, 51 (1960), 271–77, is sound, the reference at the end of this section is probably to the "Dialogue" (*Opere*, I, 405.27–406.25) rather than to the "Essay" (*Opere*, I, 319.30–320.7). (Emphasis added here and in 9, above.)
8. Cf. *Opere*, I, 313.22–314.2.

12.*a* Burley, *Commentary on* [Aristotle's] *Physics* 8.76, and Contarenus, *On the Elements,* Book 1, assign the cause of acceleration toward the end of natural motion to the parts of air, both those that precede and those that follow [the body].[9]

b See Aristotle, *De Caelo* 1.88 [277a.27–33], where he says that the speed of natural motion always becomes greater, and, if the motion continued without limit, the speed would also become greater without limit.[10]

c See also Section 89 where Aristotle says [277b.4–8] that a larger piece of earth moves [in natural motion] more swiftly than a smaller piece, and that the cause of the acceleration of natural motion is not the extruding action of the air. For, if it were, a larger piece of earth would move more slowly than a small piece, because it would be thrust out with more difficulty. Nor would its motion be accelerated, for it would be a case of forced motion, the speed of which decreases.

d Benedict Pereira writes thus, Book 14, Ch. 3 *fin.* [cf. Pererius, *De communibus omnium rerum naturalium principiis,* p. 744.8–11 of the 1595 edition; Galileo used the edition of 1579 or 1585]: "I should say without hesitation that, if the extent of the air through which a stone moves downward were endless, the motion of that stone would always become swifter and stronger and yet there would be no increase in its weight." But observe Pereira's error in what he goes on to say. For he says [cf. *De communibus,* p. 744. 14–21]: "Aristotle errs in concluding that, because in natural motion the speed keeps increasing, the weight of the moving body must also keep increasing. For if a stone moves [in natural motion] over a space that is denser and thicker in its first part, and rarer and thinner in its latter part, doubtless the motion will be swifter toward the end, but this result will not be due to an increment in weight."

13. Light bodies are not moved more swiftly by a greater force. For just as straw and tow are not heated more by the largest and strongest fire, because they do not wait for so much

---

9. See B. Pererius, *De communibus omnium rerum naturalium principiis et affectionibus libri xv* (Coloniae, 1595), p. 742. There had been previous editions of this work at Paris in 1579 and 1585. Editions of Walter Burley's commentary on the *Physics* of Aristotle had appeared in 1476 and 1501, and an edition of Gasparo Contarini's *De elementis . . . libri v* in 1548, but it is doubtful that Galileo used them directly.

10. Cf. *Opere,* I, 328.15–23.

heat, but are set afire sooner by a smaller fire, so light bodies do not resist [motion] until a large force is impressed on them, but are set in motion sooner.

14. Julius Scaliger in his work against Cardano, Exercise 28, adduces certain arguments against those who say that projectiles are moved by the air.[11]

15. Themistius on *Physics* 4.74 [see Themistius, *In Aristotelis Physica Paraphrasis,* ed. Schenkl (Berlin, 1900), pp. 132.32–133.1, on *Physics* 216a.11–21] says: "Thus, since the void gives way uniformly, yet in fact does not give way at all (for since the void is nothing, it is the mark of an oversubtle mind to suppose that it gives way)—the result is that differences of heavy and light, i.e., [differences in] the motive forces [momenta] of things, are removed [in the void]. As a consequence, all things which move [in natural motion in the void] have the same, undifferentiated speed." So Themistius. But not only is this false; indeed, the very opposite is true. For it is only in the void that the weights of things differ exactly and naturally; and it is only in the void that the speeds of their [natural] motion differ most [precisely]. A chapter should therefore be written in which this is demonstrated.[12]

16. A mover can impress contrary qualities in the projectile, namely, upward and downward. For the beginning of motion depends on the will, which has the power of moving the arm either upward or downward. And the force that impels an object upward is not different from that which impels it downward. There is the example of the iron spring in a clock which moves [the hands of] the clock up or down, or forward or back, depending on how the clock is turned. Its function is to unwind and straighten itself, just as it is the function of the arm to move the stone away from it.[13]

17. There will be many who, when they read my writings, will turn their minds not to reflecting whether what I have written is true, but solely to seeking how they can, justly or unjustly, undermine my arguments.

18. It is better to say that objects moving naturally are moved by the medium than to say this of bodies moving in forced motion.

11. Cf. Julius Caesar Scaliger, *Exotericarum exercitationum liber xv De subtilitate ad Hieronymum Cardanum* (Francofurti, 1592), pp. 126–33. There had been previous editions in 1557 and 1576.

12. Ch. 13 (*Opere,* I, 294–96) of the "Essay" is obviously the chapter that Galileo here plans; note that it begins (294.18–22) with the same quotation from Themistius.

13. The portion of this note from "And the force that impels" to the end is added in the margin.

19.   Aristotle says, *De Caelo* 1.89 [277b.1–2]: "That which moves[14] does not move because of something else, as by extrusion." But the opposite view could be maintained.

20.*a*   That a contrary force is more strongly impressed in heavier bodies is clear from objects which are suspended by a thread and allowed to move back and forth. For the heavier these objects are, the longer will they move.[15]

b   Heavier, denser, and more solid bodies conserve all opposite qualities for a longer time, more strongly, and more readily [than do lighter bodies]: as stones, which in winter become much colder than air, and in summer much warmer.

21.   Aristotle says: "Bodies that move in natural motion do not move by extrusion. For in that case the motion would be forced motion and would become slower toward the end, whereas actually we see that natural motion is accelerated." The answer to this is that forced motion becomes slower only when the body has left the hand of the mover; but so long as the body is in contact with the mover—and we may speak of a body which moves [through the air] toward its natural place as moved by the air by extrusion—in that case the motion need not necessarily become weaker toward the end.

22.   The definition of heavy and light bodies given on the basis of their motion is not a good one. For while the heavy or the light body is moving, it is neither heavy nor light. That is, the heavy is that which presses down on something. But that which presses down on another body is resisted by that other body. Hence the heavy body, while exerting weight, does not move. This is clear if you have a stone in your hand. The stone will then press down so long as the hand resists its weight; but if the hand moves down with the stone, the stone will no longer press down in the hand. Therefore a better definition will be "the heavier is that which remains at rest under the lighter." For, if we said "the heavy is that which remains at rest down and the light is that which remains at rest up," our definition would be faulty, since up and down are differentiated not in fact but only in theory.

23.   We must consider whether, if there were a void above a body of water, the things moving on the water would move more slowly or more swiftly; and whether different moving bodies would keep the same ratio in [the speeds of] their

14. [The word "naturally" is needed; cf. 39, p. 387. —*S. D.*]

15. Ch. 22 of the "Essay" (*Opere*, I, 333–37) employs and illustrates the principle here set forth. Of course, "heavier" refers here not to total weight but to specific weight, as the examples in the "Essay" on p. 334 (*Opere*, vol. I) indicate. The example of the pendulum is given on p. 335.22–26 of the "Essay"; that of the stones on p. 336.3–6.

motions [as that which they have in the absence of such a void].

24.   Downward motion is far more "natural" than upward motion. For upward motion depends entirely on the weight of the medium, which confers an accidental lightness on the moving body. But downward motion is caused by the intrinsic heaviness of the body. Apart from any consideration of the medium, all things will move downward. Upward motion is caused by the extruding action of a heavy medium. Just as in a balance the less heavy is forcibly moved upward by the more heavy, so a body is forcibly extruded upward by a heavier medium.

25.   It is clear that the difficulty of cleaving the medium is not the reason why wood does not sink in water. For [if it were], then, if we overcame this difficulty by choosing a suitable shape, the wood would sink, e.g., if it were made in the shape of a cone or an arrow. But the fact is that such a form of wood floats no less than does a flat board.

26.   A fragment of Euclid attests that the treatise on light and heavy is mathematical.

27.   Telesio says that the cause of the acceleration of [natural downward] motion toward the end is that the substance, having had its fill of descending, speeds up its motion.[16]

28.   The proof that upward motion is not natural motion is as follows. That which moves naturally moves, so long as it is unimpeded, toward a terminus in which it is naturally at rest and from which it cannot depart except by force. If, therefore, wood rises naturally in water, it moves to a place from which it will not depart except by force. That is, wood, if unimpeded, moves toward a terminus which is next to the surface of the water. Hence it will not depart from it except by force. But this is false; for, if the water is removed, the wood will move from its position and descend in natural motion. Now you cannot argue that the terminus of the natural upward motion of the wood is the surface of the water itself and that therefore, if the terminus moves, that which was in it also moves. Such an argument is ridiculous. For the terminus of a natural motion is not an actual body but must be something indivisible and immobile. And only the center [of the universe] is such a thing. Hence only toward this center does anything move naturally,

16. The reference is to Bernardino Telesio, *De rerum natura* (Naples, 1570). See P. Duhem, *Études sur Léonard de Vinci* (Paris, 1955), III, 193–95.

and only there does it remain at rest naturally, unable to move away except by force. But a body does not approach the center except by moving downward.

Furthermore, that which moves naturally must move toward some determinate goal. But in the upward direction there is no point of which we can say: "This is the upper terminus." On the contrary, there might be an infinite number of upper termini, and [the motion] may be extended upward without limit. Therefore there can be no natural motion that is without limit, to an indeterminate goal, that is to say, upward. But the same cannot be said of downward motion. For in the downward direction there is a terminus that is unique, determinate, and in fact indivisible, from which any object in motion, as it moves downward, cannot diverge even a nail's breadth. And that terminus is the center.

Nor can one say that there is a terminus of upward motion, namely, the sphere of the moon.[17] This would be erroneous. For the terminus of a given motion must be such that whatever diverges from that terminus no longer moves with the motion that had that terminus. But the sphere of the moon is not of this kind; for it is not a terminus of upward motion in the sense that whatever diverges from it can no longer move upward. On the other hand, the center is a terminus of downward motion in the sense that a body that diverges from it can no longer be said to be moving downward.

Distance from the center is without limit, but nearness to that center is finite and determinate. If, then, something exists that has the property of receding from the center, it will certainly be prone to move without end. But what could be more absurd than this?

That motion cannot be natural to which a terminus cannot be assigned. But a terminus cannot be assigned to upward motion. Therefore upward motion is not natural. The minor premise is proved as follows. The terminus of a motion is a place such that there can be no receding from that place in the course of the same motion. (For if the same motion could proceed farther, the terminus would not be in the place assumed.) But nowhere can a terminus of an upward motion be so assigned that in the same upward motion the body would not recede from that terminus. Therefore nowhere can there be a terminus of upward motion. That is, upward motion is without a terminus, and therefore it cannot be natural. And it

17. I.e., the surface of the sphere containing the moon's orbit.

is clear that nowhere can a terminus of an upward motion be so assigned that in the same motion the body would not recede from that terminus. For no matter what place is assigned, the body as it rises can recede from it, and another place [in the path of the motion] can be assigned more distant than the first from the center.

29. There is the greatest difference between upward and downward motion. For in addition to what has just been said, there is this difference: it happens that downward motion is never aided by the medium but is always hindered by it. For, since the medium lessens the weight of the moving body, it hinders its motion; on the other hand, upward motion can never take place unless it is aided by the medium.

30. The cause of a positive effect must be positive. Therefore lightness, which is a deficiency, cannot be the cause of motion. The conclusion remains, therefore, that heaviness is the cause and that things which move upward are moved by heaviness.

31. We call locomotion that motion in which the center of gravity of the body moves. Hence we shall not speak of the locomotion of the celestial spheres since in their case the center of gravity, which is also the center of magnitude, always remains unmoved.

32. If the thickness and density of water were the cause why wood does not sink, then surely, after the wood had been immersed by another agency, the same cause would prevent the wood from moving up again.

33. Aristotle, *On Generation and Corruption* Book 1, argued against Plato for his excessive concern with geometry.[18]

34. Aristotle in Book 3, Section 8, *Divinorum,* writes as follows: "Perceptible lines are not of the same kind as those that the geometer assumes. For no perceptible object is so straight or so curved. That is, the [perceptible] circle is not tangent to a line at a single point, but as Protagoras said in refuting the geometers."[19]

35. Aristotle says (*Physics* 7.10) that for naturalness of motion an internal, not an external, cause of motion is required.[20]

36. Aristotle says (*De Caelo* 3.72): "If fire had a heating action because of triangles, it would follow that mathematical bodies have a heating action."[21]

18. Cf. *On Generation and Corruption,* 315b.30-316a.4.
19. The passage is from *Metaphysics,* 997b.35-998a.4.
20. Cf. *Physics,* 243a.12-15.
21. Cf. *De Caelo,* 307a.13-22.

37.   Aristotle says (*De Caelo* 1.44 [273a.9–13]) that, if one of two contraries is determinate, the other must also be determinate. And since the center, which is the terminus of downward motion, is determinate, then upward motion must also be determinate [i.e., limited]. The same may be gathered from Section 58 [274b.11–14].[22] Read the text.

38.   Aristotle says (*De Caelo* 1.51): "Speed is to speed as weight is to weight."[23]

39.   Aristotle shows (*De Caelo* 1.89 [277a.33–277b.2]) that what moves naturally does not move by force and by extrusion.[24]

40.   Aristotle writes (*De Caelo* 1.96 [278b.14–15]): "It is our custom to call the outermost and uppermost region heaven." And he says in Section 21 [270b.6–8]: "And the place which is above is assigned to God."

41.   Bodies that move upward rise by force rather than naturally. For upward motion has an external, and downward motion an internal, cause.

42.   Aristotle in *De Caelo* 1.5 [268b.21–22]: "I say that upward [motion] is away from the center, and downward toward the center."

[There follows (*Opere*, I, 418–19) what is virtually an outline for a work on motion. This outline does not really continue the "Memoranda" (*Opere*, I, 409–17). It is on a sheet separate from the other notes and is preserved separately from them. I have translated the outline in *Galileo Galilei On Motion and On Mechanics*, pp. 130–31, and have there indicated, by reference to the "Essay" *De motu*, the connection between the "Essay" and this outline.]

22.   The reference to Section 58 is a marginal addition.
23.   *De Caelo*, 273b.32–274a.2 ("As weight is to weight, so time is to time, inversely").
24.   [Reading *ut* for *et* we would have "as by extrusion," which would better convey Aristotle's intention. —*S. D.*]

Reference
Matter

Bibliography

The following list of books attempts to provide a broad view
of the materials available to Italian writers of the sixteenth
century, their own productions in motion and mechanics, and
materials useful in the study of those works. The first section
includes printed sources of the earlier traditions. Books
therein are listed in order of year of publication, with a sub-
section being devoted to each of six traditions described in the
introduction. The second section deals with sixteenth-century
books, those written by Italians or published in Italy being
separated from foreign productions. A few books published
after 1601 are included; they are presumed to have been composed before that date. Works on astronomy are omitted,
except for that of Copernicus. The third section comprises
books and articles relating generally to matters that may have
affected Italian mechanics of the sixteenth century. The final
section includes specialized material pertaining to particular
sixteenth-century Italian authors.

Traditions of Earlier Centuries

The "Aristotelian" tradition of the *Questions of Mechanics*

Greek text, in the edition by Aldus Manutius of the collected
  works of Aristotle, in five volumes. Venice, 1497.
*Aristotelis Mechanica Victoris Fausti . . . restituta ac latinate
  donata.* Paris, 1517.
Latin translation by Niccolo Leonico Tomeo in *Opuscula
  nuper in lucem aedita,* with commentary by the translator.
  Venice, 1525; Paris, 1530.
Alessandro Piccolomini. *In mechanicas quaestiones Aris-
  totelis paraphrasis.* Rome, 1547; reprinted Venice, 1565.
Ἀριστοτέλους Μηχανικά. Paris, 1566.
Antonio Guarino. *Le Mechaniche d'Aristotile.* Modena, 1573.
Oreste Vannocci Biringuccio. *Parafrasi di Monsignor Ales-
  sandro Piccolomini sopra le Mechaniche de Aristotile.*
  Rome, 1582.
Henri de Monantheuil. *Aristotelis mechanica graeca, emen-
  data, latina facta, et commentariis illustrata.* Paris, 1599.

The Archimedean tradition

Tetragonismus id est circuli quadratura . . . . Venice, 1503. Comprises translations of three mathematical works of Archimedes, made by William of Moerbeke about 1269 and published by Luca Guarico.

Opera Archimedis . . . per Nicolaum Tartaleam . . . in luce posita. Venice, 1543. Contains the previous texts with the addition of the two books on centers of gravity and the first book on bodies in water, all in the Moerbeke translation.

Archimedis . . . opera, quae quidem extant, . . . et graece et latine . . . edita. Adjecta sunt Eutocii . . . Commentaria . . . . Basel, 1544. Greek text with Latin translations made by Jacobus Cremonensis about 1450; omits the work on bodies in water.

Archimedis opera non nulla a Federico Commandino . . . nuper in latinum conversa et commentariis illustrata. Venice, 1558. Based on the translations of Moerbeke, but without the two works relating to mechanics.

Archimedis de iis quae vehuntur in aqua libri duo. Bologna, 1565. Edited by F. Commandino from the Moerbeke translation.

Archimedis de insidentibus aquae. Liber primus et secundus. Venice, 1565. Published by Curtius Troianus from papers left to him by Tartaglia, comprising a defective version of the Moerbeke translation.

In duos Archimedis aequeponderantium libros paraphrasis . . . . Regarding this work, see the bibliography of Guido Ubaldo del Monte, below.

The tradition of Hero of Alexandria

Heronis Alexandrini Spiritalium liber . . . . Venice, 1575; Paris, 1583. Latin translation by F. Commandino.

Gli artificiosi et curiosi moti spiritali di Herone. Ferrara, 1589. Italian translation by G. B. Aleotti, with added theorems by the translator.

Pappi Alexandrini mathematicae collectiones . . . . Pesaro, 1588. Latin translation with commentaries by F. Commandino.

Spiritali di Herone . . . . Urbino, 1592. Italian translation by Alessandro Giorgi, with a "Life of Hero."

The tradition of Jordanus; medieval science of weights

Jordanus Nemorarius (13th c.). *Liber . . . de ponderibus propositiones xiii . . . cum aliud commentum,* ed. Petrus Apianus. Nürnberg, 1533.

Pseudo-Euclid. *Liber Euclidis de ponderoso et levi . . . .* Basel, 1537, in Latin version of Euclid; included in editions after 1565 of Tartaglia's translation of Euclid (see p. 400, below).

Jordanus Nemorarius (13th c.). *Opusculum de ponderositate . . . .* Venice, 1565. Published by Curtius Troianus from papers left to him by Tartaglia.

Pseudo-Archimedes. *Le livre d'Archimede des pois qui aussi est dict des choses tombantes en l'humide.* Paris, 1565. Latin text included in Jordanus, *Opusculum,* above; the present book includes a commentary by the translator, Pierre Forcadel.

The technological tradition

Roberto Valturio (15th c.). *De re militari lib. xii.* Verona, 1472; Paris, 1532.

Idem, in Italian translation by Paulo Ramusio. Verona, 1483.

Leon Battista Alberti (1404–72). *De re aedificatoria.* Florence, 1485; reprinted Paris, 1512.

Marcus Vitruvius Pollio (1st c. B.C.). *De architectura.* Rome, 1486.

Idem, in Italian translation by Cesare Cesariano, Buono Mario, and Benedetto Giovio. Como, 1521.

Idem, trans. F. L. Durantino. Venice, 1524.

Idem, trans. G. B. Caporali. Perugia, 1536.

Magnus Aurelius Cassiodorus (479?–575). *De quatuor mathematicis disciplinis . . . .* Paris, 1540, 1550, 1580.

Walther H. Ryff (Rivius) (mid-16th c.). *Der fürnembsten . . . Architectür . . . .* Nürnberg, 1547, 1558.

———. *Vitruvius teutsch.* Nürnberg, 1548. Contains German translation of *De architectura.*

Alberti, *De re aedificatoria,* in French translation. Paris, 1553.

Valturio, *De re militari,* in French translation by Louis Meigret. Paris, 1555.

Diego de Sagredo. *Raison d'architecture antique: extraicte de Vitruve et autres anciens architecteurs.* Paris, 1555.

Vitruvius, *De architectura*, in Italian translation by Daniel Barbaro. Venice, 1556, 1567.

Oribasius (4th c.). *Opera*, ed. J. B. Rasario. Basel, 1557.

Petrus Peregrinus (13th c.). *De magnete* . . . . Augsburg, 1558.

W. H. Ryff, *Der Architectür* . . . . Basel, 1575, 1582. Original editions listed above.

Vitruvius, *De architectura*, in French translation by Jean Martin, *L'Architecture ou l'art de bien bastir*. Paris, 1572.

Heronis mecanici liber de machinis bellicis . . . . Venice, 1572. Latin translation by Francesco Barozzi; this work should not be confused with the *Mechanics* of Hero as later discovered to exist in an Arabic translation.

Vitruvius, *De architectura*, in Spanish translation by Miguel de Urrea. Alcala, 1582.

Di Herone Alessandrino de gli automati . . . . Venice, 1589, 1601. Italian translation by Bernardino Baldi, with historical preface.

Onosander (1st c.). *Strategicus*, ed. N. Rigault. Paris, 1599.

Heronis Ctesibii Belopoeeca hoc est telifactiva. Augsburg, 1616. Latin translation by B. Baldi, with his "Life of Hero."

The Peripatetic tradition: medieval philosophy of motion of authors who died before 1501

This listing, while far from complete, is intended to indicate authors and titles available in print at the opening of the sixteenth century, and the trend of interest during that century as reflected in the printing history of older works.

Aristotle. *Physica,* Greek text with commentaries of Averroes and of Michael Scot. Venice, 1474?.

Walter Burley (1275–1344). *Expositio in libros octo de physico auditu*. Padua, 1476; Venice, 1491.

Paul of Venice (d. 1429). *Expositio librorum naturalium* . . . . Venice, 1476, 1486, 1491, 1494.

———. *Summule naturalium* . . . . Milan, 1476.

Gaetano of Thien (1387–1465). *Commentaria in libros metheororum Aristotelis*. Padua, 1476.

———. *Recollecte super octo libros physicorum* . . . . (n.p.), 1477; Venice, 1496.

Richard Swineshead (14th c.). . . . *Calculationum liber* . . . . Padua, 1477?.

Albert of Saxony (13th c.). *Tractatus proportionum* . . . .
Venice, 1480?.

Giovanni Marliani (d. 1483). . . . *De proportione motuum* . . . .
Pavia, 1482.

———. *De motu locali* . . . . Pavia, 1483?. Commentary on
Swineshead.

Raffaele Francesco (15th c.). *Verificatio universalis* . . . *de
motu* . . . ; and Bernardo Torni (d. 1500). *In capitulum de
motu locali Hentisberi* . . . . Pisa, 1484.

John of Jandun (d. 1328). *Quaestiones in* . . . *libros physicorum* . . . . [?], 1485; Venice, 1488.

Nicole Oresme (14th c.). *Tractatus de latitudinibus formarum* . . . ; and Blasius of Parma (Pelacani, 15th c.). *Quaestiones super tractatum* . . . *magistri* . . . *Horen* [*Oresme*] . . . .
Padua, 1486.

Marsilius of Inghen (14th c.). *Abbreviationes libri physicorum* . . . . Pavia, 1490?.

Gaetano of Thien (1387–1465). *De intensione et remissione
formarum* . . . . Venice, 1491.

Thomas Aquinas (1225–74). *Commentaria super libros
physicorum* . . . . Venice, 1492. Often reprinted.

Theophilus of Ferrara (15th c.). *Propositiones ex omnibus
Aristotelis libris* . . . . Venice, 1493.

Angelo Fosinfronte (15th c.). *De motu locali* . . . . Venice,
1494.

William Heytesbury (14th c.). *De sensu* . . . , *Regulae* . . . ,
with commentaries of Gaetano, Torni, *et al.* Venice, 1494.

Albertus Magnus (12th c.). *Physica.* Venice, 1494.

William of Ockham (1284?–1349). *Summule librorum physicorum.* Bologna, 1494, 1496.

Gregory Reisch (15th c.). *Margarita philosophica* . . . .
Freiburg, 1496. The first of many editions, chiefly in the 16th
century.

Giovanni Tollentini (15th c.) . . . . *Suiseth calculationum liber
emendatus* . . . . Pavia, 1498. *Suiseth* = Swineshead.

Paul of Venice (d. 1429). *Expositio* . . . and *Super commento
Averois* . . . . Venice, 1499.

Walter Burley (1275–1344). *Super octo libros physicorum.*
Venice, 1501.

Egidio Colonna (13th c.). *In libros de physico* . . . *commentaria.* Venice, 1502.

Jean Buridan (d. 1358). *Quaestiones super* . . . *phisicorum
libros.* Paris, 1503. Revised by Johannes Dullaert, Paris,
1509.

Nicolo Verna (Vernia) (1420–99). *Contra . . . Averroys opinionem . . .* and *De gravibus et levibus.* Venice, 1504.

Blasius of Parma, Thomas Bradwardine, Nicole Oresme, *et al. Tractatus proportionum; Tractatus de latitudinibus formarum; Sex inconvenientium . . . .* Venice, 1505.

Giovanni di Casali (14th c.). *De velocitate motus alterationis.* Venice, 1505.

Giovanni Tollentini (15th c.). *Introductorium ad calculationem.* Venice, 1505.

William of Ockham (1284?–1349). *Summule . . . physicorum.* Venice, 1506.

Benedetto Vittori (15th c.). *Commentaria in tractatum proportionum Alberti de Saxonia.* Bologna, 1506.

Walter Burley (1275–1344). *In physicam . . . quaestiones.* Venice, 1508.

Paul of Venice (d. 1429). *Summa naturalium.* Venice, 1512.

Nicholas of Cusa (1401–64). *Opera omnia.* Paris, 1514.

John of Jandun (d. 1328). *. . . Librorum metheororum Aristotelis . . . expositio et questiones.* Paris, 1514.

Averroës (12th c.) and Michael Scot (12th c.). Commentaries on the *Physics.* Venice, 1516.

Albert of Saxony (1316–90). *Quaestiones super . . . physicam . . . ;* and Nicolo Verna (1420–99). *De gravibus et levibus . . . .* Venice, 1516.

Giovanni Tollentini (15th c). *Introductorium . . . ;* and Vittorio Trincavalli (15th c.). *Quaestio de reactione . . . .* Venice, 1520.

Marsilius of Inghen (14th c.). *Abbreviationes . . . .* Venice, 1521.

Gaetano of Thien (1387–1465). *In libros metheororum Aristotelis commentaria.* Venice, 1522.

Simplicius (6th c.). Greek commentary on the *Physics.* Venice, 1526.

Themistius (4th c.). *Paraphrasis libri . . . in physica . . . .* Florence, 1530.

Philoponus (6th c.). Greek commentary on *Physics,* Bks. I–IV. Venice, 1535.

Idem, in Latin translation by Dorotheo. Venice, 1539.

Simplicius (6th c.). *Commentaria . . . de physico auditu . . .* Venice, 1546, 1566.

Ermolao Barbaro (1454–93). *Compendium scientiae naturalis . . . .* Paris, 1547.

Thomas Aquinas (1225–74). *In octo physicorum libros commentaria.* Venice, 1551.

Marsilius of Inghen (14th c.). *Quaestiones . . . super . . . libros physicorum*. Lyons, 1553.

Philoponus (6th c.). *Physicorum libri 4 commentarii*. Venice, 1558.

John of Jandun (d. 1328). *Super octo libros . . . de physico . . . .* Venice, 1560.

Nicholas of Cusa (1401–64). *Opera*. Basel, 1565.

Philoponus (6th c.). *Physicorum libri 4 comm.* Venice, 1569, 1581.

Walter Burley (1275–1344) *. . . . De physica auscultatione . . . commentaria . . . .* Venice, 1589.

Richard of Middleton (13th c.). *Super . . . Petri Lombardi questiones . . . .* Brescia, 1591.

Appendix: Works of the traditions in modern
translations or discussions

Following are modern language translations or detailed discussions of some works comprising the foregoing traditions. They are grouped by their respective traditions in order corresponding to that above.

Aristotle. *Mechanica,* trans. E. S. Forster in *Works of Aristotle* (W. D. Ross, ed.), Vol. 6. Oxford, 1913.

Aristotle. *Mechanical Problems,* trans. W. S. Hett in *Aristotle: Minor Works* (Loeb Classical Library), Vol. I. Harvard, 1936.

The Works of Archimedes, ed. and trans. T. L. Heath. Original edition Cambridge, 1897; reprinted with *The Method of Archimedes,* New York: Dover, various dates.

Drachmann, A. G. *The Mechanical Technology of Greek and Roman Antiquity.* Copenhagen, 1963. Includes translations of Hero.

Herons von Alexandria Mechanik . . . , trans. L. Nix. Leipzig, 1900.

Pappus d'Alexandrie. La Collection mathématique, trans. Paul ver Eecke. Paris, 1933.

Woodcroft, Bennet, ed. *The Pneumatics of Hero of Alexandria,* trans. J. G. Greenwood. London, 1851.

Moody, E. A., and M. Clagett, eds. and trans. *The Medieval Science of Weights.* Madison, 1952, 1960. Contains texts and English translations of all the principal works in the tradition of Jordanus.

Hero of Alexandria. *Opera quae supersunt omnia,* Greek, Arabic, and Latin texts with German translation by W. Schmidt and L. Nix. 5 volumes. Leipzig, 1899–1914.

Peter Peregrinus of Maricourt. *Epistle . . . Concerning the Magnet,* trans. S. P. Thompson. London, 1903.

————. *The Letter of Peter Peregrinus on the Magnet, A.D. 1269,* trans. Brother Arnold. New York, 1904.

Vitruvius. *The Ten Books on Architecture,* trans. M. H. Morgan. Original edition Harvard, 1914; reprinted New York: Dover, 1960.

————. *De architectura,* Latin text with English translation by F. Granger. 2 volumes. London: Loeb Classical Library, 1931–34.

Aquinas, Thomas. *Commentary on Aristotle's* Physics, ed. W. Stark, London, 1963.

Clagett, Marshall. *The Science of Mechanics in the Middle Ages.* Madison, 1959. Contains most of the significant texts relating to the Peripatetic tradition.

Crosby, H. L. *Thomas of Bradwardine, His* Tractatus de proportionibus: *Its Significance for the Development of Mathematical Physics.* Madison, 1955.

Grant, Edward. *Nicole Oresme,* De proportionibus proportionum *and* Ad pauca respicientes. Madison, 1966.

Nicholas of Cusa. *Of Learned Ignorance,* trans. Germain Heron. London, 1954.

————. *De staticis experimentis,* trans. Henry Viets in *Annals of Medical History,* Vol. 4 (1922), 115–35.

Wilson, Curtis. *William Heytesbury: Medieval Logic and the Rise of Mathematical Physics.* Madison, 1957.

Sixteenth-Century Philosophy of Motion in Italy

The ensuing list comprises books by authors living in the sixteenth century who continued the Peripatetic tradition, as that term is used in this book.

Achillini, Alessandro (1463–1512). *De distributionibus ac de proportione motuum . . . .* Bologna, 1494.

————. *Opera.* Venice, 1545.

Borro, Girolamo (1512–92). *De motu gravium et levium* . . . . Florence, 1576.

Brucioli, Antonio (16th c.). *Dialogi* . . . . Venice, 1537–38; reprinted 1545.

Buonamico, Francesco (d. 1603). *De motu libri x.* Pisa, 1591.

Cagni, Francesco (16th c.). *Theoremata physicorum Aristotelis.* Venice, 1571.

Campanella, Tommaso (1568–1639). *Philosophia sensibus demonstrata* . . . . Naples, 1591.

Canini, Marco (16th c.). *Libri duo de soliditate, fluiditate, gravitate et levitate corporum terrestrium.* Venice, 1566.

Cesalpino, Andrea (1519–1603). *Peripateticarum quaestionum: libri quinque.* Florence, 1569.

Cremonini, Cesare (1550–1631). *Explanatio prooemii . . . de physico auditu.* Padua, 1596.

Crippa, Bernardino (16th c.). *In Aristotelis librum de animalium motu* . . . . Venice, 1566.

Girolamo, Fra (d. 1570). *De quantitatibus* . . . . Rome, 1570.

Isolani, Isidoro (16th c.). *De velocitate motuum* . . . . Pavia, 1521.

Landi, Bassiano (d. 1563). *Opuscula de motu* . . . . Padua, 1550.

Leone, Ambrogio (16th c.). *Opus questionum* . . . . Venice, 1523.

Manuzio, Paolo (1512–74). *De gli elementi* . . . . Venice, 1557.

Mellini, Domenico (16th c.). *Discorso . . . contra . . . un movimento . . . perpetuo* . . . . Florence, 1583.

Michele, Agostino (16th c.). *Trattato della grandezza dell' acqua e della terra* . . . . Venice, 1583.

Morelli, Gregorio (16th c.). *Scala di tutte le scienze* . . . . Venice, 1567.

Nifo, Agostino (1473–1546). *Expositio super octo libros de physico auditu,* with commentary of Averroës added. Venice, 1508, 1552, 1569.

Patrici [Patrizzi], Francesco (1529–?). *De rerum natura* . . . . Ferrara, 1587.

Pomponazzi, Pietro (1462–1525). *Tractatus, in quo disputatur penitus quid intensio et remissio formarum intendantur* . . . . Bologna, 1514.

———. *De reactione.* Venice, 1525.

Raggi, Giambattista (16th c.). *Physica.* 1559.

Rosso, Paolo del (16th c.). *La fisica* . . . . Paris, 1578.

Scozio, Antonio (1469–?). *De potissima demonstratione* . . . . Venice?, 1516.

Telesio, Bernardino (1507?–88). *De rerum natura* . . . . Rome, 1565; Naples, 1587.

———. *Varii de naturalibus rebus libelli.* Venice, 1590.

Unicorno, Giuseppe (1523–1610). *De admiranda vi proportionis* . . . . Venice, 1552.

———. *Liber de utilitate mathematicarum artium.* Venice, 1561.

———. *De mathematicarum artium utilitate* . . . . Bergamo, 1583.

Vimercato, Francesco (mid-16th c.). *In* . . . *libros* . . . *de naturali auscultatione.* Venice, 1564.

———. *In* . . . *libros meteorologicorum commentarii* . . . . Paris, 1556; reprinted Venice, 1565.

Zabarella, Giacomo (1533–1589). *De rebus naturalibus.* Padua, 1589.

———. *In libros Aristotelis physicorum commentaria.* Venice, 1601.

Zimara, Marcantonio (1460–1532). *Questio de movente et moto* . . . . Venice, 1504.

———. *Solutiones contradictionum in dictis Averroys.* Lyons, 1542.

Sixteenth-Century Mechanics in Italy

A general listing is preceded by complete bibliographies for Tartaglia, Benedetti, and Guido Ubaldo del Monte, and several works composed by Galileo during the sixteenth century.

Niccolò Tartaglia. *La Nova Scientia* . . . . Venice, 1537. Reprinted at Venice 1546?, 1550, 1551, 1558, 1562, 1583, and in the *Opere* of 1606. Additions in 1550 and later editions. German translation of Bks. I–II (without Tartaglia's name) in W. H. Ryff, *Der Architectür;* see "The technological tradition," above. French translation of Bks. I–II by Rieffel in *La Balistique de Nicolas Tartaglia,* Paris, 1845–46, said to have appeared first in *Journal des armes speciales.*

———. *Euclide* . . . *di Latino in volgar tradotto* . . . . Venice, 1543. Reprinted at Venice, 1544 and 1545; then, with the added fragment given as pseudo-Euclid (in "The tradition of Jordanus"), above, in 1565, 1569, 1585, and 1586.

———. *Opera Archimedis* . . . *per Nicolaum Tartaleam* . . . *multis erroribus emendata, expurgata, ac in luce posita* . . . . Venice, 1543.

———. *Quesiti, et inventioni diverse* . . . . Venice, 1546.

Reprinted, with additions to Bk. VI, at Venice, 1554, 1562, and in the *Opere,* 1606. Reported editions of 1550 and 1551 are spurious. German translations of several parts in Ryff (see "The technological tradition," above) with names of interlocutors changed and no mention of Tartaglia. French translation of Bk. VI by unknown translator, Rheims, 1556. English translation of Bks. I–III in Cyprian Lucar, *Three Bookes of Colloquies concerning the Arte of Shooting . . .* , London, 1588. German translation of Bk. VI by Andrea Böhm in *Magazin für ingenieure Artilleristen,* Vol. 4 (1778). French translation of Bks. I–III by Rieffel in *La Balistique de Nicolas Tartaglia,* Paris, 1845–46. Reprinted in facsimile of 1554 edition, with introduction by Arnaldo Masotti, Brescia, 1959.

————. *Risposta data . . . a Messer Lodovico Ferraro . . . .* Milan and Venice, 1547. Followed by a second, third, and fourth *Risposta* published at Venice in 1547, and a fifth and sixth in 1548. Published with the six corresponding letters of Ferrari to Tartaglia in Silvestro Gherardi, *Di alcuni materiali per la storia della facoltà matematica di Bologna,* Bologna, 1846.

————. *Regola generale da sulevare . . . ogni affondata nave . . . intitolata la travagliata inventione, con . . . li ragionamenti con Riccardo Wentworth . . . .* Venice, 1551. The *Ragionamenti,* also issued separately, contains Tartaglia's Italian translation of the first book of Archimedes' *On Bodies in Water.* Reprinted at Venice, 1558, 1562, and in the *Opere,* 1606. English translation by Thomas Salusbury, *Mathematical Collections and Translations, The Second Tome,* London, 1665, reprinted 1967.

————. *General trattato di numeri, et misure* (in six parts). Venice, 1556–60. Reprinted as *Tutte l'opere d'aritmetica . . . ,* Venice, 1592–93. French translation of Pts. I–II by Guillaume Gosselin, Paris, 1578.

————. *Archimedis de insidentibus aquae* (see "The Archimedean tradition," above). Venice, 1565.

————. *Jordani opusculum de ponderositate* (see "The tradition of Jordanus," above). Venice, 1565.

————. *Opere* (containing four works as designated above). Venice, 1606.

Giovanni Battista Benedetti. *Resolutio omnium Euclidis problematum aliorumque ad hoc necessario inventorum una tantummodo circini data apertura . . . .* Venice, 1553.

————. *Demonstratio proportionum motuum localium* . . . .
Venice, 1554. Two issues, the second dated "Ides of February." First issue reprinted as his own by Joannes Taisnier in *Opusculum perpetua memoria dignissimum* . . . , Cologne, 1562; and translated into English as Taisnier's in *A Very Necessary and Profitable Booke concerning Navigation* . . . , by Richard Eden, London, [1578].

————. *De gnomonum umbrarumque solarium usu liber.* Turin, 1573.

————. *De temporum emendatione opinio.* Turin, 1578; reprinted in *Diversarum speculationum,* below.

————. *Consideratione* . . . *d'intorno al discorso* . . . *del Ecc. Signor Antonio Berga* . . . . Turin, 1579. Latin translation reported in A. Berga, *Disputatio de magnitudine terrae et aquae,* trans. F. M. Vialardi, Turin, 1580.

————. *Lettera* . . . *all'Ill. Sig. Bernardo Trotto* . . . *contra alli calculatore delle effemeridi.* Turin, 1581. Latin version in *Diversarum speculationum,* below.

————. *Diversarum speculationum mathematicarum, et physicarum liber.* Turin, 1585. Reissued, Venice, 1586 and 1599. A Venice edition of 1585 is also reported.

Guido Ubaldo del Monte. *Mechanicorum liber.* Pesaro, 1577. Italian translation by F. Pigafetta, *Le Mechaniche,* Venice, 1581. Both versions reprinted, Venice, 1615.

————. *Planisphaeriorum universalium theorica.* Pesaro, 1579.

————. *De ecclesiastici calendarii restitutione opusculum.* Pesaro, 1580.

————. *In duos Archimedis aequeponderantium libros paraphrasis.* Pesaro, 1588.

————. *Perspectivae libri sex.* Pesaro, 1600.

————. *Problematum astronomicorum libri septem.* Venice, 1609.

————. *De cochlea libri quatuor.* Venice, 1615.

Galileo Galilei. *Juvenilia* (1584), in *Opere di Galileo Galilei,* ed. Antonio Favaro, Vol. I. (Ed. Naz., Florence, 1890; reprinted, 1929), pp. 15–177. Comprising lecture notes or student compositions.

————. *La Bilancetta* (1586), in *Opere,* I, 215–20. First published in G. B. Hodierna, *Archimede redivivo* . . . , Palermo, 1644. Translation in L. Fermi and G. Bernardini,

Galileo and the Scientific Revolution, New York, 1961.
———. Theorems on centers of gravity (1588), *Opere,* I, 187–208. First published in Galileo, *Discorsi e dimostrazione* . . . , Leyden, 1638.
———. *Sermones de motu* (ca. 1590), in *Opere,* I, 251–366. English translation by I. E. Drabkin in *Galileo On Motion and On Mechanics,* Madison, 1960.
———. *Le meccaniche* (1593–1600), in *Opere,* II, 155–90. First published in French translation by Marin Mersenne, *Les Méchaniques de Galilée* . . . , Paris, 1634; reprinted, ed. B. Rochot, Paris, 1966. First Italian publication by Luca Danesi, *Della scienza meccanica, e delle utilità che si traggono dagli strumenti di quella,* Ravenna, 1649. English translation by Thomas Salusbury in *Mathematical Collections and Translations, The Second Tome,* London, 1665; reprinted, 1967. Recent English translation by S. Drake in *Galileo On Motion and On Mechanics,* Madison, 1960. Early manuscript versions first published in A. Favaro, "Delle meccaniche lette in Padova dal Sr. Galileo Galilei l'anno 1594," *Memorie del R. Ist. Veneto di Scienze, Lettere ed Arti,* Vol. 26 (1899); and in S. Drake, "The Earliest Version of Galileo's Mechanics," *Osiris,* Vol. 13 (1958).
———. *Cosmografia* (ca. 1596), in *Opere,* II, 211–55. Early variant manuscript readings in S. Drake, "An Unrecorded Manuscript Copy of Galileo's Cosmography," *Physis,* Vol. I, No. 4 (1959).

Affaitati, Fortunio (16th c.). *Physicae ac astronomicae considerationes.* Venice, 1549.
Agricola (Georg Bauer, 1494–1555). *Opera . . . de l'arte de metalli* . . . , trans. Michelangelo Florio. Basel, 1563.
Agrippa, Camillo (16th c.). *Trattato di scientia d'arme* . . . . Rome, 1553; reprinted Venice, 1568.
———. *Modo di comporre il moto* . . . . Rome, 1575.
———. *Trattato di trasportar la guglia in su* . . . . Rome, 1583.
———. *Nuove inventioni* . . . . Rome, 1595.
———. *La virtù, dialogo . . . della causa de' moti* . . . . Rome, 1597.
Baldi, Bernardino (1553–1617). *In mechanica Aristotelis problemata exercitationes.* Mainz, 1621.
Berga, Antonio (16th c.). *In prohemium physicorum Aristotelis.* Turin, 1573.

————. *Discorso . . . della grandezza dell'acqua e della terra* . . . . Turin, 1579. Reported Latin translation under Benedetti, above.

Biringuccio, Vannocci (1480–1539). *Pirotechnia* . . . . Venice, 1540; reprinted, 1552, 1558, and 1559.

Bozio, Tommaso (1548–1610). *De signis ecclesiae Dei*. Rome, 1591.

Bruno, Giordano (1548–1600). *La cena delle ceneri*. Paris [London], 1584.

————. *De la causa, principio, et uno*. Venice [London], 1584.

————. *De l'infinito universo et mondi*. Venice [London], 1584.

[Campi, Bartolomeo?] (16th c.). *Descrizione dell'artificiosa macchina fatta per cavar il galeone* . . . . Venice, 1560.

Cardano, Girolamo (1501–76). *De subtilitate libri xxi*. Nürnberg, 1550; reprinted Lyons, 1551, 1558, 1580; Basel, 1554, 1560, 1582. French translation by Richard Le Blanc, Paris, 1556; reprinted 1566, 1578, 1584.

————. *De rerum varietate libri xvii*. Basel, 1557; reprinted Avignon, 1558; Basel, 1571.

————. *Opus novum de proportionibus* . . . . Basel, 1570.

Cataneo, Girolamo (16th c.). *Opera nuova di fortificare* . . . . Brescia, 1564; reprinted various times under varying titles. French translation, Lyons, 1574.

Cellini, Benvenuto (1500–1571). *Due trattati . . . dell'oreficeria . . . della scultura*. Florence, 1568.

Ceredi, Giuseppe (16th c.). *Tre discorsi sopra il modo d'alzar acque da' luoghi bassi*. Parma, 1567.

Commandino, Federico (1509–75). *Liber de centro gravitatis*. Bologna, 1565.

Delfino, Domenico (16th c.). *Sommario di tutte le scienze* . . . . Venice, 1556; reprinted many times.

Fioravanti, Leonardo (d. 1588). *Dello specchio di scientia* . . . . Venice, 1564; often reprinted. French translation by G. Chappuys, Paris, 1584.

Fontana, Domenico (1543–1607). *Della trasportazione dell' obelisco Vaticano* . . . . Rome, 1590.

Ghetaldi, Marino (1566–1627). *Promotus Archimedis* . . . . Rome, 1603.

Isaachi, Giambattista (16th c.). *Inventioni* . . . . Parma, 1579.

————. *Repertorio de' secreti*. Reggio, 1573.

Lorini, Bonajuto (1540–1611). *Delle fortificationi* . . . . Venice, 1596.

Marinati, Aurelio (16th c.). . . . *Somma di tutte le scienze* . . . .
Rome, 1587.

Masini, Francesco (16th c.). *Discorso . . . di trasportar . . .
la guglia* . . . . Cesena, 1586.

Maurolico, Francesco (1494–1575). *Opuscula mathematica.*
Venice, 1575.

——. *Problemata mechanica.* Messina, 1613.

Montemerlini, Francesco (16th c.). *Discorsi del modo di
fortificare le città* . . . . Venice, 1548.

Parisio, Attilo (16th c.). *Discorso sopra la sua nuova inventione d'horologi* . . . . Venice, 1598.

Paschali, Giulio (16th c.). *Il theatro de gl'instrumenti* . . . .
Lyons, 1582. Also published in Latin, Lyons, 1582.

Piccolomini, Alessandro (1508–78). *In mechanicas* . . . . (see
"The 'Aristotelian' tradition," above).

——. *La prima (seconda) parte della filosofia naturale* . . . .
Venice, 1551–54.

Pifferi, Francesco (d. *ca.* 1615). *Sfera . . . tradotta e dichiarata.*
Siena, 1604.

Pigafetta, Filippo (1533–1604). *Le meccaniche* . . . . (see
Guido Ubaldo, above).

——. *Discorso intorno all'istoria dell'aguglia* . . . . Rome,
1586.

Porta, Giovanni Battista della (1535?–1615). *Magia naturalis
libri iiii.* Naples, 1558; reprinted and translated many times;
expanded to *libri xx* in 1589 (1569?) and thereafter again
reprinted and translated into Italian, French, and English.

——. *Pneumaticorum libri tres.* Naples, 1601. Italian translation *I tre libri de'spiritali* . . . . Naples, 1606.

Prado, Geronimo (1547–95), and Juan Bautista Villalpando
(1552?–1608). *In Ezechielem explanationes* . . . (3 vols.;
Vol. III by Villalpando). Rome, 1596–1604.

Raimondi, Luigi (16th c.). *Animadversiones in librum F.
Commandini de centro gravitatis* . . . . [?], 1597.

Ramelli, Agostino (1531–1605?). *Le diverse et artificiose
machine* . . . (editions in Italian and French). Paris, 1588.

Rinaldi, Orazio (16th c.). *Specchio di scienze* . . . . Venice,
1583.

Ringhieri, Innocenzo (16th c.). *Cento giuochi liberali* . . . .
Bologna, 1551. French translation by Hubert Philippe de
Villiers, Lyons, 1555.

Stelliola, Niccolò Antonio (1555–1624). *De gli elementi me-
chanici.* Naples, 1597.

——. *Encyclopedia Pythagorea.* Naples, 1616.

Valerio, Luca (1552–1618). *De centro gravitatis solido-*
rum . . . . Bologna, 1604.

Valla, Giorgio (1447–1500). *De expetendis et fugiendis re-*
bus. Venice, 1501.

Valle, Battista della (d. 1550). *Vallo; libro continente apperti-*
nentie a capitanii . . . . Venice, 1524. Often reprinted.

Veranzio, Fausto (1551–1617). *Machinae novae.* Florence,
1615.

Villalpando, Juan Bautista. See Prado, Geronimo, above.

Vinci, Leonardo da (1452–1519). *Del moto e misura dell'-*
acqua, in *Raccolta d'autori italiani che trattano del moto*
dell'acqua, Vol. 10. Bologna, 1826. Another ms. ed. E.
Carusi and A. Favaro. Bologna, 1923.

———. *Les manuscrits . . . de la Bibliothèque de l'Institut,*
ed. Ch. Ravaisson-Mollien. Paris, 1881–91.

———. *Il codice atlantico,* ed. R. Acc. dei Lincei. Milan,
1894–1904.

Zanchi, Giovanni Battista (16th c.). *Del modo di fortificar* . . . .
Venice, 1556.

Zonca, Vittorio (1568–1602). *Novo teatro di machine.*
Padua, 1607.

Sixteenth-Century Mechanics and Philosophy
of Motion Outside Italy

Agricola (Georg Bauer, 1494–1555). *De re metallica.* Basel,
1530. German translation by Phillip Bech, Frankfurt, 1580.

Astudillo, Diego de (d. 16th c.). *Questiones super octo libros*
phisicorum . . . . Valladolid, 1532.

Bayfius, Lazarus (16th c.). . . . *De re navali* . . . . Basel,
1537.

Besson, Jacques (1540?–76). *Théâtre des instrumens mathé-*
matiques et méchaniques . . . . Lyons, 1579; Geneva, 1594.
The first edition appeared about 1570. German version,
Mümbelgart, 1595. Spanish version, Cardon, 1602.

Bourne, William. *Inventions or Devises.* London, 1578.

Buteo, Johannes (Jean Borrel, 1492–1572). *Opera geometrica.*
Paris, 1554.

———. *Logistica.* Paris, 1559.

Celaya, Juan (d. 16th c.). *Expositio in octo libros phisico-*
rum . . . . Paris, 1517.

College of Coimbra (authors of 16th c.). *In octo libros physi-*
corum and *Commentarii in meteorologicorum libros.*
Coimbra, 1592.

Copernicus, Nicholas (1473–1543). *De revolutionibus orbium coelestium*. Nürnberg, 1543; 2nd ed. Basel, 1562.

Coronel, Luis (d. 16th c.). *Physicae perscrutationes*. Paris, 1511.

Diest, Diego de (d. 16th c.). *Quaestiones phisicales*. Saragossa, 1511.

Digges, Thomas (1546–95). *An Arithmeticall Militare Treatise, named Stratioticos*. London, 1579.

Dullaert, Joannes (1470?–1513). *Quaestiones super octo libros phisicorum* . . . . Paris, 1506; Lyons, 1512.

Gilbert, William (1544–1603). *Physiologia nova de magnete* . . . . London, 1600.

Lax, Gaspar (d. 16th c.). *Cunabula omnium fere scientiarum*. Montaubon, 1518.

Mair, Jean (John Major, 1478?–1540?). *In primum sententiarum* . . . . Paris, 1519.

———. *Octo libri physicorum cum naturali philosophia* . . . . Paris, 1526.

Margallo, Pedro (d. 16th c.). *Physices compendium*. Salamanca, 1520.

Melanchthon, Phillip (1497–1560). *Doctrinae physicae elementa*. Wittenberg, 1549.

Norman, Robert. *The Newe Attractive*. London, 1581.

Nuñez, Pedro (1492–1577). *De arte atque ratione navigandi*. Coimbra, 1573.

———. *Opera*. Basel, 1566.

Palissy, Bernard (1510–89). *Discours . . . des eaux et fontaines*. Paris, 1580.

Pereira, Bento (1535–1610). *De communibus omnium rerum naturalium principiis*. Rome, 1562.

Ramus, Petrus (Pierre de la Ramée, 1515–72). *Aristotelicae animadversiones*. Lyons, 1543.

———. *Scholarum mathematicarum libri xxxi*. Basel, 1569.

———. *Scholae in liberales artes* . . . . Basel, 1578.

———. *Scholarum physicarum libri viii*. Frankfurt, 1583.

Ryff, W. H. See under "The technological tradition," above.

Scaliger, Julius Caesar (1484–1558). *Exotericarum exercitationum libri xv; de subtilitate ad Cardanum*. Paris, 1557; often reprinted.

Schoffer, Hartman (16th c.). *Panoplia omnium*. Frankfurt, 1568.

Soto, Domingo de (1494–1560). *Super octo libros physicorum Aristotelis subtilissime quaestiones*. Salamanca, 1545; seven editions to 1572; Venice, 1588.

————. *Super octo libros physicorum Aristotelis commentaria.* Salamanca, 1545? six editions to 1582; Venice, 1582.

Stevin, Simon (1548–1620). *De Beghinselen der Weeghconst.* Leyden, 1586. Latin translation, *Hypomnemata mathematica,* Leyden, 1605–08. French translation, *Les Oeuvres mathématiques,* Leyden, 1634.

Strada, Jacob de (1523–88). *Künstlicher Abris allerhand Wasser, Wind, Ross, und Handmühlen beneben . . . nutzlichen Pompen . . . .* Frankfurt, 1618, 1629.

Suarez, Francisco (1548–1617). *Disputationes metaphysicae.* Salamanca, 1587.

Taisnier, Joannes (mid-16th c.). *Opusculum . . . .* Cologne, 1562. See Benedetti, above.

Thomaz, Alvarez (fl. 1510). *Liber de triplici motu . . .* and *Suiseth calculationes . . . .* Paris, 1509.

Titelmans, Francis (1502–37). *Compendium naturalis philosophiae.* Paris, 1542; often reprinted.

Toledo, Francisco (16th c.). *Commentaria . . . in Aristotelis de physica . . . .* Venice, 1570, 1580; Cologne, 1585.

Varro, Michel (16th c.). *Tractatus de motu.* Geneva, 1584.

Velcurio, Johannes (16th c.). *Commentariorum . . . in universam Aristotelis physicen liber.* London, 1588.

Verro, Sebastian (16th c.). *Physicorum libri x.* London, 1581.

Vives, Juan Luis (1492–1540). *Opera . . . .* Basel, 1555.

Appendix: Sixteenth-century works in modern English
translation

Agricola, Georgius. *De re metallica,* trans. H. C. and L. H. Hoover. London, 1912; reprinted New York: Dover, 1950.

Biringuccio, Vannoccio. *The Pirotechnia,* trans. C. S. Smith and M. T. Gnudi, 1942; reprinted New York: Basic Books, 1959.

Bruno, Giordano. "On the Infinite Universe and Worlds," trans. D. W. Singer in *Giordano Bruno, His Life and Thought.* New York, 1950. Contains bibliography and summaries of other works by Bruno.

————. *Cause, Principle and Unity,* trans. J. Lindsay. Essex, 1962.

Cardano, Girolamo. *De subtilitate.* Bk. I, trans. M. M. Cass, Williamsport, Pa., 1934.

Cellini, Benvenuto. *The Treatises on Goldsmithing and Sculpture,* trans. C. R. Ashbee, London, 1898.

Copernicus, Nicholas. *On the Revolutions of the Heavenly Spheres,* trans. C. G. Wallis in *Great Books of the Western World,* Vol. 16. Chicago, 1952.

Gilbert, William. *On the Magnet,* trans. S. P. Thompson. London, 1900; reprinted New York: Basic Books, 1958. Another translation by P. F. Mottelay, New York, 1893; reprinted in *Great Books of the Western World,* Vol. 28, Chicago, 1952.

Porta, Giovanni Battista della. *Natural Magick,* anonymous trans. London, 1658; reprinted New York: Basic Books, 1957.

Stevin, Simon. "Writings on mechanics," trans. C. Dijkshoorn in *The Principal Works of Simon Stevin,* Vol. I. Amsterdam, 1955.

Vinci, Leonardo da. *The Notebooks of Leonardo da Vinci,* ed. and trans. Edward MacCurdy. New York, 1939.

Works relating to Sixteenth-Century Mechanics in Italy

Baldi, Bernardino. *Cronica de' matematici* . . . . Urbino, 1707.
———. "Vite inedite di matematici Italiani," ed. E. Narducci in *Bullettino di bibliografia e di storia delle scienze matematiche e fisiche,* Vol. 19 (1886); and separately Rome, 1887.

Beaujouan, G. *L'Interdépendence entre la science scolastique et les techniques utilitaires.* Paris, 1957.

Beck, T. *Beiträge zur Geschichte des Maschinenbaues.* Berlin, 1900.

Boas (-Hall), Marie. "The Establishment of the Mechanical Philosophy," *Osiris,* Vol. 10 (1952), 412–51.
———. "Hero's *Pneumatica,* a Study of Its Transmission and Influence," *Isis,* Vol. 40 (1949), 38–48.
———. *The Scientific Renaissance, 1450–1630.* London and New York, 1962.

Boll, Marcel. *Histoire de la mécanique.* Paris, 1961.

Bortolotti, Ettore. *Studi sulla storia delle matematiche in Italia nei secoli XVI et XVII.* Bologna, 1928.
———. *Studi e ricerche sulla storia della matematica in Italia.* Bologna, 1944.

Burtt, E. A. *The Metaphysical Foundations of Modern Science.* 1924; revised edition, 1932; reprinted New York: Anchor Books, 1954.

Cantor, Moritz. *Vorlesungen über die Geschichte der Mathematik.* Leipzig, 1892; reprinted New York, 1965. See especially Vol. 2.

Caverni, Raffaello. *Storia del metodo sperimentale in Italia.* Florence, 1895–98. See especially Vols. 4 and 5.

Clagett, Marshall. *The Science of Mechanics in the Middle Ages.* Madison, 1961.

———, ed. *Critical Problems in the History of Science.* Madison, 1959.

Cooper, L. *Aristotle, Galileo, and the Tower of Pisa.* Ithaca, 1935.

Crombie, A. C. *Medieval and Early Modern Science.* New York, 1959.

———, ed. *Scientific Change.* New York, 1963.

Dainville, F. de. "L'Enseignement des mathématiques dans les Collèges Jésuites de France du XVIe au XVIIIe siècle," *Revue d'histoire des sciences et de leurs applications,* Vol. 7 (1954), 6–21.

DeCamp, L. S. *The Ancient Engineers.* Norwalk, Conn., 1966.

Deshayes, M. *La Découverte de l'inertie. Essai sur les lois générales du mouvement de Platon à Galilée.* Paris, 1930.

Dibner, Bern. *Moving the Obelisks.* Norwalk, Conn., 1952.

Dijksterhuis, E. J. *The Mechanization of the World Picture.* Oxford, 1961. English translation by C. Dikshoorn of *De Mechanisering van het Wereldbeeld,* Amsterdam, 1950.

———. "The Origins of Classical Mechanics from Aristotle to Newton," in Marshall Clagett, ed., *Critical Problems in the History of Science* (Madison, 1959), pp. 163–96.

———. *Val en Worp. Een Bijdrage tot de Geschiedenis der Mechanica van Aristoteles tot Newton.* Gröningen, 1924.

Dingler, Hugo. *Das Experiment; sein Wesen und seine Geschichte.* Munich, 1928.

Drabkin, I. E. "Aristotle's Wheel: Notes on the History of a Paradox," *Osiris,* Vol. 9 (1940), 162–98.

Drew, Katherine F., and F. S. Lear, eds. *Perspective in Medieval History.* Chicago, 1963.

Duehring, E. *Kritische Geschichte der allgemeinen Principien der Mechanik* (2nd ed.). Leipzig, 1887.

Dugas, René. *A History of Mechanics.* New York, 1955. English translation by J. R. Maddox of *Histoire de la mécanique,* Paris, 1950.

———. *Mechanics in the Seventeenth Century.* New York,

1958. English translation by Freda Jacquot of *La Mecanique au XVIIe siècle,* Paris, 1954.

Duhem, Pierre. "De l'Accélération produite par une force constante: Notes pour servir à l'histoire de la dynamique," *Comptes rendus du IIe Congrès international de philosophie* (Geneva, 1904), pp. 895–915.

———. *Études sur Léonard de Vinci.* 3 Vols. Paris, 1906–13; reprinted Paris, 1955.

———. *L'Evolution de la mécanique.* Paris, 1903.

———. "Le Mouvement absolu et le mouvement relatif," *Revue de philosophie,* Vols. 11–14 (1907–09); reprinted, Montligeon, 1909.

———. *Les Origines de la statique.* 2 Vols. Paris, 1905–06.

———. Σώζειν τὰ φαινόμενα; *Essai sur la notion de théorie physique de Platon à Galilée.* Paris, 1908.

———. *Le Système du monde.* 10 Vols. Paris, 1913–59.

Durand, D. B. "Nicole Oresme and the Medieval Origins of Modern Science," *Speculum,* Vol. 16 (1941), 167–85.

Favaro, A. "Se e quale influenza abbia Leonardo da Vinci esercitata su Galileo e sulla scuola galileiana," *Scientia,* Vol. 20 (1916), 417–34.

Febvre, L. P. V. *Le Problème de l'incroyance au XVIe siècle.* Paris, 1947.

Fierz, Markus. "Ueber den Ursprung und Bedeutung von Newtons Lehre vom absolutem Raum," *Gesnerus,* Vol. 11, No. 3/4 (1954).

Fink, Eugen. *Zur ontologischen Frühgeschichte von Raum, Zeit, Bewegung.* The Hague, 1957.

Forbes, R. J., and E. J. Dijksterhuis. *Ancient Times to the Seventeenth Century. (A History of Science and Technology,* Vol. 1.) Baltimore, 1963.

Forti, U. *Storia della tecnica italiana.* Florence, 1940.

———. *Storia della tecnica dal Medioevo al Rinascimento.* Florence, 1957.

Gentile, Giovanni. *Studi sul Rinascimento* (2nd ed.). Florence, 1936.

Gerland, E., and F. Trantmüller. *Geschichte der physicalischen Experimentierkunst.* Leipzig, 1899.

Gille, Bertrand. *Les Ingénieurs de la Renaissance.* Paris, 1964.

Gilson, E. *Le Rôle de la pensée médiévale dans la formation du système cartésienne* (2nd ed.). Paris, 1951.

Ginzburg, B. "Duhem and Jordanus Nemorarius," *Isis,* Vol. 25 (1936), 341–62.

Grant, Edward. "Aristotle, Philoponus, Avempace, and Galileo's Pisan Dynamics," *Centaurus,* Vol. 11 (1965), 79–95.

———. "Later Medieval Thought, Copernicus, and the Scientific Revolution," *Journal of the History of Ideas,* Vol. 23 (1962), 197–220.

Haas, A. E. *Die Grundgleichung der Mechanik, dargestellt auf Grund der geschichtlichen Entwickelung.* Leipzig, 1914.

Hall, A. R. *The Scientific Revolution, 1500–1800* (2nd ed.). London, 1962.

———. *Ballistics in the Seventeenth Century.* Cambridge, England, 1952.

———. "Early Modern Technology, to 1600," in M. Kranzberg and C. W. Pursell, eds., *Technology in Western Civilization* (New York, 1967), Vol. 1, 79–103.

Holton, G. "Science and the Changing Allegory of Motion," *Scientia,* Vol. 98 (1963), 191–200.

Hooykas, R. *Das Verhältnis von Physik und Mechanik in historischer Hinsicht.* Wiesbaden, 1963.

Jammer, Max. *Concepts of Space.* Cambridge, Mass., 1954.

———. *Concepts of Force.* Cambridge, Mass., 1957.

———. *Concepts of Mass in Classical and Modern Physics.* Cambridge, Mass., 1961.

Keller, A. G. *A Theatre of Machines.* New York, 1964.

Koyré, A. *Études Galiléennes.* Paris, 1939. Reprinted, 1968.

———. *From the Closed World to the Infinite Universe.* Baltimore, 1957.

Koyré, A., *et al. La Science au seizième siècle.* Paris, 1960.

———. *Études d'histoire de la pensée scientifique.* Paris, 1966.

———. *Metaphysics and Measurement.* Cambridge, 1968.

Kristeller, P. O. "Renaissance Aristotelianism," *Greek, Roman, and Byzantine Studies,* Vol. 6 (1965), 157–74.

Lange, Heinrich. *Geschichte der Grundlagen der Physik,* Vol. 1. Freiburg and Munich, 1954.

Lange, Ludwig. "Die geschichtliche Entwickelung des Bewegungsbegriffs und ihr voraussichtliches Endergebniss," *Philosophische Studien,* Vol. 3 (1886), 337–419, 463–91.

Lasswitz, Kurd. *Geschichte der Atomistik vom Mittelalter bis Newton,* Vol. 1. Hamburg and Leipzig, 1890; reprinted Hildesheim, 1963.

Libri, G. *Histoire des sciences mathématiques en Italie,*

depuis la renaissance des lettres jusqu'à la fin du XVIIe siècle. 4 volumes. Paris, 1838–41.

Mach, E. *Die Mechanik in ihrer Entwickelung historisch-kritisch dargestellt.* Leipzig, 1883. English translation by T. J. McCormack, *The Science of Mechanics,* La Salle, Ill., 1960.

MacLean, J. *De historische Ontwikkeling der Stootwetten van Aristoteles tot Huygens.* Rotterdam, 1959; Amsterdam, 1960.

Maier, A. *An der Grenze von Scholastik und Naturwissenschaft.* Rome, 1952.

———. *Die Vorläufer Galileis in 14. Jahrhundert.* Rome, 1949.

———. *Zwischen Philosophie und Mechanik.* Rome, 1958.

Marcolongo, R. "Lo Sviluppo della meccanicá sino ai discepoli di Galileo," *Memorie della Reale Accademia dei Lincei,* Vol. 13 (1919).

Mieli, A. *La Ciencia del Renacimiento.* Buenos Aires, 1948(?).

Moscovici, S. "Notes sur le *De motu tractatus* de Michel Varro," *Revue d'histoire des sciences,* Vol. 11 (1958), 108–29.

Nardi, B. *Saggi sull'Aristotelismo padovano dal secolo XIV al XVI.* Florence, 1958(?).

Olschki, L. *Die Literatur der Technik und der angewandten Wissenschaften vom Mittelalter bis zur Renaissance. (Geschichte der neusprachlichen wissenschaftlichen Literatur,* Vol. 1.) Heidelberg, 1919; reprinted Vaduz, 1965.

———. *Bildung und Wissenschaft im Zeitalter der Renaissance in Italien. (Geschichte der neusprachlichen wissenschaftlichen Literatur,* Vol. 2.) Leipzig, 1922; reprinted Vaduz, 1965.

Papp, D., and J. Babini. *Astronomia, fisica, y biologia. (Panorama general de historia de la ciencia,* Vol. 6 [pertaining to the Renaissance].) Buenos Aires, 1952.

Parsons, W. B. *Engineers and Engineering in the Renaissance.* Baltimore, 1939.

Poni, C. *Economia, scienza, tecnologia e controriforma.* Bologna, 1966.

Randall, J. H. "The Development of Scientific Method in the School of Padua," *Journal of the History of Ideas,* Vol. 1 (1940), 177–206.

———. *The Making of the Modern Mind.* Boston, 1940.

————. *The School of Padua and the Emergence of Modern Science.* New York, 1961.

Rhys, H. H., ed. *Seventeenth Century Science and the Arts.* Princeton, 1961.

Rossi, P. *I Filosofi e le macchine.* Milan, 1962.

Rouse, H., and S. Ince. *History of Hydraulics.* Ames, Iowa, 1957; reprinted New York: Dover.

Russo, F. "Deux ingénieurs de la Renaissance, Besson et Ramelli," *Thalès,* Vol. 15 (1948), 108–12.

Saitta, G. *La Scolastica del secolo XVI.* Turin, 1911.

Sarton, G. *The Appreciation of Ancient and Medieval Science during the Renaissance (1450–1600).* Philadelphia, 1955.

————. *Six Wings. Men of Science in the Renaissance.* Bloomington, 1957.

Singer, C., et al. *From the Renaissance to the Industrial Revolution, 1500–1750. (A History of Technology,* Vol. 3.) New York, 1957.

Singleton, C. S., ed. *Art, Science, and History in the Renaissance.* Baltimore, 1968.

Strong, E. W. *Procedures and Metaphysics. A Study in the Philosophy of Mathematical-Physical Science in the Sixteenth and Seventeenth Centuries.* Berkeley, 1936.

Tannery, Paul. *Mémoires scientifiques,* Vol. 6. Paris, 1926.

Taton, René, ed. *La Science moderne (de 1450 à 1800). (Histoire générale des sciences,* Vol. 2.) Paris, 1958. English translation by A. J. Pomerans, *History of Science,* Vol. 2, New York, 1964.

Taylor, H. O. *Thought and Expression in the Sixteenth Century.* 2 volumes. New York, 1920.

Thorndike, Lynn. *A History of Magic and Experimental Science,* Vols. 5 and 6. New York, 1941.

Thurot, C. "Recherches historiques sur le principe d'Archimède," *Revue Archéologique,* N.S., Vol. 19 (1869), 111–23, 284–99.

Uccelli, A., et al. *Storia della tecnica dal medio evo ai giorni nostri. Enciclopedia storica delle scienze e delle loro aplicazione,* Vol. 2. Milan, 1943.

Usher, A. P. *A History of Mechanical Inventions* (revised ed.). Cambridge, Mass., 1954.

Vailati, G. *Scritti.* Leipzig, 1911.

Villoslada, R. G. *La Universidad de Paris durante los estudios de Francisco de Vitoria O. P. (1507–1522). (Analecta Gregoriana,* Vol. 14.) Rome, 1938.

Wallace, W. A. "The Concept of Motion in the Sixteenth

Century," *Proceedings of the American Catholic Associa-tion of the Catholic University of America*, Section 6 (1967), 184–95.

———. "The Enigma of Domingo De Soto," *Isis* (in press).

Walsh, J. J. *The Popes and Science*. New York, 1908.

Weisheipl, J. A. *The Development of Physical Theory in the Middle Ages*. London and New York, 1959.

Wiener, P., ed. *Roots of Scientific Thought*. New York, 1959.

Wightman, W. P. D. *Science and the Renaissance*. Edinburgh and New York, 1962.

Wohlwill, E. "Die Entdeckung des Beharrungsgesetzes," *Zeitschrift für Völkerpsychologie und Sprachwissenschaft*, Vol. 14 (1883), 365–410; Vol. 15 (1884), 70–135; 337–87.

———. *Galilei und sein Kampf für die Copernikanischen Lehre*, Vol. 1. Hamburg and Leipzig, 1909.

Wolf, Abraham. *A History of Science, Technology and Phi-losophy in the 16th and 17th Centuries*, Vol. 1 (revised ed.). London, 1951; reprinted New York: Harper Torchbooks, 1959.

Zilsel, E. "The Sociological Roots of Science," *American Journal of Sociology*, Vol. 47 (1941–42), 544–62.

Zoubov, Vassily P. "Expérience scientifique et expérience technique à l'époque de la Renaissance," *Proceedings of the Tenth International Congress of the History of Science, 1962*, Vol. 1 (1964), 65–80.

Works Relating to Particular Authors

In this section, some biographies of sixteenth-century scientists who did not write specifically on mechanics are included as source material for the general state of science at that epoch.

Bernardino Baldi

Affò. I. *Vita di Monsignor Bernardino Baldi da Urbino . . . .* Parma, 1783.

Giovanni Battista Benedetti

Bordiga, G. "Giovanni Battista Benedetti filosofo e mate-matico veneziano del secolo XVI," *Atti del Reale Istituto Veneto di Scienze, Lettere ed Arti*, Vol. 85 (1925–26), 585–754.

Drabkin, I. E. "G. B. Benedetti and Galileo's *De motu*,"

Proceedings of the Tenth International Congress of the History of Science, 1962, Vol. 1 (1964), 627–30.

––––––. "Two Versions of G. B. Benedetti's *Demonstratio proportionum motuum localium,*" *Isis,* Vol. 54 (1963), 259–62.

Koyré, A. "Jean-Baptiste Benedetti, critique d'Aristote," in *Mélanges Gilson* (Paris, 1959), pp. 351–72. English translation in E. McMullin, ed., *Galileo, Man of Science,* New York, 1967.

Maccagni, C. "Contributi alla biobibliografia di Giovanni Battista Benedetti," *Physis,* Vol. 9 (1967), 337–64.

––––––. *Le speculazioni giovanili "de motu" di Giovanni Battista Benedetti.* Pisa, 1967. Includes texts of *Resolutio* and *Demonstratio.*

Tycho Brahe

Dreyer, J. L. E. *Tycho Brahe, A Picture of Scientific Life and Work in the Sixteenth Century.* Edinburgh, 1890; reprinted New York: Dover, 1963.

Gade, J. A. *The Life and Times of Tycho Brahe.* New York, 1947.

Giordano Bruno

Berti, D. *Giordano Bruno da Nola, su vita e sua dottrina.* Turin, 1889.

Michel, P. H. *La Cosmologie de Giordano Bruno.* Paris, 1962.

––––––. "L'Atomisme de Giordano Bruno," *La Science au seizième siècle.* Paris, 1960.

Yates, F. A. *Giordano Bruno and the Hermetic Tradition.* London, 1964.

Girolamo Cardano

Bellini, A. *Girolamo Cardano e il suo tempo.* Milan, 1947.

Eckman, J. *Jerome Cardan.* Baltimore, 1946.

Morley, H. *The Life of Girolamo Cardano of Milan . . . .* 2 volumes. London, 1854.

Öre, O. *Cardano, the Gambling Scholar.* Princeton, 1953.

Rivari, E. *La Mente di Girolamo Cardano.* Bologna, 1906.

Waters, W. G. *Jerome Cardan, a Biographical Study.* London, 1898.

Federico Commandino

Baldi, B. "Vita di Federigo Commandino," *Giornal dei Letterati,* Vol. 19, Venice, 1714; reprinted in F. Ugolini and F. L. Polidori, eds., *Versi e prose scelte di Bernardino Baldi,* Florence, 1859.

Nicholas Copernicus

Armitage, A. *Copernicus, the Founder of Modern Astronomy.* New York and London, 1947. Also titled *Sun, Stand Thou Still* and *The World of Copernicus.*

Kesten, H. *Copernicus and His World,* trans. E. Ashton and N. Guterman. New York, 1945.

Galileo Galilei (work before 1600)

Busulini, B. "Componente Archimedea e componente medioevale nel *De motu* di Galileo," *Physis,* Vol. 6 (1964), 303–21.

Clagett, M. "Galileo and Medieval Kinematics," *History of Science* (Voice of America). Washington, 1964.

Drabkin, I. E. "A Note on Galileo's *De motu,*" *Isis,* Vol. 51 (1960), 271–77.

Drake, S. "Galileo and the Law of Inertia," *American Journal of Physics,* Vol. 32 (1964), 601–08.

Favaro, A. "Galileo Galilei e i Doctores Parisienses," *Reale Accademia dei Lincei, Rendiconti della classe di scienze morali, storiche e filologiche,* Vol. 27 (1918), 139–50.

Freiesleben, H. C. *Galileo Galilei; Physik und Glaube an der Wende der Neuzeit.* Stuttgart, 1956.

Giacomelli, R. *Galileo Galilei giovane e il suo "De motu."* Pisa, 1949.

Golino, C. L., ed. *Galileo Reappraised.* Berkeley and Los Angeles, 1966.

Hessen, S. "Die Entwickelung der Physik Galileis und ihr Verhältnis zum physikalischen System von Aristoteles," *Logos,* Vol. 18 (1929), 339–61.

Marcolongo, R. "La meccanica di Galileo," in *Nel terzo centenario della morte di Galileo Galilei* (Milan: Università Cattolica del S. Cuore, 1942), 36–57.

Olschki, L. *Galilei und seine Zeit.* (*Geschichte der neusprachlichen wissenschaftlichen Literatur,* Vol. 3.) Halle, 1927; reprinted Vaduz, 1965.

————. "Galileo's Literary Formation," E. McMullin, ed., *Galileo, Man of Science.* New York, 1967.

Strong, E. W. "Galileo on Measurement," in M. F. Kaplon, ed., *Homage to Galileo.* Cambridge, 1965.

Johannes Kepler (work before 1600)

Armitage, A. *John Kepler.* London, 1966.

Baumgardt, C. *Johannes Kepler: Life and Letters.* New York, 1951.

Caspar, Max. *Kepler,* trans. C. D. Hellman. New York, 1959.

Francesco Maurolico

Rossi, G. *F. Maurolico e il risorgimento filosofico e scientifico nel secolo XVI.* Messina, 1888.

Guido Ubaldo del Monte

Mamiani, G. *Memoria . . . su la vita e gli scritti di Guid' Ubaldo del Monte . . . .* Senigallia, 1821.

Bernard Palissy

Morley, H. *Palissy the Potter.* 2 volumes. London, 1852.

Alessandro Piccolomini

Ceretti, F. *Alessandro Piccolomini, letterato e filosofo senese del cinquecento.* Siena, 1960.

Pietro Pomponazzi

Fiorentino, F. *Pietro Pomponazzi.* Florence, 1868.

Wilson, C. "Pomponazzi's Criticism of Calculator," *Isis,* Vol. 44 (1953), 355–62.

Giovanni Battista [della] Porta

Parascandolo, G. *Notizie autentiche sulla famiglia e sulla patria di Gio. Battista della Porta . . . .* Naples, 1903.

Pierre de la Ramée

Graves, F. P. *Peter Ramus and the Educational Reformation of the Sixteenth Century.* New York, 1912.

Hooykas, R. "Pierre de la Ramée et l'empirisme scientifique au xvie siècle," *La Science au seizième siècle.* Paris, 1960.

Waddington, C. *Ramus, sa vie, ses écrits, et ses opinions.* Paris, 1855.

Simon Stevin

Depau, R. *Simon Stevin.* Brussels, 1942.
Dijksterhuis, E. J. *Simon Stevin.* Gravenhage, 1943.
Sarton, G. "Simon Stevin of Bruges," *Isis,* Vol. 21 (1934), 234–300.
Van de Velde, A. T. *Simon Stevin.* Brussels, 1948.

Niccolò Tartaglia

Favaro, A. *Intorno al testamento inedito di Niccolò Tartaglia.* Padua, 1882.
———. *Per la biografia di Niccolò Tartaglia.* Rome, 1913.
———. "Niccolò Tartaglia e la determinazione dei pesi specifici," *Commentarii dell'Ateneo di Brescia per l'anno 1916.* Brescia, 1916.
———. "A proposito della famiglia di Niccolò Tartaglia," *Commentarii dell' Ateneo di Brescia anno 1919.* Brescia, 1919.
Koyré, A. "La dynamique de Nicolo Tartaglia," *La Science au seizième siècle.* Paris, 1960.
Masotti, A. "Commemorazione di Niccolò Tartaglia," *Commentarii dell'Ateneo di Brescia per il 1957.* Brescia, 1958.
———. "Sui 'Cartelli . . . ' scambiati fra Lodovico Ferrari e Niccolò Tartaglia," *Rendiconti Istituto Lombardo — Accademia di Scienze e Lettere, classe di scienze* (A), Vol. 94 (1960).
———, ed. *Quarto Centenario della Morte di Niccolò Tartaglia: Atti del Convegno* [*di Storia delle Matematiche,* May, 1959]. Brescia, 1962. The *Atti* contain the following papers: B. Finzi, "La Meccanica dal Tartaglia ai nostri giorni," pp. 91–111; A. Masotti, "Tartaglia e i suoi *Quesiti,*" pp. 17–55, and "Rarità tartagliane—Spigolature bibliografiche e archivistiche," pp. 119–59; L. Tenca, "Niccolò Tartaglia e la balistica esterna," pp. 115–18; M. Villa, "La Matematica dal Tartaglia ai nostri giorni," pp. 59–87.

Bernardino Telesio

Gentile, G. *Bernardino Telesio, con appendice bibliografica.* Bari, 1911.
Troilo, E. *Bernardino Telesio.* Modena, 1910.
———. *La Filosofia di Bernardino Telesio.* Bari, 1914.

Van Deusen, N. C. *Telesio the First of the Moderns.* New York, 1932.

Leonardo da Vinci

Dibner, Bern. "Leonardo da Vinci: Military Engineer," in M. F. Ashley Montagu, ed., *Studies and Essays in the History of Science and Learning Offered . . . to George Sarton . . .* (New York, 1946), pp. 85–111.

Giacomelli, R. "Leonardo da Vinci, aerodinamico, aerologo, aerotecnico, e osservatore del volo degli ucelli," *Atti Convegno Studi Vinciani.* Florence, 1953.

Gille, Bertrand. "Léonard de Vinci et son temps," *Techniques et Civilisations,* Vol. 2 (1952), 69–84.

Hart, I. B. *The World of Leonardo da Vinci.* New York, 1961.

———. *The Mechanical Investigations of Leonardo da Vinci* (2nd ed.) Berkeley and Los Angeles, 1963.

International Colloquium. *Léonard de Vinci et l'expérience scientifique au XVIe siècle.* Paris: Centre National de la Recherche Scientifique, 1953.

Lilley, S. "Leonardo da Vinci and the Experimental Method," *Atti Convegno Studi Vinciani.* Florence, 1953.

Luporini, C. *La mente di Leonardo.* Florence, 1953.

Marcolongo, R. *Memorie sulla geometria e la meccanica di Leonardo da Vinci.* Naples, 1937.

Orellana, L. E. *La filosofía del movimento en Leonardo de Vinci; el influjo de su teoría escolástica del impetú en la formulación de la mecánica moderna.* Quito, 1960.

Pedretti, C., ed. *Documenti e Memorie Riguardanti Leonardo da Vinci a Bologna in Emilia.* Bologna, 1953.

Reti, L. "Die wiedergefunden Leonardo-Manuskripte der Biblioteca Nacional in Madrid," *Technikgeschichte,* Vol. 34 (1967), 193–225.

Ronna, A. M. *Léonard de Vinci—Peintre, Ingénieur, Hydraulicien.* Paris, 1902.

Schmidt, W. "Leonardo da Vinci und Hero von Alexandria," *Bibliotheca Mathematica,* Ser. 3, Vol. 3 (1902), 180–87.

Schuster, F. *Zur Mechanik Leonardo da Vincis.* Erlangen, 1915.

Solmi, E. "Leonardo da Vinci ed il metodo sperimentale nelle ricerche fisiche," *Atti e memorie della Reale Accademia Vergiliana di Mantova* (1904–05), pp. 8–218.

Somenzi, V. "Leonardo ed i principi della dinamica," in *Leonardo: Saggi e ricerche* (Rome, 1953), pp. 145–58.

Index